AF730568

Advances in Experimental Medicine and Biology

Volume 1475

Series Editors

Haidong Dong, Departments of Urology and Immunology
Mayo Clinic, Rochester, USA
Nima Rezaei, Research Center for Immunodeficiencies, Children's Medical Center, Tehran University of Medical Sciences
Tehran, Iran
Ortrud Steinlein, Institute of Human Genetics, LMU University Hospital
Munich, Germany
Junjie Xiao, Cardiac Regeneration and Ageing Lab, Institute of Cardiovascular Sciences, School of Life Science, Shanghai University
Shanghai, China
Avia Rosenhouse-Dantsker, Department of Chemistry
University of Illinois at Chicago, Chicago, IL, USA
Robert Gerlai, Department of Psychology, University of Toronto
Mississauga, ON, Canada

Advances in Experimental Medicine and Biology provides a platform for scientific contributions in the main disciplines of the biomedicine and the life sciences. This series publishes thematic volumes on contemporary research in the areas of microbiology, immunology, neurosciences, biochemistry, biomedical engineering, genetics, physiology, and cancer research. Covering emerging topics and techniques in basic and clinical science, it brings together clinicians and researchers from various fields.

Advances in Experimental Medicine and Biology has been publishing exceptional works in the field for over 40 years, and is indexed in SCOPUS, Medline (PubMed), EMBASE, Reaxys, EMBiology, the Chemical Abstracts Service (CAS), and Pathway Studio.

2023 CiteScore: 5.9

Karen Ballen
Editor

Clinical Management of Challenges in Allogeneic Stem Cell Transplant

Springer

Editor
Karen Ballen
University of Virginia
Charlottesville, VA, USA

ISSN 0065-2598 ISSN 2214-8019 (electronic)
Advances in Experimental Medicine and Biology
ISBN 978-3-031-84987-9 ISBN 978-3-031-84988-6 (eBook)
https://doi.org/10.1007/978-3-031-84988-6

© The Editor(s) (if applicable) and The Author(s), under exclusive license to Springer Nature Switzerland AG 2025

This work is subject to copyright. All rights are solely and exclusively licensed by the Publisher, whether the whole or part of the material is concerned, specifically the rights of translation, reprinting, reuse of illustrations, recitation, broadcasting, reproduction on microfilms or in any other physical way, and transmission or information storage and retrieval, electronic adaptation, computer software, or by similar or dissimilar methodology now known or hereafter developed.
The use of general descriptive names, registered names, trademarks, service marks, etc. in this publication does not imply, even in the absence of a specific statement, that such names are exempt from the relevant protective laws and regulations and therefore free for general use.
The publisher, the authors and the editors are safe to assume that the advice and information in this book are believed to be true and accurate at the date of publication. Neither the publisher nor the authors or the editors give a warranty, expressed or implied, with respect to the material contained herein or for any errors or omissions that may have been made. The publisher remains neutral with regard to jurisdictional claims in published maps and institutional affiliations.

This Springer imprint is published by the registered company Springer Nature Switzerland AG
The registered company address is: Gewerbestrasse 11, 6330 Cham, Switzerland

If disposing of this product, please recycle the paper.

Preface

Allogeneic stem cell transplant is a curative therapy for many patients with hematologic malignancies, bone marrow failure syndromes, and genetic disorders. Over 1 million stem cell transplants have been performed worldwide. Improvements in transplant access, Human Leukocyte Antigen (HLA) typing, Graft vs Host Disease (GVHD) prevention and management, and prevention and treatment of infections have all contributed to improved survival. Yet, challenges remain, especially as we treat older and sicker patients.

In this volume, we explore the etiology of common transplant complications, and best practices and management. We discuss the importance of social determinants of health in access to transplant care. We review the unique needs of older patients and the importance of the pre-transplant evaluation. There are many new options for conditioning regimens, of varying intensity, based on the patient population and underlying health. We review the role of expansion and the number of different graft choices, including peripheral blood stem cells, umbilical cord blood, and bone marrow grafts. The optimal prevention and management strategies for GVHD will be summarized, including the use of post-transplant cyclophosphamide.

The use of newer techniques to document, prevent, and treat measurable residual disease will be reviewed. We explore the changing paradigm for transplant for acute lymphoid leukemia, and how transplant aligns with other therapies. We discuss the unique features of myelofibrosis, and the challenges of transplant timing and splenomegaly. Finally, with many long-term survivors, we focus on post-transplant strategies including post-transplant maintenance and long-term survivorship.

Much has been accomplished in allogeneic stem cell transplant, and the next 5 years should be exciting indeed!

Charlottesville, VA, USA Karen Ballen

Contents

1 Access to Transplant Care and Social Determinants of Health

Krithika Shanmugasundaram and Karen Ballen

Abstract

Stem cell transplant is a potentially curative treatment for many hematologic conditions and is often one of the only curative options for many otherwise fatal hematologic malignancies. However, the ability to offer allogeneic transplant for patients who need this treatment is often limited by availability of an HLA-matched donor. Recently, developments in graft-vs-host disease prevention and the advent of PTCy have expanded the donor pool to include HLA-mismatched and HLA-haploidentical donors, but challenges in optimizing GVHD risk and post-transplant complications remain in these settings and can limit access to patients who have comorbidities or are of older age.

Outside of medically optimizing patients for transplant, the success of transplant is largely determined by a patient's social support and access to dedicated caregivers for extended periods of time. Ability to recover post-transplant is also determined by their ability to fund adjunctive therapies including nutritional support, physical therapy and rehabilitation, and the resources required to obtain necessary medications and medical attention in a timely manner. Thus, social determinants of health such as food insecurity, transportation insecurity, housing insecurity, and social support become large determinants of transplant success as well.

Keywords

Social determinants of health · Transplant access · Stem cell transplant · Post-transplantation cyclophosphamide · Transplant donors

1.1 Introduction

Hematopoietic stem cell transplant (HSCT) is a potentially curative treatment for many hematologic conditions and is often the only curative option for many otherwise fatal hematologic malignancies. However, the ability to offer an allogeneic transplant to patients who need this treatm ent is often limited by the availability of an HLA (Human Leukocyte Antigen)-matched donor. Recently, developments in graft-versus-host disease (GVHD) prevention and the advent of post-transplantation cyclophosphamide (PTCy) have expanded the donor pool to include HLA-mismatched and HLA-haploidentical donors, but challenges in optimizing GVHD and post-transplant complications remain in these settings.

K. Shanmugasundaram (✉) · K. Ballen
University of Virginia, Charlottesville, VA, USA
e-mail: kme5gp@uvahealth.org

© The Author(s), under exclusive license to Springer Nature Switzerland AG 2025

K. Ballen (ed.), *Clinical Management of Challenges in Allogeneic Stem Cell Transplant*, Advances in Experimental Medicine and Biology 1475, https://doi.org/10.1007/978-3-031-84988-6_1

Outside of medically optimizing patients for transplant, the success of a transplant is largely determined by a patient's social support and access to dedicated caregivers for extended periods of time. The ability to recover post-transplant is also determined by their ability to fund adjunctive therapies, including nutritional support, physical therapy, and rehabilitation, and the resources required to obtain necessary medications and medical attention in a timely manner. Thus, social determinants of health such as food insecurity, housing insecurity, and social support become large determinants of transplant success as well.

1.2 Availability of HLA-Matched Donors

The ability to proceed with an allogeneic stem cell transplant depends on the availability of a donor. The preferred donor, if available, is an 8/8 HLA-matched sibling donor. While availability varies based on region (Shamsi et al. 2008; Elbjeirami et al. 2013; Wang et al. 2010), in the United States, only 20–30% of patients eligible for transplant have a fully HLA-matched sibling donor available (D'Souza et al. 2020; Hsieh et al. 2009). Additionally, there are many barriers that may prevent potential sibling donors from being eligible, such as a potential genetic predisposition to the condition that the recipient has (Acevedo et al. 2019), age, and comorbidities.

Thus, many patients rely on finding an unrelated donor that is fully matched. However, the likelihood of finding an 8/8 HLA-matched unrelated donor is widely discrepant based on ethnic background, largely due to differences in genetic variance among different populations. Gragert et al. described this large discrepancy among donor availability in the United States, based on ethnicity, showcasing the highest likelihood of finding a fully matched unrelated donor, who is matched at the major histocompatibility (MHC) gene loci of HLA-A, HLA-B, HLA-C, and HLA-DRB1, is 75% among those of European descent, whereas all other ethnicities have a likelihood of less than 50% of identifying a fully matched donor, and the lowest likelihood being among those of African descent, with only a 16–19% chance of identifying an 8/8 HLA-matched donor (Gragert et al. 2014).

Among those that do have an available fully HLA-matched donor, there are other barriers that may arise such as timing of transplant in relation to donor scheduling, donor age, and location and resources of international donor centers that may impact the ability to complete timely evaluations of donors. The scarcity of international donor evaluation and collection centers may increase the burden of travel for many potential donors, which may limit their ability to complete the process of donation.

1.3 Alternative Donors: Challenges, Limitations, Safety

While using a fully HLA-matched donor remains the standard of care, alternative donor sources have been increasingly used for patients with no available matched sibling donor (MSD) or matched unrelated donor (MUD). Alternative donor sources include the following: mismatched unrelated donors (mMUD), which carry at least one allele difference in the four HLA gene loci used to determine a match, haploidentical donors, which are related donors who share one haplotype with the recipient, and umbilical cord blood (UCB).

Transplants using a mMUD or haploidentical donor do carry a higher risk of graft rejection and graft-versus-host disease (GVHD) if using a T-cell replete graft. Thus, initially haploidentical grafts were T-cell depleted, either by direct manipulation of the graft or in vivo T-cell depletion with antithymocyte globulin or alemtuzumab (Ho and Soiffer 2001; Saad and Lamb 2017). These transplantation platforms carried a much higher risk of severe infections and relapse (Goldman et al. 1988; Soiffer et al. 2011). However, with the advent of post-transplantation cyclophosphamide (PTCy) in 2008 (Luznik et al. 2008), acute GVHD and relapse outcomes in haploidentical transplantation are essentially comparable to matched donor transplantation (Jain et al. 2024; McCurdy and Fuchs 2015). Mismatched unrelated donor transplantation has also exhibited lower rates of GVHD with the use of PTCy-based platforms (Battipaglia et al. 2019). Thus, we have seen a large increase in mMUD and haploidentical transplantation in the last decade and ultimately an increase in the total number of transplants we have been able to offer and perform for patients (D'Souza et al. 2020; Ballen et al. 2008). Challenges with these PTCy-based platforms, however, are the noted increased risk of infections, particularly viral reactivation, which can contribute significantly to morbidity and, in some cases, can be fatal (McCurdy and Fuchs 2015).

Umbilical cord blood is rich in hematopoietic stem cells and often requires less stringent HLA-matching (Barker et al. 2002, 2005; Wagner et al. 1996; Kurtzberg et al. 1996) when considered for use in transplantation. Retrospective data suggests comparable outcomes in relapse as Matched sibling donor (MSD) or matched related donor (MRD) transplantation, but there can be considerable morbidity regarding infection as engraftment is often delayed compared to peripheral blood or marrow grafts. These delays in engraftment can be overcome with FDA-approved expansion techniques for cord blood products (Horwitz et al. 2021) or by using double-unit cord blood (Anastasia et al. 2011). However, recent randomized controlled trial data in reduced intensity transplants has also shown that double-unit cord blood transplantation has less favorable overall survival when compared to haploidentical transplantation in patients who do not have an available MSD or MRD (Fuchs et al. 2021).

1.4 Patient Factors: Social Determinants of Health

Although there may be an available and acceptable donor for transplantation, many patients still face significant barriers to receiving the recommended allogeneic transplant (Fig. 1.2). We will discuss social factors, such as race, income, socioeconomic status (SES), marital status, food insecurity, housing insecurity, and transportation insecurity, along with medical factors, primarily age.

1.4.1 Health Insurance

In the United States, one of the earliest barriers that patients face in pursuing transplantation is a lack of health insurance coverage. Many patients, even if medically appropriate, are less likely to be referred for transplant if they do not have active insurance (Pidala et al. 2013; Getta et al. 2017). Still, many patients who are referred are unable to proceed with transplant because of variations in coverage with Medicare/Medicaid or lack of insurance and inability to obtain it (Getta et al. 2017; Bhatt et al. 2018; Flannelly et al. 2020; Majhail et al. 2010). Patients who eventually do obtain coverage may often be ineligible for transplant after this delay because of relapse or evolution of other medical comorbidities.

1.4.2 Caregiver Access

A large aspect of the rigorous evaluation patients undergo prior to transplant is evaluating their social environment and caregiver support. This is due to the long and demanding process of recovery post-transplant. Patients remain vulnerable to infection and organ toxicity from post-transplant medications for many months, and thus they are expected to be able to commit to frequent healthcare appointments, reliably take prescribed medications, and have 24/7 caregiver support for a minimum of 3 months post-transplant. Thus, food insecurity, housing insecurity, transportation insecurity, education, and broad socioeconomic status may all have an impact on the patient's ability to commit to the rigors of post-transplant care.

Having a reliable caregiver plan is paramount to the success of a transplant. This can be difficult for many patients. Those that are unmarried are less likely to undergo hematopoietic transplantation (Jabo et al. 2017; Mock et al. 2021). Families with lower incomes and socioeconomic status are less likely as well (Jabo et al. 2017; Ashok Kumar et al. 2024), which may be in part due to financial and logistical challenges of finding acceptable childcare or elder care while the family resources are utilized for the patient during the post-transplant period (Preussler et al. 2019). Among surveyed caregivers and patients, notable barriers include lack of income for both during the post-transplant period, additional responsibilities of caring for other family members, and lack of transportation for both (Preussler et al. 2019). Accordingly, distance from the transplant center has also been shown to be a negative predictor for receipt of HSCT (Mitchell and Conklin 2015).

1.4.3 Racial Disparity in Access

There is profound racial disparity among those referred for and ultimately receiving an HSCT. Patients of racial and ethnic diversity are at a much lower likelihood of receiving HSCT as compared to their non-Hispanic White counterparts. While some of this disparity is due to donor availability as discussed previously, there are multiple other factors that contribute to lower rates of stem cell transplantation among minorities. In fact, a report from 2010 demonstrated that Black patients who have a matched sibling donor available are still approximately two-thirds less likely than White patients to receive a transplant (Joshua et al. 2010). Since then, some limited retrospective studies have actually reported no difference in transplant utilization between Black and White patients, but these studies do remark that more time to transplant was substantially increased for Black patients and that death due to relapse before achieving transplant was higher in the Black patient population (Shah et al. 2022; Bashey et al. 2023). More recently, when looking at all patients who were eligible for transplant, Black patients were much less likely to be referred for transplant (Landry 2021; Pidala et al. 2013), which creates one of the earliest barriers to care.

Hispanic patients and Asian patients were also found to have decreased access to HSCT, as they were less likely to receive a transplant compared to non-Hispanic White patients (Jabo et al. 2017). One potential aspect to this discrepancy is that Hispanic donors were found to be less likely to donate and reported being discouraged from donating (Landry 2021; Switzer et al. 2013). The same study found a similar, but less extreme, ambivalence among potential Asian donors. These patterns appear to exist outside of the United States as well, with notable attrition rates in potential donors of diverse ethnicities, with quoted reasons including medical mistrust, religious objection, family concerns, and receiving discouragement to donate (Anthias et al. 2020). While lower socioeconomic status (SES) has been shown to be a predictor of access to transplant, average household income in Asian families within the United States is higher than that of non-Hispanic White families, yet we still see an underutilization of transplant for Asian patients, indicating other barriers to care independent of SES (Ashok Kumar et al. 2024).

Systemic barriers that minorities face within healthcare at large do still apply to patients undergoing evaluation for transplant, and these issues can, unfortunately, translate to the underutilization of this potentially life-saving therapy.

1.4.4 Socioeconomic Status

Socioeconomic status (SES) is a large determinant of health in multiple arenas. Transplant care can offer considerable financial toxicity, and thus, SES affects access to HSCT for many patients. Many aspects of social determinants of health are affected by a patient's SES, such as transportation insecurity, housing insecurity, and food insecurity. While nearly two-thirds of US residents are within 1 hour of a transplant center, there are notable disparities in access to care based on location and distance from a transplant center (Mock et al. 2021; Mitchell and Conklin 2015), with patients living further away being less likely to proceed with transplant. Patients and caregivers often have to relocate, and while there are programs that can help deter some of the costs, patients often a bear significant financial burden from this alone (Ashok Kumar et al. 2024). Additionally, patients also incur the cost of some necessary post-transplant medications that their insurance will not cover, despite medical necessity.

Patients with higher SES also have a higher likelihood of having received higher levels of formal education. Patient education levels also seem to correlate with the likelihood of receiving HSCT. This may be due to baseline medical literacy prior to transplant, which can present a significant discrepancy in patient hesitancy, medical trust, and decision-making prior to transplant, given the high medical complexity of HSCT (Flannelly et al. 2020; Mock et al. 2021).

While many aspects of social determinants of health appear to be important in optimizing access to HSCT, these aspects also appear to have some interdependence. A broader understanding of how these aspects play into transplant access was determined using the social vulnerability index (SVI), which takes into account SES, household characteristics, racial and ethnic minority status, and housing type and transportation type. Describing regional and population differences in access to care using SVI has helped evaluate barriers to both autologous and allogeneic transplants (Fig. 1.1) (Salafian et al. 2024; Bhandari et al. 2022). In allogeneic transplants, a high SVI increases baseline nonrelapse mortality (NRM) for medically high-risk patients from 15% to 21%. The applicability of SVI extends

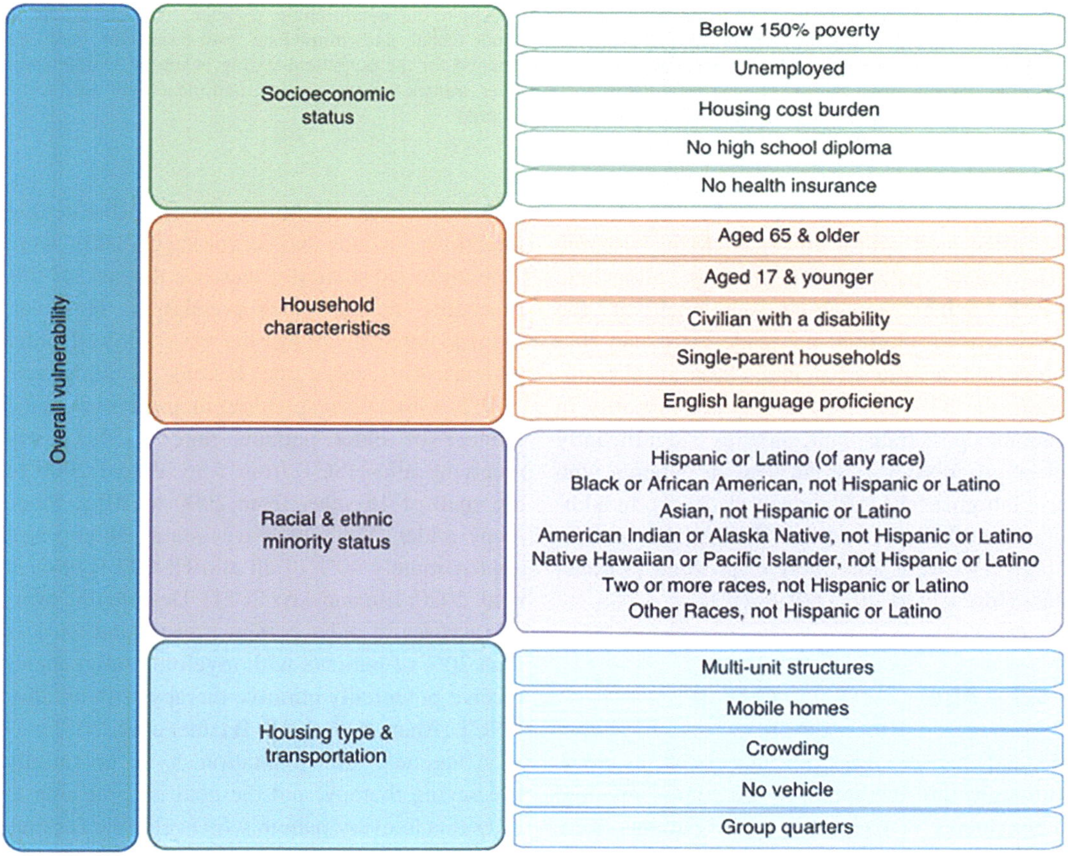

Fig. 1.1 Social vulnerability index, comprising 16 subfactors. (From Salafian et al, *Bone Marrow Transplantation, 2024*)

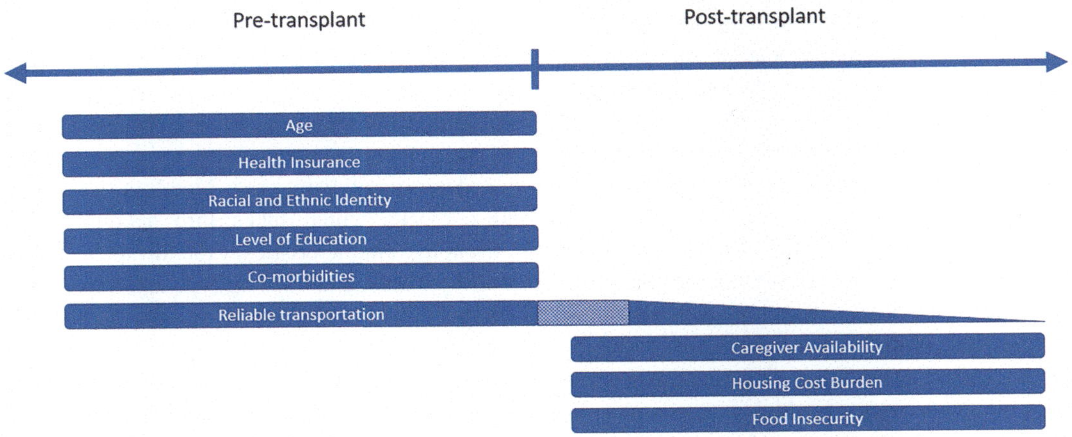

Fig. 1.2 Factors that impact access to care are crucial to the transplant process at different times. Here, age, access to health insurance, racial and ethnic identity, level of education, and preexisting comorbidities may delay or prevent a patient's access to a transplant. Lack of reliable transportation also affects pretransplant evaluations and appointments to optimize factors that improve eligibility for transplant, and the assurance of this ongoing access to transportation post-transplant is needed in order to proceed with the transplant. Similarly, patients need to confirm a reliable caregiver and access to nutrition for post-transplant recovery and secure a housing plan prior to proceeding with the transplant. Most institutions still plan for inpatient immediate peri-transplant care, which does alleviate some need for a caregiver, transportation, housing, and food security during admission. However, some institutions do offer an outpatient-based transplant, in which the conditioning regimen, transplantation of donor cells, and immediate post-transplant care are planned for the outpatient setting, where a reliable caregiver, transportation, housing, and meals will need to be secured

across racial demographics as well, with a higher SVI associated with a higher NRM in Hispanic and Asian patients undergoing allogeneic HSCT. When describing outcomes at the University of Virginia, an academic center that serves both a large urban and a large rural population, the SVI was able to capture disparity in outcomes post-transplant, showing a significantly lower progression-free survival in patients who have a higher SVI (Salafian et al. 2024). In addition, acute leukemia patients living in areas with a high SVI were less likely to proceed to transplant (Mock et al. 2021) (Fig. 1.2).

1.4.5 Age

Medical barriers to transplant exist as well. Although the median age of many myeloid malignancies is late 60 s, older patients were rarely considered for transplant historically. The severity of symptoms and high risk of organ toxicity patients faced in the post-transplant setting (mucositis, prolonged cytopenias, life-threatening infections, acute and chronic GVHD) were thought to substantially increase the risk of this procedure in the older population. However, reduced intensity platforms have allowed older patients to receive the benefit of allogeneic HSCT, with a corresponding increase now in the number of older patients (age > 55) yearly receiving allo-HSCT from 5 to almost 2000 in the span of 15 years from 2000 to 2015. Since then, older patients have stably comprised approximately 40% of all allo-HSCTs (D'Souza et al. 2020; Lin and Artz 2021). Despite this overall increase in older patient transplantation, less than 20% of patients with myeloid malignancies receive potentially curative therapy with an allo-HSCT (Bhatt et al. 2018; Bazinet et al. 2023).

Allogeneic transplantation is a potentially life-saving therapy, and the goal is to be able to offer this to every patient who is eligible for one. While some barriers, such as preexisting health conditions, may be difficult to overcome by the time a transplant is necessary, there are many factors, particularly within the SVI, which would make transplants more accessible and more successful for patients. Patients who lack access to health insurance are sometimes eligible for transplantation in clinical trials, which can offer a range of financial support. Patients can even receive complete travel, housing, and meal compensation when on clinical trials at the National Institutes of Health. However, most nonfederal transplant programs work within the means of what is available in social support. To better serve patients who currently face barriers to transplant, work is needed to build robust programs that readily offer hygienic local housing, financial support for hired caregiving for those that do not have available family, and dependable transportation that will enable patients to make every appointment with ease, ensuring smooth and reliable care before and after transplantation.

References

Acevedo MJ et al (2019) Outcomes of related and unrelated donor searches among patients with primary immunodeficiency diseases referred for allogeneic hematopoietic cell transplantation. Biol Blood Marrow Transplant 25(8):1666–1673. (1523-6536 (Electronic))

Anastasia S et al (2011) An overview of the progress on double umbilical cord blood transplantation. Haematologica 96(8):1213–1220

Anthias C et al (2020) Role of race/ethnicity in donor decisions about unrelated hematopoietic progenitor cell donation: exploring reasons for higher attrition among racial/ethnic minorities. Biol Blood Marrow Transplant 26(3):593–599

Ashok Kumar P et al (2024) Utilization and outcome disparities in allogeneic hematopoietic stem cell transplant in the United States. Eur J Haematol 112(3):328–338

Ballen KK et al (2008) The national marrow donor program 20 years of unrelated donor hematopoietic cell transplantation. Biol Blood Marrow Transplant 14(9 Suppl):2–7

Barker JN et al (2002) Searching for unrelated donor hematopoietic stem cells: availability and speed of umbilical cord blood versus bone marrow. Biol Blood Marrow Transplant 8(5):257–260

Barker JN et al (2005) Transplantation of 2 partially HLA-matched umbilical cord blood units to enhance engraftment in adults with hematologic malignancy. Blood 105(3):1343–1347

Bashey A et al (2023) Improved access to HCT with reduced racial disparities through integration with leukemia care and haploidentical donors. Blood Adv 7(15):3816–3823

Battipaglia G et al (2019) Posttransplant cyclophosphamide vs antithymocyte globulin in HLA-mismatched unrelated donor transplantation. Blood 134(11):892–899

Bazinet A et al (2023) Evolving trends and outcomes in older patients with acute myeloid leukemia including allogeneic stem cell transplantation. Am J Hematol 98(9):1383–1393

Bhandari R et al (2022) Social vulnerability and risk of nonrelapse mortality after allogeneic hematopoietic cell transplantation. J Natl Cancer Inst 114(11):1484–1491

Bhatt VR et al (2018) Socioeconomic and health system factors associated with lower utilization of hematopoietic cell transplantation in older patients with acute myeloid leukemia. Bone Marrow Transplant 53(10):1288–1294

D'Souza A et al (2020) Current Use of and Trends in Hematopoietic Cell Transplantation in the United States. Biol Blood Marrow Transplant 26(8):e177–e182

Elbjeirami WM, Abdel-Rahman F, Hussein AA (2013) Probability of finding an HLA-matched donor in immediate and extended families: the Jordanian experience. Biol Blood Marrow Transplant 19(2):221–226

Flannelly C et al (2020) Barriers to hematopoietic cell transplantation for adults in the United States: A systematic review with a focus on age. Biol Blood Marrow Transplant 26(12):2335–2345

Fuchs EJ et al (2021) Double unrelated umbilical cord blood vs HLA-haploidentical bone marrow transplantation: the BMT CTN 1101 trial. Blood 137(3):420–428

Getta BM et al (2017) Allogeneic hematopoietic stem cell transplantation is underutilized in older patients with myelodysplastic syndromes. Biol Blood Marrow Transplant 23(7):1078–1086

Goldman JM et al (1988) Bone marrow transplantation for chronic myelogenous leukemia in chronic phase: increased risk for relapse associated with T-cell depletion. Ann Intern Med 108(6):806–814

Gragert L et al (2014) HLA Match Likelihoods for Hematopoietic Stem-Cell Grafts in the U.S. Registry. N Engl J Med 371(4):339–348

Ho VT, Soiffer RJ (2001) The history and future of T-cell depletion as graft-versus-host disease prophylaxis for allogeneic hematopoietic stem cell transplantation. Blood 98(12):3192–3204

Horwitz ME et al (2021) Omidubicel vs standard myeloablative umbilical cord blood transplantation: results of a phase 3 randomized study. Blood 138(16):1429–1440

Hsieh MM, Kang EM, Fitzhugh CD, Link MB, Bolan CD, Kurlander R, Childs RW, Rodgers GP, Powell JD, Tisdale JF. Allogeneic hematopoietic stem-cell transplantation for sickle cell disease. N Engl J Med. 2009 Dec 10;361(24):2309-17. doi: 10.1056/NEJMoa0904971. PMID: 20007560; PMCID: PMC3627532

Jabo B et al (2017) Sociodemographic disparities in chemotherapy and hematopoietic cell transplantation utilization among adult acute lymphoblastic and acute myeloid leukemia patients. PLoS One 12(4):e0174760

Jain TA-OX et al (2024) Donor types and outcomes of transplantation in myelofibrosis: a CIBMTR study. Blood Adv 8(16):4281–4293. (2473-9537 (Electronic))

Joshua TV et al (2010) Access to hematopoietic stem cell transplantation. Cancer 116(14):3469–3476

Kurtzberg J et al (1996) Placental blood as a source of hematopoietic stem cells for transplantation into unrelated recipients. N Engl J Med 335(3):157–166

Landry I (2021) Racial disparities in hematopoietic stem cell transplant: a systematic review of the literature. Stem Cell Investig 8:24

Lin RJ, Artz AS (2021) Allogeneic hematopoietic cell transplantation for older patients. Hematology Am Soc Hematol Educ Program 2021(1):254–263

Luznik L et al (2008) HLA-haploidentical bone marrow transplantation for hematologic malignancies using nonmyeloablative conditioning and high-dose, posttransplantation cyclophosphamide. Biol Blood Marrow Transplant 14(6):641–650

Majhail NS et al (2010) Access to hematopoietic cell transplantation in the United States. Biol Blood Marrow Transplant 16(8):1070–1075

McCurdy SR, Fuchs EJ (2015) Comparable outcomes for hematologic malignancies after HLA-haploidentical transplantation with posttransplantation cyclophosphamide and HLA-matched transplantation. Adv Hematol. 2015:431923. (1687-9104 (Print))

Mitchell JM, Conklin EA (2015) Factors affecting receipt of expensive cancer treatments and mortality: evidence from stem cell transplantation for leukemia and lymphoma. Health Serv Res 50(1):197–216

Mock J et al (2021) Barriers to access to hematopoietic cell transplantation among patients with acute myeloid leukemia in virginia. Transplant Cell Ther 27(10):869.e1–869.e9

Pidala J et al (2013) Practice variation in physician referral for allogeneic hematopoietic cell transplantation. Bone Marrow Transplant 48(1):63–67

Preussler JM et al (2019) Caregiver availability and patient access to hematopoietic cell transplantation: social worker perspectives inform practice. Support Care Cancer 27(11):4253–4264

Saad A, Lamb LS (2017) Ex vivo T-cell depletion in allogeneic hematopoietic stem cell transplant: past, present and future. Bone Marrow Transplant 52(9):1241–1248

Salafian K et al (2024) The impact of social vulnerability index on survival following autologous stem cell transplant for multiple myeloma. Bone Marrow Transplant 59(4):459–465

Shah H et al (2022) Racial disparities in time to hematopoietic cell transplant among patients with hematologic malignancies at a large urban academic center. Bone Marrow Transplant 57(7):1213–1215

Shamsi TS et al (2008) The stem cell transplant program in Pakistan—the first decade. Bone Marrow Transplant 42(1):S114–S117

Soiffer RJ et al (2011) Impact of immune modulation with anti–T-cell antibodies on the outcome of reduced-intensity allogeneic hematopoietic stem cell transplantation for hematologic malignancies. Blood 117(25):6963–6970

Switzer GE et al (2013) Race and ethnicity in decisions about unrelated hematopoietic stem cell donation. Blood 121(8):1469–1476

Wagner JE et al (1996) Successful transplantation of HLA-matched and HLA-mismatched umbilical cord blood from unrelated donors: analysis of engraftment and acute graft-versus-host disease. Blood 88(3):795–802

Wang X et al (2010) Transplantation related donors: experience of 1355 Chinese patients. Arch Med Res 41(6):483–486

2 Allogeneic Transplantation for Older Adults

Ashley D. Hadjis and Shannon R. McCurdy

Abstract

Allogeneic hematopoietic stem cell transplant (HCT) is a potentially curative treatment for acute myeloid leukemia (AML) and myelodysplastic syndrome (MDS)—diseases that disproportionately affect older adults. Yet historically, older adults were excluded from HCT given its high morbidity and mortality and their increased susceptibility to these outcomes. This created a vast disparity: Those most often affected by AML and MDS were not considered eligible for its cure. Fortunately, clinicians have increasingly recognized that geriatric, functional, and frailty assessments, rather than chronological age alone, may help to identify older adults who are "fit" for HCT. This change in practice accompanied by the advent of azacitidine/venetoclax, which induces disease remission in a greater number of older adults, has led to the implementation of HCT in a wider population of patients than ever before. Moreover, improvements to HCT platforms, including reduced-intensity conditioning (RIC), post-transplantation cyclophosphamide (PTCy), abatacept, and the increasing use of unrelated and HLA-mismatched donors, have all contributed to the increased adoption and safety of HCT in adults up to 75 years old. In fact, select centers have extended the age of HCT eligibility up to the eighth decade. Herein, we review the data regarding the safety of existing platforms, the unique challenges inherent to HCT in older adults, and the efforts to optimize outcomes. Ultimately, in the modern era of HCT, age may, in fact, just be a number.

Keywords

Older adults · Elderly · Allogeneic hematopoietic cell transplant · Geriatric

2.1 Introduction

Among adults, the most common indications for allogeneic (allo) hematopoietic cell transplant (HCT) are acute leukemia, myelodysplastic syndrome (MDS), myeloproliferative neoplasm (MPN), and non-Hodgkin lymphoma (NHL) (Cusatis et al. 2024). Approximately 3000 transplants for a diagnosis of acute myeloid leukemia (AML), 2000 transplants for a diagnosis of either MDS or MPN, and 1000 transplants for acute

A. D. Hadjis
Department of Internal Medicine, Hospital of the University of Pennsylvania, Philadelphia, PA, USA
e-mail: ashley.hadjis@pennmedicine.upenn.edu

S. R. McCurdy (✉)
Abramson Cancer Center and the Division of Hematology and Oncology, Hospital of the University of Pennsylvania, Philadelphia, PA, USA
e-mail: shannon.mccurdy@pennmedicine.upenn.edu

© The Author(s), under exclusive license to Springer Nature Switzerland AG 2025

K. Ballen (ed.), *Clinical Management of Challenges in Allogeneic Stem Cell Transplant*, Advances in Experimental Medicine and Biology 1475, https://doi.org/10.1007/978-3-031-84988-6_2

lymphoblastic leukemia (ALL) were performed in 2022 in the United States (Cusatis et al. 2024). HCT in first complete remission (CR1) improves survival for AML patients with European LeukemiaNet (ELN) 2017 of intermediate or higher-risk disease. In the absence of HCT, only 16.6% of patients ≤60 years old and 2.4% of patients aged ≥60 years old with AML are disease-free at 10 years after treatment with chemotherapy alone (Vasu et al. 2018; Döhner et al. 2017). In MDS, patients with higher-risk disease, as classified by the International Prognostic Scoring System (IPSS), benefit from expeditious HCT (DeFilipp et al. 2023a). This is also true for older adults with MDS, specifically those between the ages of 50 and 75 years old with intermediate- or higher-risk disease have improved overall survival (OS) with reduced-intensity conditioning (RIC) HCT when compared to hypomethylating agents (HMA) alone or best supportive care (Nakamura et al. 2021). As such, HCT is the preferred consolidation therapy for HCT-eligible older adults with AML or MDS.

Fortunately, older adults are benefiting more than ever before from HCT. Among first-time HCT recipients in 2020, 38% of patients were ≥60 years old, a 10% increase from the previous decade (Auletta et al. 2023). Patients 65 years and older made up 28% of HCT recipients in 2022, a significant increase from 12% in 2012 (Cusatis et al. 2024). Adults over 70 years old have also benefited from this trend; from 2014 to 2022, the percentage of HCTs performed in the United States for adults ≥75 increased from <1% to 3% (Muffly et al. 2017). Not only is the number of older adults undergoing HCT increasing, but the number of transplant programs performing allogeneic HCT in patients ≥70 years increased from 65 centers in 2000–2007 to 93 centers in 2008–2013 (Muffly et al. 2017). Importantly, despite the increasing use of HCT in high-risk populations, HCT outcomes continue to improve. For instance, a European Society for Blood and Marrow Transplantation (EBMT) retrospective analysis examining outcomes in patients with AML aged ≥65 years who underwent HCT in 2000–2009 versus 2015–2021 found that nonrelapse mortality (NRM) significantly decreased from 31% to 27%, whereas 2-year leukemia-free survival (LFS) and OS improved from 32% to 44% and 37% to 49%, respectively (Bazarbachi et al. 2023).

2.2 Advances in Remission-Inducing Chemotherapy for Older Adults with AML and MDS

Given the limited efficacy of HCT for active leukemia or aggressive uncontrolled MDS, achieving disease remission or response is often a prerequisite for consolidative HCT. Previously, AML patients deemed "unfit" for induction therapy, which included all adults over the age of 75, would receive less-intensive initial chemotherapy, often single agent HMA or low dose-cytarabine (LDAC) (DiNardo et al. 2020; Dombret et al. 2015; Cashen et al. 2010; Kantarjian et al. 2012; Al-Ali et al. 2012). These therapies are less likely to achieve a durable complete response (CR) when compared with traditional 7 + 3 induction chemotherapy, which is reserved for "fit" younger patients. For instance, in AML patients over 65 years old treated with HMA alone, the CR rate is ≤30% and the median survival is less than 1 year (Dombret et al. 2015). In newly diagnosed older adults treated with LDAC, a randomized multicenter trial demonstrated a CR rate of only 20% (Tilly et al. 1990). In a more recent multicenter randomized study, only 15.6% and 7.9% of older adults achieved CR with decitabine and LDAC, respectively (Kantarjian et al. 2012). A phase II randomized study by Cortes et al. of the combination of glasdegib, a hedgehog pathway inhibitor, with LDAC led to a significantly improved median OS of 8.8 months when compared with 4.9 months for LDAC alone with corresponding CR rates of 17% and 2.3% (Cortes et al. 2019). Given the limited efficacy of these less-intensive induction strategies, older adults were less likely to achieve CR and thus much less likely to undergo consolidative HCT than their younger counterparts. The advent of more effective "less-intensive" remission-inducing chemotherapy, specifically azacitidine/venetoclax (aza/ven), has changed that paradigm.

The VIALE-A trial, published in 2020, randomized older adults, who were considered ineligible for traditional induction due to age ≥75 years and/or comorbidities, to either aza/ven or aza/placebo. The study enrolled patients from age 49 to 91 years, with a median age of 76 years, and demonstrated a CR/CR with incomplete hematologic recovery (CR/CRi) rate of 66.4% and a median OS of 14.7 months with aza/ven (DiNardo et al. 2018, 2020; Pollyea et al. 2018). These results led to the widespread adoption of aza/ven as the initial therapy for older adults with AML and as a salvage option for patients who were refractory to traditional induction. However, the vast majority of patients treated with aza/ven will eventually progress, and median survival is only 2–3 months after progression (Gangat et al. 2023; Maiti et al. 2021). Although controversial at first, this led many HCT physicians to consider consolidative HCT after aza/ven in an effort to capitalize on remission and derive the best chance at long-term survival. HCT after aza/ven has now become routine, in part due to publications by Pratz et al. and others, demonstrating a 69% 1-year OS and 29.9 month median OS after HCT in patients treated with aza/ven on VIALE-A (Pratz et al. 2023).

Another potential advantage of less-intensive remission-inducing strategies such as aza/ven is the ability to achieve CR without impacting performance status, which may be worsened by intensive induction. Aza/ven may be less cardiotoxic than 7 + 3 and is often selected as an induction strategy in patients ineligible for 7 + 3 due to prior lifetime anthracycline exposure or due to baseline reduced ejection fraction (Cusatis et al. 2024). However, 20% of patients receiving azacitidine/venetoclax experience a cardiac event, the most frequent of which is atrial fibrillation often cooccurring at the time of neutropenic fever, with 7.6% experiencing major cardiac complications (Johnson et al. 2022). Furthermore, given that aza/ven is typically given as an outpatient for a median of 4 months (range 0.9–31.8 months) prior to HCT (Pratz et al. 2023), physicians have more time to optimize comorbidities and performance, intervening in any deficits that could impact HCT outcomes. In addition to aza/ven, other disease-modifying agents such as isocitrate dehydrogenase (IDH) and fms-like tyrosine kinase 3 (FLT3) inhibitors are increasing our armamentarium against AML and allowing a greater number of older adults to become eligible for HCT.

2.3 Conditioning Regimens in Older Adults

The development of RIC paved the way for HCT to be used safely in the older adult population (McSweeney et al. 2001). RIC is associated with less NRM, but more relapse when compared with myeloablative conditioning (MAC) regimens (Shimoni et al. 2006). The question of balancing the risk of NRM with that of relapse has been studied extensively in both older and younger populations (for a summary of studies comparing conditioning intensity, see Table 2.1).

One such retrospective study by the EBMT demonstrated a higher incidence of relapse after RIC in patients <50 years old (Saraceni et al. 2021). However, in patients ≥50 years old, there was no difference in relapse or LFS, but less NRM after RIC compared with MAC HCT (Saraceni et al. 2021). In a prospective phase II study of RIC HLA-matched HCT with methotrexate and tacrolimus, the majority also received antithymocyte globulin (ATG); the 2-year relapse rate was 44% in AML patients aged 60–74 years (Devine et al. 2015). In a study comparing various treatment intensities, Shimoni et al. also showed that patients in CR did better with RIC, while those with active disease did better with MAC due to improved protection from relapse (Shimoni et al. 2006).

Two randomized studies have explored the question of conditioning intensity. First, the RICMAC Trial by the EBMT found that in patients <65 years old with MDS, the cumulative incidence (CI) of relapse and 2-year relapse-free survival (RFS) were not significantly different when comparing RIC and MAC HCT (Kröger et al. 2017). However, due to less NRM in the

Table 2.1 Summary of major studies examining the role of conditioning intensity

Publication	Study type	Disease type	*N*	Age median (range)	Results
de Lima et al. (2004)	Retrospective single center	AML or MDS	94	58 (22–75)	NMA vs. RIC: RIC a/w higher engraftment and lower relapse, but higher TRM
Blaise et al. (2005)	Prospective 2-center phase II	High-risk AML	33	51 (26–60)	RIC safe and effective 18-month OS: 79% CuI TRM: 9% 2-year PFS: 76%
Tauro et al. (2005)	Prospective multicenter	High-risk AML or MDS	76	52 (18–71)	FluMel RIC + alemtuzumab a/w 3-yr OS: 41% 3-yr DFS 37% Day 100 TRM: 9%
van Besien et al. (2005)	Prospective phase II single center	AML or MDS	52	52 (17–71)	FluMel RIC + alemtuzumab a/w 1-yr OS: 48% 1-yr PFS: 38% CuI TRM: 33% CuI relapse: 27%
Shimoni et al. (2006)	Prospective single center	AML or MDS	112	50 (17–70)	MAC and RIC: similar when leukemia in remission; MAC: better OS for active disease
Valcárcel et al. (2008)	Prospective multicenter	AML or MDS	93	53 (21–70)	RIC (Flu 150 mg/m^2 and oral busulfan 8–10 mg/kg) 4-yr OS: 45% 1-yr relapse: 29% 4-yr NRM: 20%
Kohrt et al. (2009)	Prospective phase II single center	Lymphoid or myeloid malignancies	111	54 (21–67)	RIC with TLI + ATG Durable chimerism: 97% 3-yr OS 60% 3-yr DFS 40%
Gyurkocza et al. (2010)	Retrospective single center	AML	274	60 (5–74)	NMA safe and effective in MSD and MUD HCT 5-yr OS: 33% 5-yr NRM: 26% 5-yr relapse: 42%
McClune et al. (2010)	Retrospective CIBMTR	AML and MDS	545 535	59.5 (40–79) 59 (40–78)	40–54 vs. 55–59 vs. 60–64 vs. ≥65 yrs. old: no significant difference in NRM, aGVHD, cGVHD, or relapse or OS after RIC HCT Pre-HCT performance status predicted 2-yr OS
Baron et al. (2015)	Retrospective EBMT	AML	394	56 (21–76)	FB vs. FM 2-yr relapse: 31% vs. 20%; similar OS and LFS
Devine et al. (2015)	Prospective multicenter phase II	AML	114	65 (60–74)	RIC MUD with ATG 2-yr relapse: 44% 2-yr NRM: 15% 2-yr OS: 48%

(continued)

Table 2.1 (continued)

Publication	Study type	Disease type	*N*	Age median (range)	Results
Damlaj et al. (2016)	Retrospective single center	AML, MDS, MPN	134	61 (18–72)	FB vs. FM 2-yr relapse: 35.6% vs. 17.3% 2-yr PFS: 60.5% vs. 48.7% NRM and OS: similar
Kawamura et al. (2017)	Retrospective multicenter	AML	1607	61 (50–74)	FB4 and FM140 higher NRM, but lower relapse compared to FB2 FB4, FM140, and FB2 no difference in OS 3-yr OS in high-risk AML and MDS superior in FM140 (37.0% and 60.2%), vs. either FB2 (24.4% and 45.5%) or FB4 (24.6% and 40.6%)
Kröger et al. (2017)	Phase III randomized multicenter EBMT	MDS or secondary AML	129	50.5 (19–63)	RIC vs. MAC 1-yr NRM: 17% vs. 25% 2-yr relapse: 17% vs. 15% 2-yr OS: 76% vs. 63%
Scott et al. (2017)	Phase III randomized multicenter (BMT CTN 0901)	AML or MDS	272	54.8 (18–65)	Study terminated early due to higher relapse rates in RIC vs. MAC: 48.3% vs. 13.5% 18-month OS: 67.7% vs. 77.5% TRM: 4.4% vs. 15.8% MDS subgroup had no difference in OS in RIC vs. MAC
Ustun et al. (2019)	Retrospective multicenter	AML	431	67.5 (60–75)	HCT vs. chemo consolidation on clinical trials: OS and NRM better after the first 9 months in the HCT arm: 5-yr OS 29% vs. 13.8%
Ciurea et al. (2020)	Retrospective single center	AML	404	65 (60–79)	FM100 significantly better PFS and GRFS compared to FM140, Bu ≥ 20,000, and Bu1600 due to lower NRM Benefit of FM100 more pronounced in older adults with poor PS
Hourigan et al. (2020)	Retrospective ultra-deep sequencing of samples from Scott et al. (2017)	AML or MDS	190	54.8 (18–65)	Using samples from Scott et al. randomized study: Patients with MRD had better OS due to less relapse with MAC vs. RIC Patients without MRD had no difference in OS with MAC vs. RIC
Oran et al. (2021)	Retrospective CIBMTR	MDS	1145	67 (60–83)	FM is associated with superior DFS, reduced RI, but higher TRM compared with FluBu
Scott et al. (2021)	Long-term follow-up of BMT CTN 0901	AML or MDS	272	54.8 (18–65)	4-yr TRM significantly higher with MAC vs. RIC, but OS superior due to higher rates of relapse in the RIC group

(continued)

Table 2.1 (continued)

Publication	Study type	Disease type	*N*	Age median (range)	Results
Hirschbühl et al. (2023)	Retrospective, EBMT	B- (Ph- and Ph+) and T-ALL	501	57 (45–76)	Addition of TBI to IIC regimens for ALL safe in adults ≥45 years old and a/w lower relapse
Niederwieser et al. (2024)	Long-term follow-up of RICMAC	AML, CMML, MDS	129	50.5 (19–63)	Median follow-up of 6.2 years, no significant difference in 10-yr OS or RFS between RIC and MAC

Abbreviations: *NMA* nonmyeloablative conditioning, *yr.* year, *AML* acute myeloid leukemia, *RIC* reduced-intensity conditioning, *HCT* hematopoietic cell transplant, *MDS* myelodysplastic syndrome, *a/w* associated with, *CuI* cumulative incidence, *MAC* myeloablative conditioning, *TRM* transplant-related mortality, *OS* overall survival, *MRD* minimal residual disease, *RFS* relapse-free survival, *FluMel* fludarabine-melphalan, *DFS* disease-free survival, *RI* relapse-incidence, *FluBu* fludarabine-busulfan, *FM100* fludarabine+melphalan 100 mg/m^2, *FM140* fludarabine+melphalan 140 mg/m^2, *Bu* ≥ 20,000 fludarabine+IV busulfan AUC ≥ 5000/dx4d, *Bu16000* fludarabine+IV busulfan AUC 4000/dx4d, *PS* performance status, *FB* fludarabine+busulfan, *FM* fludarabine+melphalan, *CIBMTR* Center for International Blood and Marrow Transplant Research, *EBMT* European Society for Blood and Marrow Transplantation, *NRM* non-relapse mortality, *LFS*, leukemia-free survival, *CIR* cumulative incidence of relapse, *PFS* progression-free survival, *FB4* busulfan at 12.8 mg/kg iv, *FB2* busulfan at 6.4 mg/kg iv, *TLI* total lymphocyte irradiation, *ATG* antithymocyte globulin, *ALWP* acute leukemia working party, *CMML* chronic myelomonocytic leukemia, *IIC* intermediate conditioning intensity, *TBI* total

[a]Included AML, CML, MDS, NHL, multiple myeloma, ALL, Gaucher's disease, Fanconi's anemia, Blackfan Diamond anemia, beta thalassemia major, non-Hodgkin lymphoma

RIC arm, there was significantly improved OS when compared with the MAC cohort (Kröger et al. 2017). The long-term results of this study (median follow-up of 6.2 years) demonstrated comparable long-term outcomes with no difference in 10-year OS, NRM, or RFS after RIC and MAC HCT (Niederwieser et al. 2024). However, multivariate analysis did show that cytogenetic low-risk patients had better OS, RFS, and NRM after RIC compared to MAC (Niederwieser et al. 2024). Another randomized trial of RIC versus MAC (BMT CTN 0901) in AML and MDS patients was terminated early due to a significant increase in relapse in the RIC arm in patients with AML, although notably there was no increase in relapse in patients with MDS (Scott et al. 2017). However, long-term follow-up and ultra-deep sequencing showed that patients without minimal residual disease (MRD) prior to HCT had similar OS regardless of conditioning arm, while those with MRD who underwent RIC had a higher incidence of relapse relative to MAC HCT (Hourigan et al. 2020). The BMT CTN 0901 restricted enrollment to patients aged 18–65 (median age of 54) with Hematopoietic Cell Transplantation-specific Comorbidity Index (HCT-CI) scores ≤4, thus enrolling a relatively young and healthy population (Scott et al. 2021). Given age restrictions at enrollment, limits on comorbidity burden, and low enrollment of patients with Eastern Cooperative Oncology Group (ECOG) scores >1, these randomized studies do not accurately reflect the real-world population of older adults with hematologic malignancies (Kröger et al. 2017; Scott et al. 2021).

While MRD status at HCT plays a huge role in outcomes after induction with intensive chemotherapy, it may play less of a role after aza/ven therapy. For instance, patients with MRD after aza/ven had no significant difference in RFS ($p = 0.39$) or OS ($p = 0.23$) when compared to those who were MRD-negative at the time of HCT (Winters et al. 2022). In that same study, MRD-positive patients who received induction chemotherapy had inferior RFS ($p = 0.0028$) and OS ($p = 0.0045$) when compared to those who were MRD-negative at HCT (Winters et al. 2022). The above studies suggest that both the older adult in CR without MRD or the older adult in CR after aza/ven agnostic of MRD would likely benefit from an RIC regimen. However, larger studies are needed to confirm the impact of MRD at HCT after less-intensive induction strategies.

Despite the utility of RIC in the older adult population, RIC platforms may not be interchangeable, and current studies aim to perfect it. For instance, the use of melphalan instead of busulfan has been heralded to reduce relapse. A Center for International Blood and Marrow Transplant Research (CIBMTR) study showed that there were higher rates of NRM but less relapse and better disease-free survival (DFS), which resulted in improved OS at 3 years in patients with MDS who received fludarabine-melphalan (FluMel) compared to those who received fludarabine-busulfan (FluBu) conditioning (Oran et al. 2021). While FluMel improved long-term survival for older adults, there was a higher incidence of NRM within the first 3 months after HCT. Given the increase in early NRM associated with FluMel, there have been efforts to optimize dosing to minimize toxicity, while still maintaining post-HCT disease control. In one study, four RIC regimens were compared: fludarabine+melphalan 100 mg/m^2 (FM100), fludarabine+melphalan 140 mg/m^2 (FM140), fludarabine+IV busulfan area under the curve (AUC) ≥ 5000/d x4d (Bu ≥ 20,000), and fludarabine+IV busulfan AUC 4000/dx4d (Bu16000) (Ciurea et al. 2020). Among these four regimens, the FM100 group had superior 5-year PFS due to lower NRM (Ciurea et al. 2020). Importantly, there was no significant difference in outcomes between the four conditioning regimens in patients with high-risk AML (defined as secondary AML, induction failure, high-risk cytogenetics, or adverse ELN 2017) (Ciurea et al. 2020). Several studies have demonstrated that the FM140 regimen achieves better disease control; however, the increase in NRM suggests that this platform should be reserved for those patients with the highest-risk disease (Baron et al. 2015; Damlaj et al. 2016; Kawamura et al. 2017). All in all, these data have led to an increase in the use of FM100 at many centers (Saraceni et al. 2021; Ciurea et al. 2020).

Two alternative strategies to improve outcomes in older adults include the use of myeloablative fractionated doses of busulfan given over a few weeks pre-HCT and the integration of venetoclax into conditioning regimens. A phase II trial investigated an extended fractionated MAC Bu dosing schedule with outpatient Bu administered on days -20 and -13 and Flu administered on days -6 to -3 (Popat et al. 2023). This was followed by Bu dosed to achieve a taret AUC of 20,000 mol/min for the whole course (Popat et al. 2023). This regimen was safe and effective with 100-day, 1-year, and 3-year NRM of 3.8%, 8%, and 9.3%, respectively, and 3-year OS of 80% (Popat et al. 2023). Cao et al. first showed that the integration of venetoclax into a MAC platform was safe and effective in patients aged 3–58 years (median age 25) (Cao et al. 2023). However, when this regimen was studied in an older population (median enrollment age of 64), there were high rates of grades II–IV and III–IV aGVHD (Zheng et al. 2023). In the RIC setting, the addition of venetoclax to FluBu was both safe and effective, without dose-limiting toxicities (Garcia et al. 2021). Nevertheless, the rate of morphologic relapse rate before day 100 was 41%, likely due to the high-risk features of the population (33% of patients with MDS and MDS/MPN had 5–10% marrow blasts and 82% of patients had persistent molecular MRD at time of HCT) (Garcia et al. 2021). Regardless, the addition of venetoclax or other novel agents to RIC conditioning may be an attractive option to more effectively control disease without increasing the dose of cytotoxic chemotherapy and warrants further study.

2.4 HCT in Older Adults with Myeloid Malignancies

As previously outlined, the majority of myeloid malignancies are diseases of older adults that require HCT for cure (National Cancer Institute 2024; Sekeres and Taylor 2022). The decision to pursue HCT in AML is often based upon stratification into favorable-, intermediate-, and unfavorable-risk diseases based on cytogenetic and molecular features. Patients with intermediate- or adverse-risk disease by ELN have improved OS with remission-inducing chemotherapy followed by HCT. For instance, a time-dependent retrospective analysis of 715 patients who underwent HCT in CR1 demonstrated improved OS with HCT for ELN 2017 intermediate- (HR 0.46; 95% CI, 0.29–0.71) and high-risk (HR, 0.45; 95% CI, 0.30–0.69) disease (Boddu et al. 2019). Ustun et al. found that among older patients with AML (ages 60–77), relatively younger patients with higher-risk disease were more likely to undergo HCT (Ustun et al. 2019). Despite higher-risk disease, patients who underwent HCT had better OS and DFS when compared to patients who were treated on prospective National Clinical Trials Network trials for induction and non-HCT chemotherapy consolidation (Ustun et al. 2019). As with AML, older adults with MDS who undergo HCT have better OS and LFS compared with those who receive HMA or the best supportive care (Nakamura et al. 2021).

Myelofibrosis (MF), another disease that requires HCT for cure, represents one of the most challenging diagnoses for HCT physicians. Advanced age, transfusion-dependency, iron overload, splenomegaly, and portal hypertension are inherent features of advanced MF, which increase the risk of graft failure, sinusoidal occlusive syndrome (SOS), GVHD, and infection post-HCT (Devlin and Gupta 2016). The American Society of Transplantation and Cellular Therapy (ASTCT) recommends HCT for Dynamic International Prognostic Scoring System (DIPSS) intermediate- and high-risk disease and increasingly emphasizes HCT for patients with high-risk mutations (Kanate et al. 2020). Using the Mutation-Enhanced International Prognostic Score System (MIPSS-70+), which incorporates cytogenetic and molecular data into risk prediction, patients with scores ≥3 should consider early HCT (Tefferi et al. 2018). MAC does not improve relapse rates in MF, and thus RIC is usually preferred, especially in older adults with MF, given high rates of patient comorbidity (Patriarca et al. 2008; McLornan et al. 2019). To overcome issues with engraftment in MF using RIC, low doses of total body irradiation (TBI) are often added to chemotherapy conditioning (e.g., flu/bu/TBI), which may mitigate graft failure (Freyer et al. 2022). An EBMT study of predominantly RIC HCT in adults ≥65 years with MF demonstrated a 5-yr OS, NRM, and relapse rate of 40%, 37%, and 25%, respectively (Hernández-Boluda et al. 2021). Notably, mortality was higher in the HCT group compared to the non-HCT group in the first year, but survival was improved during years 4–8 in patients who underwent HCT (Hernández-Boluda et al. 2021). This was also true in another study by Gowin et al., which showed HCT was associated with inferior OS compared with non-HCT therapies in both intermediate and low-risk DIPSS disease due to the upfront risk of TRM (Gowin et al. 2020). However, after 1-year OS was not significantly different and was in fact improved with HCT in intermediate and higher DIPSS score patients (Gowin et al. 2020). All in all, this suggests that, although HCT in this population comes with inherent risks, the early post-HCT risk of increased TRM may be outweighed by the long-term survival benefit. What is more, early HCT outcomes for MF may improve with the addition of novel agents. For instance, JAK inhibitors are increasingly being recognized as important in the pre-, peri-, and post-HCT care of MF patients. For instance, ruxolitinib improved OS in patients with high-risk MF when taken in the peri-transplant setting until engraftment, decreasing both NRM and aGVHD (Kröger et al. 2018; Ali et al. 2022).

2.5 HCT in Older Adults with Lymphoid Malignancies

Before the development of modern chemotherapy regimens and tyrosine kinase inhibitors, HCT improved OS compared with consolidative chemotherapy in adults with Philadelphia negative (Ph^-) B-cell ALL (Goldstone et al. 2008) and with Philadelphia positive (Ph^+) B-cell ALL (Yanada et al. 2006). More recently, several large retrospective studies in Ph^- B-ALL have demonstrated that a pediatric-based regimen can improve survival for adolescents and young adults and thus the paradigm has shifted to avoiding HCT in this population (Leonard and Hayes-Lattin 2018; Huguet et al. 2009; DeAngelo et al.

2015; Stock et al. 2014). Several of the original studies of pediatric regimens included patients up to the age of 60, with one group reporting that regimens that include peg asparaginase were tolerated up to age 50 (Leonard and Hayes-Lattin 2018; Stock et al. 2014). However, other studies demonstrated high rates of toxicity when using pediatric regimens in patients over 35–45 years old (Leonard and Hayes-Lattin 2018). These data have largely limited the use of intensive curative chemotherapy to younger patients or very fit adults. Yet, in adult patients who do not receive ALL regimens that include peg asparaginase, HCT may be considered in CR1 (Leonard and Hayes-Lattin 2018).

The TKI era and B-cell-directed therapies such as blinatumomab (Gökbuget et al. 2018, 2020; Topp et al. 2014, 2015, 2021; Kantarjian et al. 2017; Dombret et al. 2019; Martinelli et al. 2017; Hodder et al. 2024; Boissel et al. 2021) and inotuzumab (Kantarjian et al. 2016; Kantarjian et al. 2013; Stelljes et al. 2024; Goekbuget et al. 2022) have also led to a paradigm shift in the treatment of Ph$^+$ B-cell ALL and have led to fewer older adults undergoing HCT in CR1 (Mathisen et al. 2011). HCT for Ph$^+$ ALL is discussed in more detail in another section. Although treatment options have flourished for B-cell ALL, therapy choices are unfortunately more limited for T-cell ALL and HCT is recommended for patients in CR2 or high-risk patients in CR1 (Hamilton et al. 2017).

Also, contributing to the delay of HCT for ALL to CR2 is the validation of MRD-guided pursuit of HCT. For instance, Ghobadi et al. showed that Ph$^+$ ALL patients who were molecularly MRD-negative within 90 days of induction, when treated with consolidation chemotherapy and TKI, had comparable outcomes to those who went on to HCT (Ghobadi et al. 2022). Short et al. also showed similar results, with a 4-year OS rate of 66% and median OS of more than 10 years in Ph$^+$ ALL patients who achieved complete molecular response (CMR) at 3 months and did not undergo HCT in CR1 (Short et al. 2016). Similarly to Ph$^+$ B-cell ALL, MRD status can guide HCT decision-making in Ph$^-$ B-cell ALL (Hay et al. 2019; Frey et al. 2020).

In contrast to B-ALL, many patients with T-ALL are MRD-positive at the end of induction, but children have still been shown to have favorable outcomes if they have undetectable MRD at the end of consolidation, as demonstrated in the AIEOP-BFM 2000 trial (Schrappe et al. 2011). The German Multicenter Study Group for Adult Acute Lymphoblastic Leukemia (GMALL) study showed that patients with ALL (both B cell and T cell) who failed to achieve MRD-negative status after consolidation had a higher probability of achieving durable CR when proceeding to HCT in CR1 compared to those treated with chemotherapy alone (Gökbuget et al. 2012).

Just as in AML and MDS, when considering HCT for patients with ALL, the selection of conditioning is of critical importance. The cumulative effects of prior cycles of chemotherapy often make adults with ALL more susceptible to mortality from MAC HCT, as such many HCT physicians avoid MAC for ALL patients over 45 years old, a much younger cutoff than traditionally used for patients with AML. Instead, many opt for either RIC or intermediate intensity conditioning (IIC), which removes the TBI portion of the MAC regimen commonly employed in younger adults and children with ALL (Roth-Guepin et al. 2017; Mohty et al. 2010). However, the addition of lower doses of TBI to IIC regimens may provide an additional survival advantage. A retrospective study comparing fludarabine/TBI 8 Gy (FluTBI8), fludarabine/busulfan 6.4 mg/kg (FluBu6.4), and fludarabine/busulfan 9.6 mg/kg (FluBu9.6) in ALL patients ≥45 showed a significant reduction in the incidence of relapse at 2 years in the FluTBI8 group compared to the groups without TBI (37.3% vs. 30.9%, $p = 0.014$), but no difference in NRM, CI of aGVHD, or OS (Hirschbühl et al. 2023). Multivariate analysis did show that age was associated with higher NRM and lower OS, LFS, and GVHD-free relapse-free survival (GRFS), but this variable did not interact with the conditioning regimen (Hirschbühl et al. 2023). Therefore, the addition of TBI to currently used IIC regimens may be a feasible option for older adults to reduce relapse, but whether this improves OS for older adults with

ALL remains to be seen, and many institutions use identical RIC conditioning for ALL and AML patients.

Several studies have compared the efficacy of MAC and RIC regimens for adults and demonstrated no significant difference in NRM or relapse after HLA-matched HCT (Marks et al. 2010). Notably, the RIC group in this study tended to be older and have worse Karnofsky Performance Scores (KPS), so similar outcomes were encouraging for the efficacy of RIC in disease control (Marks et al. 2010). Another study, in patients ≥45 years old with ALL (including both B cell [Ph^+ and Ph^-] and T cell), demonstrated that while there was no difference in 1-year relapse risk, there was a significantly increased risk of relapse at 3-year post-HCT at 26% versus 15%, with RIC compared to MAC, respectively (Tanaka et al. 2013). However, this did not translate to a difference in 3-year OS, possibly due to the relatively greater efficacy of salvage therapies for late relapses (Tanaka et al. 2013). A retrospective EBMT analysis of patients ≥45 years old (median age of 50 and 56 in the MAC and RIC groups, respectively) showed an increase in 2-year NRM in MAC, but a lower risk of relapse, which again did not impact OS (Mohty et al. 2010). In multivariable analysis, patients transplanted in ≥CR2 or with poor KPS had worse NRM, but without an interaction with conditioning intensity (Marks et al. 2010).

Reassuringly, there are signals that both treatments of and HCT for ALL have improved in recent decades. A single-center retrospective study at Stanford showed that patients who underwent HCT from 2014 to 2019 had superior OS when compared with those transplanted in the proceeding 5 years (2-year OS of 73% compared to 55%), owing to improvements in NRM and better treatments for post-HCT relapse (Liang et al. 2022). This study also noted an increase in the number of HCTs performed, an expansion in the use of alternative donors, a rise in racially diverse HCT recipients, and finally a higher percentage of recipients over 40 years old (56% in 2008–2013 and 84% in 2014–2019) (Liang et al. 2022).

2.6 Donor Options in Older Adults

Lack of suitable donors previously represented a major limiting factor to HCT candidacy for older adults. When compared to their younger counterparts, matched sibling donors (MSDs) for older adults tend to also be of older age, have already passed away, or have comorbidities precluding them from stem cell donation. Even if an older MSD or haplo sibling donor is available, many donor centers also tend to avoid using older donors due to fear of inferior outcomes. For instance, older donors have been shown to have poor graft cell collection (Richa et al. 2009), higher incidence of clonal hematopoiesis of indeterminate potential (CHIP) (Frick et al. 2019; Newell et al. 2021a), donor-derived leukemia (Gondek et al. 2016), GVHD (Flowers et al. 2011; Kollman et al. 2001; Nademanee et al. 1995), relapse (Abid et al. 2023), and inferior OS.

In light of multiple recent studies showing comparable outcomes between MUD and MSD HCT (Gupta et al. 2010; Schlenk et al. 2010; Walter et al. 2010) and between MUD and haplo donors (Mushtaq et al. 2024) in the general HCT population, some centers pursue younger MUD or haplo donors over older MSD. However, a retrospective CIBMTR study using tacrolimus or cyclosporine-based GVHD prophylaxis examined older (≥ 50 years old) MSDs versus younger (<50 years old) MUDs in recipients ≥50 years old and found that there were higher rates of grades II–IV and III–IV aGVHD and cGVHD after MUD HCT (Alousi et al. 2013). Interestingly, the effect of donor age on NRM and OS was dependent on the performance status of the recipient (Alousi et al. 2013). Patients with higher KPS of 90 or 100 had worse NRM and OS with MUDs, while those with lower KPS had comparable NRM, relapse, and OS with MUDs compared to MSDs (Alousi et al. 2013). In contrast, another group compared younger MUDs to older MSDs in recipients ≥50 years old, demonstrating better OS, less relapse, and lower CI of cGVHD with MUDs compared to older MSDs (Schetelig et al. 2023). Notably, there was no difference in

NRM, event-free survival (EFS), or aGVHD between the groups (Schetelig et al. 2023).

The advent of post-transplantation cyclophosphamide (PTCy) has now broadened the field of eligible donors for older adults beyond MUDs and MSDs to include haplo donors (Kasamon et al. 2015). Many studies have demonstrated that the safety and efficacy of haplo HCT with PTCy is similar to that of MUD HCT (Ciurea et al. 2015; Luznik et al. 2010; Mehta et al. 2022; Saliba et al. 2020; Brissot et al. 2020). For instance, a retrospective analysis comparing patients ≥60 with AML who underwent haplo or MUD HCT showed no significant difference in aGVHD, relapse, NRM, LFS, GRFS, or OS between the two groups (Santoro et al. 2018). Extensive cGVHD was higher in the MUD group, which did not receive PTCy, and engraftment was worse in the haplo group (Santoro et al. 2018).

Haplo HCT has also been shown to be safe in older adults with ALL. A single-center retrospective study examining patients ≥55 years old who underwent HCT with PTCy showed that 5-year RFS and OS were 46% and 49%, respectively (Webster et al. 2023). Additionally, grades III–IV aGVHD and chronic GVHD occurred in only 3% and 13% of patients, respectively (Webster et al. 2023). Several studies have even demonstrated less relapse with haplo versus HLA-matched HCT for ALL (Dwabe et al. 2021; Al Malki et al. 2020).

Due to concerns over graft failure with mismatched grafts, haplo HCT has historically been used with caution in MPN and MF. However, haplo HCT with PTCy in MDS/MPN overlap syndromes was safe in a retrospective study of 120 adults with a median age of 62.5 years (Artz 2016). There has also been increasing use of haplo HCT in MF, with explorations of lower dose PTCy as a means to improve engraftment (Gambella et al. 2023; Kunte et al. 2022; García-Cadenas et al. 2023).

Haplo HCT is also efficacious in patients with relapse after prior HCT or with MRD at the time of HCT. A retrospective single-center study that examined patients who underwent haplo HCT for post-HCT relapse showed a 3-year PFS and OS of 53% and 60%, respectively, in patients without donor-specific antibodies and with HCT-CI < 3 (Srour et al. 2020). Additionally, in a single-center retrospective study using flu/mel conditioning, patients with MRD at the time of haplo HCT with PTCy had no significant difference in PFS when compared to patients who were MRD-negative at the time of haplo HCT, which prompted the authors to conclude the HLA-mismatch may overcome some of the poor prognosis of MRD positivity at HCT (Srour et al. 2019).

With the increasing use of HLA-mismatched donors, donor choice is becoming increasingly complex as many patients have several donor options available to them. Just as in MSD and MUD HCT, donor age may play a role when selecting a haplo donor. Mehta et al. compared younger haplo donors using PTCy to older MUDs using conventional GVHD prophylaxis for adult patients (≥18 years old) with ALL; they found that the older MUD group was associated with a significantly higher-risk of cGVHD, higher NRM that outweighed the lower relapse risk, and this equated to inferior OS (Mehta et al. 2023). When MSD versus haplo donors were compared, both using PTCy-based GVHD prophylaxis, there was similar OS [hazard ratio (HR), [95% confidence interval] 0.94, [0.78–1.14], $p = 0.54$], NRM (HR 0.98 [0.72–1.32], $p = 0.87$), and relapse (HR 0.87, [0.70–1.08], $p = 0.20$) in multivariate analysis (Mehta et al. 2024). When younger haplo donors (<35 years old) were compared with older MSD (≥50 years old) in older recipients (≥50 years old), patients with younger haplo B-leader matched donors had significantly better OS (Mehta et al. 2024).

All in all, more work is needed to improve donor selection and allow HCT physicians to prioritize among donor source, donor age, HLA-match, cytomegalovirus (CMV) serostatus, and other factors. Tools to aid decision-making when selecting a haplo donor exist (Fuchs et al. 2022), but additional models are needed that weigh the balance of donor sources, donor factors, and recipient characteristics to allow us to personalize the selection of a donor to the individual patient. Finally, given the many studies showing comparable outcomes between donor sources, factors such as convenience, ease of access, and costs should also be considered when picking a donor for the older adult recipient (Pereira et al. 2023). It is not surprising that haplo donors may be easier to access and schedule and may also be more willing to provide additional assistance later in the disease course, such as readiness of donation of cells for CD34^{+} stem cell boosts or donor-lymphocyte infusions (Pereira et al. 2023). Regardless, many donor sources, including HLA-mismatched unrelated donors, represent viable options for older adults (Kasamon et al. 2015; Shaw et al. 2021).

2.7 GVHD Prophylaxis and Treatment in Older Adults

GVHD is a major cause of morbidity and mortality post-HCT, and its prevention and effective treatment are of paramount importance to older adults. Older HCT recipients have a higher incidence of both acute (Weisdorf et al. 1991; Hahn et al. 2008; Gale et al. 1987) and chronic (Bhatt et al. 2022; Remberger et al. 2002) GVHD, which impacts quality of life and increases NRM. Additionally, patients with higher HCT-CI scores, which represent a vast majority of the older adult population, are more likely to die of GVHD (Sorror et al. 2014). Not only is the prevention and effective treatment of GVHD important in the older adult population but also the limitation of toxicity associated with the prophylaxis strategy and the GVHD treatment itself is essential.

A randomized trial of PTCy/mycophenolate mofetil (MMF)/tacrolimus (tac) versus methotrexate/tac as GVHD prophylaxis for RIC HLA-matched peripheral blood stem cell transplant (PBSCT) demonstrated improved GRFS due to less severe acute and chronic GVHD and a higher incidence of immunosuppression-free survival in the PTCy arm (Holtan et al. 2023; Bolaños-Meade et al. 2023). Moreover, PTCy was shown to lead to better patient-reported outcomes, including improved nutrition and mouth scores at days 100 and 365, respectively, with no difference in hemorrhagic cystitis symptoms compared to methotrexate/tac (Holtan et al. 2023).

Intermediate-dose PTCy is another strategy to lessen HCT toxicity, which is particularly relevant to HCT for older adults. In mouse models, intermediate-dose (25 mg/kg/day on days +3,+4) PTCy prevented fatal GVHD more effectively than high-dose (50 mg/kg/day on days +3,+4) PTCy (Wachsmuth et al. 2019; Wachsmuth et al. 2020). In clinical studies, intermediate doses of PTCy have been shown to be effective in both haplo and HLA-matched HCT (McAdams et al. 2021; Zhang et al. 2023; Ackula et al. 2021). For instance, interim results from a phase I/II clinical trial examining intermediate-dose PTCy after MAC haplo bone marrow transplant (BMT) showed excellent control of grades II–IV and III–IV aGVHD (McAdams et al. 2021). Notably, intermediate-dose PTCy was associated with faster engraftment, less mucositis, and fewer cases of CMV reactivation requiring treatment (McAdams et al. 2021). Given the findings in MAC haplo BMT, intermediate-dose PTCy was expanded to an RIC haplo BMT platform that included a high-risk older adult cohort. There were again low rates of GVHD, NRM, relapse, and faster engraftment, with results comparable across age groups (McCurdy et al. 2024). In a study of RIC HLA-matched PBSCT, three doses of PTCy (50 mg/kg/day on days +3,+4; 25 mg/kg/day on day +4; and 25 mg/kg/day on days +3,+4) were compared with no significant difference in the CI of grades II–IV aGVHD between the dose levels (Zhang et al. 2023). The intermediate-dose PTCy group also demonstrated very low rates of both grades II–IV aGVHD (CI: 28.6%) and relapse (1-year CI: 10.6%) (Zhang et al. 2023). These early findings bring the hope that deescalated doses of PTCy may lessen HCT morbidity, facilitating the expansion of HCT to an older and frailer population. Recently, the OPTIMIZE study, a large, prospective CIBMTR study, began recruitment to examine reduced-dose PTCy in MMUD and will aid in examining outcomes with reduced-dose PTCy on a large scale.

Similar to PTCy, Abatacept, a CTLA4 analog, has also been shown to be effective in reducing severe aGVHD in HLA-mismatched unrelated donor HCT (Watkins et al. 2021) and in HCT recipients ≥60 years old, but with substantial 1-year moderate–severe cGVHD rates of 57.9% and 39% in those two studies (Raghunandan et al. 2023). Longer duration abatacept is being explored as a way to decrease cGVHD rates with this platform. Several studies are also exploring combining abatacept and PTCy to gain the benefits of reduced aGVHD from abatacept and reduced cGVHD from PTCy. A phase Ib/II clinical trial with PTCy/abatacept/tacrolimus GVHD prophylaxis after haplo PBSCT resulted in CI of 17.4%, 4.4%, and 15.9% of grades II–IV aGVHD, grades III–IV aGVHD, and 1-year cGVHD, respectively, with a 1-year OS of 85.9% (Al-Homsi et al. 2023). Importantly, in those patients who did develop grades II–IV aGVHD, a secondary retrospective analysis demonstrated steroid responsiveness (Tonzi et al. 2024). Finally, abatacept is also being explored for GVHD treatment with several studies finding that it was an effective therapy for chronic GVHD, particularly bronchiolitis obliterans syndrome (Kashyap et al. 2022; Koshy et al. 2023; Wertheimer et al. 2021).

Having novel agents to treat GVHD is particularly important in the older adult population in whom the inflammatory state of GVHD and the toxicity profiles of its treatments are often poorly tolerated. Older adults are more at risk of diabetes, infection, osteoporosis, and myopathy associated with steroid use. In addition, they have a higher incidence of frailty making it more difficult for them to rebound from a GVHD event. The JAK inhibitor ruxolitinib is an effective treatment for steroid-refractory acute and chronic GVHD (Zeiser et al. 2020a; Pattipaka et al. 2024; Zeiser et al. 2021). The REACH2 study, a phase 3 trial, studied ruxolitinib versus the best-available therapy in steroid-refractory aGVHD and found a significantly improved overall response rate with the use of ruxolitinib: 62% versus 39% (Zeiser et al. 2020a). A phase 3 study of ruxolitinib versus best-available therapy (The REACH3 Study) in steroid-refractory cGVHD showed an overall response rate of 50% in the ruxolitinib group compared to 26% in the best-available therapy group (Zeiser et al. 2020b). Notably, there was also a higher rate of symptom improvement in the ruxolitinib group: 24% versus 11% (Zeiser et al. 2020b). FDA approval was granted in 2021 based on the results of this study. Ruxolitinib can also be utilized in steroid-responsive subsets with the goal of lessening the duration and/or dose of glucocorticoids (Zeiser et al. 2020a, 2021; Pattipaka et al. 2024). In the older adult population, one prospective phase II multicenter study examined prolonged administration of ruxolitinib post-HCT and found that there was no increase in relapse and there were low rates of cGVHD requiring systemic treatment, suggesting a role for ruxolitinib in cGVHD prevention (Defilipp et al. 2023b). Studies to incorporate ruxolitinib early as GVHD prophylaxis have shown promise (Zhang et al. 2024), and further studies are planned.

Like ruxolitinib, axatilimab and belumosidil were both recently approved for the treatment of cGVHD, and like ruxolitinib, these therapies may move to the prophylactic setting in future studies (Kitko et al. 2023; Cutler et al. 2021). Regardless, novel effective GVHD treatments will improve the safety of HCT and can be explored as steroid-sparing agents to decrease the morbidity and infectious mortality associated with GVHD treatment in older adults. The impact of these exciting new therapies on the outcomes of older adult HCT recipients warrants further study.

2.8 Special Considerations in Older Adult HCT

Given that comorbidities increase with aging, identifying them and the increased risk they pose for the HCT patient is critical to both risk stratification and treatment optimization. Acute kidney injury (AKI) is a common complication of HCT that increases NRM and impacts quality of life (Miyata et al. 2021). A recently published meta-analysis of reports from 1995 to 2019 found that the incidence of AKI after HCT was 55.1%, with stage III AKI occurring in 8.3% of patients (Miyata et al. 2021; Kanduri et al. 2020). A prospective study across HCT platforms in 135 patients with a median age of 38.3 years demonstrated AKI in 68.9% of patients in the first 100 days post-HCT, 8.6% of whom required renal replacement therapy (RRT) (Andronesi et al. 2022). Patients who required RRT had a mortality rate of 87.5% (Andronesi et al. 2022). In a larger study, 3% of patients required RRT, of whom 43% recovered renal function and 24% survived hospital discharge (Abramson et al. 2021). AKI after HCT is associated with increased all-cause mortality and NRM (Miyata et al. 2021), but AKIs occurring early post-HCT are associated with even higher mortality (Shingai et al. 2015). Likewise, the greater the severity of the AKI, the greater the associated mortality (Miyata et al. 2021).

Older age has been shown to be an independent risk factor for AKI after HCT (Gutiérrez-García et al. 2020). Other independent risk factors include female sex, diabetes, hypertension, higher CRP, lower albumin, CMV infection, baseline kidney dysfunction, and chronic kidney disease (CKD) (Miyata et al. 2021; Abramson et al. 2021; Cui et al. 2023; Menezes et al. 2022). A majority of these risk factors are more common in the older adult population and control of diabetes, hypertension, and nutrition prior to HCT could be a way to mitigate risk. HCT-specific risk factors include TBI, calcineurin inhibitors (CNIs) (which increase the risk of both thrombotic microangiopathy [TMA]-associated AKI (Li et al. 2018) and TMA-independent AKI), methotrexate, and the use of cord blood stem cells (Miyata et al. 2021; Cui et al. 2023). Importantly, the risk of AKI is higher with MAC compared to RIC and NMA conditioning regimens, yet another reason lower intensity conditioning regimens may be preferred in older adults (Lopes et al. 2016; Parikh et al. 2005; Parikh and Coca 2006; Lopes and Jorge 2011; Kogon and Hingorani 2010). Furthermore, PTCy, which avoids the use of methotrexate, and the use of sirolimus instead of tacrolimus may be avenues to reduce the risk of AKI in older adults. Complications from HCT including sinusoidal obstructive syndrome, sepsis, and CMV infection also increase the risk of AKI and thus strategies to reduce these outcomes may also reduce the burden of HCT-induced kidney disease (Miyata et al. 2021). Finally, many of the medications used to treat post-HCT complications such as amphotericin B, acyclovir, aminoglycosides, and foscarnet are nephrotoxic (Miyata et al. 2021). Clearly, the risk of AKI is inherent to HCT itself and the older adult population is especially vulnerable to suffering from this post-HCT complication. Therefore, studies designed to improve HCT outcomes in older adults must also focus on mitigating renal failure.

Cardiovascular complications are also common after HCT and include cardiomyopathy, arrhythmias, acute thrombosis, pulmonary hypertension, pericardial effusions, and atherosclerotic disease, among others (Vasbinder et al. 2023). A retrospective study at Michigan and Rush medical centers examined risk factors for cardiovascular complications post-HCT, not restricted to older adults (Vasbinder et al. 2023). MAC, RIC, and NMA conditioning regimens were shown to have similar rates of cardiovascular events within the first 100 days post-HCT; however, long-term cardiovascular events were increased after RIC compared to MAC, which was surprising (Vasbinder et al. 2023). Additionally, the 100-day incidence of cardiovascular events was highest in recipients who received conditioning regimens containing Flu/Mel (7.2%) and lowest in patients who received Cy/TBI (1.8%) (Vasbinder et al. 2023). The authors postulate that this unexpected increase in long-term cardiovascular events with RIC may be due to selection bias in conditioning choice. Indeed, patients who received RIC compared to MAC in this study were significantly older (mean 55.7 vs. 52.4, $p = 0.002$) and had significantly higher rates of hypertension (51.6% vs. 38.2%, $p < 0.001$), diabetes mellitus (18% vs. 11%, $p = 0.012$), CKD (5.2% vs. 2%, $p = 0.028$), and coronary artery disease (CAD) (22.2% vs. 12.5%, $p < 0.001$) (Vasbinder et al. 2023). This same study also found that CKD was associated with a greater short-term risk of cardiovascular events (Vasbinder et al. 2023). Prior history of cardiovascular disease (including history of myocardial infarction, CAD, congestive heart failure, and percutaneous coronary intervention) is also predictive of a short-term cardiovascular event after HCT (Alblooshi et al. 2021). Age, greater body mass index (BMI), lower pretransplant ejection fraction, CKD, and history of CAD were all significantly associated with long-term cardiovascular events (Vasbinder et al. 2023). Again, the older adult population is more likely to have one, or multiple, of these risk factors present prior to HCT, putting them at higher risk of developing cardiovascular complications.

A unique complication associated with HLA-mismatched HCT is cytokine release syndrome (CRS) evidenced by the onset of fever between day 0 and PTCy administration on days 3 and 4. One study demonstrated that 17% of patients undergoing haplo PBSCT developed severe CRS. Prior radiation was found to be a significant risk factor, whereas older age was associated with a trend toward increased probability of severe CRS (Imus et al. 2019). In contrast, another study showed that there was no significant impact of age on the development or severity of CRS, but one would imagine that the state of severe CRS would be tolerated poorly in the older adult contributing to subsequent morbidity (Abboud et al. 2016). Regardless, severe CRS is associated with delayed neutrophil engraftment, increased TRM (Imus et al. 2019; Abboud et al. 2016), and worse OS (Abboud et al. 2016). There are mixed results regarding the impact of CRS on the development of aGVHD, with a nonsignificant trend toward increased incidence of aGVHD in one study (Raj et al. 2018) and no difference in another (Abboud et al. 2016). While worry about CRS may cause some physicians to avoid haplo donors in the older adult, another strategy may be to use bone marrow (BM) grafts and RIC regimens, both of which are associated with a lower incidence of severe CRS (Abboud et al. 2016; McCurdy et al. 2018). Finally, the use of ruxolitinib or another JAK inhibitor, started prior to day 0 and continued through PTCy completion, may reduce CRS and be a successful strategy for older adults (Abboud et al. 2021).

2.9 The Biology of Aging and the Domains of the Older Adult

The nine cellular hallmarks of aging are as follows: "genomic instability, telomere attrition, epigenetic alterations, loss of proteostasis, deregulated nutrient sensing, mitochondrial dysfunction, cellular senescence, stem cell exhaustion, and altered intercellular communication." (Lin et al. 2021; Lopez-Otin et al. 2013) These nine hallmarks encapsulate the unique biological landscape of the older adult, which responds differently to the stress of HCT than younger physiologies. To put it simply: "aging is characterized by the progressive loss of physiologic integrity, leading to impaired function and vulnerability to death." (Lopez-Otin et al. 2013) However, understanding the underlying cellular mechanisms of aging can improve risk stratification, inform interventions to reverse any modifiable risk factors, and improve treatment decision-making in older adults. Age is more than an absolute number, but rather a complex biologic process. A heightened understanding of the genetic and cellular processes underpinning aging may provide targeted antisenescence therapies that allow us to optimize the restorative potential of HCT.

In HCT, the cellular foundations of aging are especially relevant. For example, immune reconstitution post-HCT is altered in older adults. For HCT to be successful, donor immune reconstitution must occur to provide the recipient with a marrow capable of fighting infection and to achieve a graft-versus-leukemia effect to prevent relapse. The older adult is often at a natural disadvantage when it comes to post-HCT immune reconstitution. In the aging recipient, senescent cells have been shown to accumulate in different organs and secrete proinflammatory cytokines and adhesive molecules, termed the senescence-associated secretory phenotype (SASP), which impair immune function (Lin et al. 2021). Additionally, the development of a diverse $CD4^+$ T-cell repertoire requires a functional thymus (Lin et al. 2021). Thymic atrophy is common in older adults (Lin et al. 2021). Even though older adult adults are able to employ a thymus-independent mechanism for T-cell generation, older adults lack T-cell diversity post-HCT (Lin et al. 2021). Finally, there is a natural functional decline in hematopoietic stem cells as we age, resulting in impaired self-renewal and long-term reconstitution potential, which may in part explain the benefit of utilizing younger donors as discussed previously (Lin et al. 2021; Elias et al. 2017; de Haan and Lazare 2018). On the whole, older adults often have delayed immune reconstitution and a less competent post-HCT immune system, leading to an increased risk of infection and disease relapse (Lin et al. 2021).

Survivors of cancer and HCT have both been shown to have accelerated clinical and biological features of aging, and there are several biological assessments of premature aging that have been studied in HCT. Patients who have undergone HCT have been shown to have increased levels of IL-6, a marker of accelerated aging, and $p16^{ink4A}$, a marker of cellular senescence (Uziel et al. 2020). Another measure of biological aging is cardiopulmonary exercise testing (CPET) and measurement of peak oxygen consumption, which is the standard measure of aerobic capacity and cardiovascular reserve in clinical practice (Armenian et al. 2017; Wood et al. 2013). Peak oxygen consumption has been shown to be an independent predictor of cardiovascular disease and all-cause mortality in multiple populations (Wood et al. 2013; Jones et al. 2010, 2012) Post-HCT patients had lower peak oxygen consumption on CPET than predicted for age. Additionally, MAC and active GVHD were both associated with lower peak oxygen consumption (Armenian et al. 2017).

Ultimately, the science of the biological assessment of aging is developing, but there are several promising biomarkers at both the genomic level and the patient level that may provide valuable assessment of biological age. One such method is the assessment of telomere length, as these structures are involved in the process of cellular senescence (Vaiserman and Krasnienkov 2020). Telomere length has been shown to be correlated not only with higher morbidity and all-cause mortality (Wang et al. 2018) but also is associated with multiple aging-related chronic diseases, including coronary artery disease (Xu et al. 2019), myocardial infarction (D'Mello et al. 2015), stroke (D'Mello et al. 2015; Ding et al. 2012), and diabetes. (D'Mello et al. 2015) One CIBMTR study of 1514 MDS HCT recipients showed that short telomeres pre-HCT were associated with worse OS and increased risk of NRM independent of recipient comorbidities (Myllymäki et al. 2020). However, telomere length measurement is hampered by several analytical issues that preclude its prime-time use as a measure of biological aging. This includes variations in telomere length across tissue type (Dlouha et al. 2014) and also interindividual variability of telomere lengths (Bodelon et al. 2014).

Another factor, relevant to the age of the donor and the recipient, is CHIP, which is the presence of leukemia-associated somatic mutations in individuals without a hematologic malignancy (Frick et al. 2019). CHIP increases the risk of developing hematologic malignancies and of nonhematologic adverse outcomes, including cardiovascular disease (Jaiswal et al. 2014, 2017; Gibson et al. 2022; Genovese et al. 2014). The presence of CHIP in healthy adults has also been associated with decreased OS compared to age-matched controls (Jaiswal et al. 2014). CHIP is more common in older adults and increases in frequency with age. The population as a whole has been shown to have a frequency of 2% (Xie et al. 2014), while exome sequencing of peripheral blood samples from over 30,000 patients without known hematologic malignancies demonstrated a frequency of 10% in patients over the age of 65 and 20% in patients over the age of 90 (Sperling et al. 2017). Data regarding the impact of CHIP in HCT, in particular, is limited. However, early data suggests that recipient CHIP may impact HCT outcomes, with particular mutations, such as p53, being markers of poor survival (Bejar et al. 2014; Imus et al. 2024). In a retrospective HCT cohort of lymphoma patients ≥60 years old at Johns Hopkins, recipients with pre-HCT CHIP had worse 3-year OS compared to those without pre-HCT CHIP (47% vs. 78%, $p < 0.001$) (Imus et al. 2024). Given that older adults are more likely to harbor CHIP, more studies are needed to understand how recipient CHIP impacts HCT survival.

Older adults are also more likely to have older donors, who are therefore more likely to have CHIP at the time of stem cell donation. The presence of donor CHIP has been hypothesized to increase the development of donor-derived malignancies, enhancing the risk of utilizing older donors (DeZern and Gondek 2020). The post-HCT immune milieu may be a particularly hospitable environment for the expansion of a pre-HCT clone given the associated proliferative stress and decreased immune surveillance in the immunosuppressed host (DeZern and Gondek 2020). At least one report showed a disproportionate expansion of engrafted CHIP mutations (Frick et al. 2019). In fact, several reports have shown increased rates of donor-derived leukemia in older donors (Gondek et al. 2016; Xiao et al. 2011; Diamond et al. 2011; Hirsch et al. 2016; Suárez-González et al. 2018; Kobayashi et al. 2017; Herold et al. 2017). Donor CHIP has also been associated with a high risk of cGVHD (Frick et al. 2019; Newell et al. 2021b) and unexplained post-HCT cytopenias (Gibson et al. 2017). However, donor CHIP in the form of DNMT3A mutations significantly reduced relapse risk after HCT (Gibson et al. 2022). Larger studies are needed to clarify the clinical impact of donor CHIP on outcomes in HCT and determine whether it should be utilized as part of donor selection algorithms.

2.10 Current HCT Risk and Functional Assessments in the Nongeriatric Population

In all potential HCT recipients, and especially so in older adult HCT candidates, it is important to weigh the risks and benefits of proceeding to HCT. Not only is the quantity of life important, but the quality of life must be considered, and risk-stratification tools may help us to balance these various factors. Currently, the HCT-CI, ECOG performance status, and KPS are widely used to assess risk in patients with hematologic malignancies.

The most commonly utilized tool to assess the risk of NRM after HCT is the HCT-CI, which utilizes 17 different comorbidities and organ function assessments to generate a score from 0 to 25 (Sorror et al. 2005, 2009; Broglie et al. 2023; Sung et al. 2024). Notably, in the United States, the reporting of HCT-CI is a regulatory requirement. The age-specific HCT-CI adds one point to the score for all patients ≥40 years old, giving all older adults a minimum score of one, which equates to a predicted TRM of 21%. For older adults with two or more comorbidities, the predicted TRM is 41% when using RIC HCT (Sorror et al. 2005). However, given that older adults generally have more comorbidities than younger patients, and comorbidities may be compensated for, the HCT-CI score may be less useful in differentiating risk in this population. In addition, modern HCT platforms, such as PTCy-based GVHD prophylaxis, may decrease the risk of NRM even in patients with comorbidities (Sung et al. 2024; Backhaus et al. 2023). For instance, in older AML patients who received NMA HCT, the HCT-CI score did not significantly impact NRM, relapse, or OS (Backhaus et al. 2023). Given the interest in designing a specific risk prediction tool for older adults, the BMT CTN1704 (aka the CHARM study) completed a multicenter study in adults over 60 undergoing HCT. Of the many factors analyzed, the combination of HCT-CI, CRP, albumin, and weight loss led to the best prediction of survival (Artz et al. 2023).

Performance measures are another area of focus in HCT risk assessment. ECOG and KPS, performance status assessments that correlate with survival in multiple oncologic settings, are widely used in research and the clinic to assess pre-HCT risk (Kelly and Shahrokni 2016). ECOG performance status uses a five-point scale that describes a patient's level of symptomology from their underlying oncologic disease, specifically the amount of time spent in bed (Oken et al. 1982). The KPS score was first developed in patients receiving chemotherapy for lung cancer and has since been used in a variety of oncologic diseases. It uses a linear score in integers of 10 with 0 being dead to 100 being no evidence of disease (Karnofsky et al. 1948). Despite their utility, both have major limitations, including their subjective nature and the frequency of interobserver disagreement between clinicians (Kelly and Shahrokni 2016; Roila et al. 1991). Discrepancies in performance status and function may also be present between physicians and patients, as patients have been shown to rate themselves lower than members of the healthcare team (Ando et al. 2001). Agreement between observers is more consistent when the patient population has a "good" performance status, but when the patient population is more disabled there is greater interobserver disagreement (Sørensen et al. 1993; Verger et al. 1992).

2.11 Functional Assessments

Frailty is the vulnerability of older adults to "adverse health outcomes in response to illness and iatrogenic stressors." (Fried et al. 2001) Importantly, age and frailty are not linearly related, making some researchers wonder whether frailty status may predict death more accurately than chronologic age. Given this, frailty may be an especially useful pre-HCT tool that more fully encompasses the functional status and risk profile of older adults. However, there are multiple tools to assess for frailty, and the research regarding the relationship between frailty and HCT is still new.

The most frequently employed frailty and functional assessments include Fried's frailty assessment, the Rockwood clinical frailty scale, the clinical frailty scale, the "Get up and Go" test, and gait speed (Klepin 2019). The seven-point Rockwood clinical frailty scale categorizes patients with a score of 1 as being very fit and 7 as being severely frail (Rockwood et al. 2005). The incorporation of the Rockwood scale into clinical assessment was shown in a prospective clinical study by the MDS registry of Canada to improve the risk stratification of the IPSS score by 30% (Buckstein et al. 2016). The clinical frailty scale evaluates specific domains, including comorbidity, function, and cognition to generate a score ranging from 1 (very fit) to 9 (terminally ill) and has been shown to be predictive of outcomes in multiple disease states (Church et al. 2020). The CFS has been shown to be significantly associated with OS in a large study of 676 patients admitted to an oncology center in the United Kingdom, with higher CFS correlated to a higher chance of death during admission in both patients with hematologic malignancies and solid tumors: 56.5% of patients with CFS of 8 died during admission, while only 4.5% of patients with CFS of 4 died ($P < 0.001$) (Welford et al. 2022). (Welford et al. 2022)

The "Get up and Go" test times the patient as they rise from an armchair, walk 3 meters, turn, walk back, and sit down again (Podsiadlo and Richardson 1991). This test has often been proposed as both a fall-risk screening and a simple test of functional mobility that is easy to perform and requires no specialized equipment. One study in Korean older adults found that the "Get up and Go test" was correlated to previously validated measures of clinical frailty (Cardiovascular Health Study [CHS] and the Short Physical Performance Battery [SPPB]) (Jung et al. 2020). However, an Irish study found that while the "Get up and Go test" was effective in identifying frailer members of the community, it was less sensitive in discriminating between frail and pre-frail adults (Savva et al. 2012).

Separate from the "Get up and Go test," gait speed can identify functional impairment and has been studied both in patients with hematologic malignancies and HCT recipients. Analysis of patients after HCT showed that gait parameters differed after HCT compared to age-matched controls who had not undergone HCT. This highlights that gait is an important functional marker of health and that HCT has a major impact on that functionality (Kneis et al. 2020). One group showed that for every 0.1 m/s decrease in gait speed, there was an increase in mortality and the odds of unplanned hospitalizations in patients who were evaluated at an initial consultation appointment for a hematologic malignancy (Liu et al. 2019). Muffy et al. found that slow walk speed was associated with worse OS and greater disease relapse in patients ≥50 years old who underwent HCT (Lori et al. 2014).

Although including frailty assessments as part of the pre-HCT visit is a relatively new concept, several studies have shown that there is a relationship between frailty and worse HCT outcomes. The most commonly used metric to measure frailty is Fried's Frailty Phenotype (FFP). FFP combines two objective measures (weakness measured by grip strength and slowness measured by gait speed) with three subjective measures (patient-reported weight loss, exhaustion, and low physical activity) to assess frailty on a five-point scale. A point is given for each measure for which the patient qualifies as frail (Fried et al. 2001). Patients with zero points are not frail, one to two points are prefrail, and three or more points are considered frail (Fried et al. 2001). FFP was associated with worse OS in patients ≥60 years old undergoing either initial chemotherapy for hematologic malignancies or HCT (Sung et al. 2024; Hadjis et al. 2023, 2024). Specifically, in older adults with myeloid malignancies, FFP at diagnosis was associated with survival: 2-year OS for those who were fit, prefrail, and frail was 57%, 42%, and 21%, respectively (Hadjis et al. 2023). Preconditioning FFP was also associated with post-HCT survival and relapse in older adults (Sung et al. 2024). Frailty is also a dynamic marker, and the majority of older adults have a decline in their FFP status from pre-HCT to 1-month post-HCT (Hadjis et al. 2024). Being frail versus prefrail at 1- and 6-month post-HCT was also associated with worse OS (Hadjis et al. 2024).

There is a growing emphasis that stratifying patients based on their frailty, rather than their age alone, may improve outcomes for older adults. For instance, the ongoing UK fitness trial randomizes treatment intensity for older adults with multiple myeloma based on their International Myeloma Working Group (IMWG) frailty score (Coulson et al. 2022). The IMWG frailty score combines age, deficits in ADLs, impairments in IADLs, and the Charlson comorbidity index to categorize patients as fit, intermediate-fit, or frail (Murillo et al. 2019). Data from ongoing trials will help to validate the utility of frailty to inform treatment decisions and platform selection. Regardless of the tool utilized, evaluating function should now be standard practice when considering the older adult for HCT.

2.12 The Comprehensive Geriatric Assessment

The oncologic geriatric assessment is an alternative or additive to the aforementioned tools. Older age has consistently been linked to low referral rates to HCT (Flannelly et al. 2020) and surveys of HCT physicians show that opinions on the upper age limit for HCT vary widely (Mishra et al. 2021; Meyer et al. 2019). While there is general agreement on the need for specialized assessments for older adults, uncertainty remains over the preferred metric to do so (Mishra et al. 2021; Meyer et al. 2019). The comprehensive geriatric assessment aims to identify and quantify age-related weaknesses to predict outcomes and ideally improve upon any identified weaknesses before HCT (Olin et al. 2020; Holmes et al. 2014; Huang et al. 2020). The American Society of Clinical Oncology (ASCO) recommends that geriatric assessments be performed on all patients 65 and older with cancer in order to identify age-related vulnerabilities that could be managed before treatment initiation (Williams et al. 2023).

There are many domains of geriatric assessment: functional status, nutrition, weight loss, social support, psychological testing, comorbidities, cognitive function, and polypharmacy (Table 2.2) (Artz 2016; Williams et al. 2023). The completion of a geriatric assessment has been shown to influence HCT decision-making and lead to impactful pre-HCT interventions. These remediations could include medication adjustment, an increase in physical and occupational therapy services, referral for neuro-psych testing, and nutritional counseling (Lin et al. 2018). Barriers to the completion of the geriatric assessment still exist, including perceived time constraints and uncertainty in how to intervene on any identified deficits (Dale 2022; Gajra et al. 2022; Dale et al. 2021). Nevertheless, geriatric assessments have been shown to improve communication, decrease severe treatment-related

Table 2.2 Select domains of the older adults and tools for assessment

Domains	Common and recommended tools
Comorbidities	HCT-CI, Carlson CI, cumulative illness rating scale
Functional assessments	
Patient-reported	PS, ADLs, IADLs, falls, exhaustion, MOS-PH, FACT-functional wellbeing, Fried's Frailty Phenotype
Performance-based	PS, 4-meter walk, timed up and go, 6-minute walk test, grip strength, Fried's Frailty Phenotype, gait speed
Social supports	ISSS, MOS social support
Cognition	MMS, Montreal cognitive assessment (MOCA), Mini-Cog, BOMC
Psychological	GDS, HADS, MHI-17
Nutritional status	Body mass index, weight loss, sarcopenia (SARC-F)
Polypharmacy	Beers Criteria, >5 medications or > 8 medications

Adapted from Artz (2016)

Abbreviations: *HCT-CI* hematopoietic cell transplantation-specific comorbidity index, *PS* performance status, *ADLs* activities of daily living, *IADLs* instrumental activities of daily living, *MOS* medical outcomes study, *MMS* mini-mental state, *BOMC* blessed orientation-memory concentration, *GDS* geriatric depression scale, *HADS* hospital anxiety and depression scale, *MHI-17* mental health inventory-17

toxicities, lower hospitalization risk, increase advanced direction completion, and improve patient/caregiver satisfaction in the general oncology population (Li et al. 2021; Mohile et al. 2020; Mohile et al. 2021; Soo et al. 2022; Hamaker et al. 2022). If choosing only a single geriatric assessment, we would recommend an objective functional assessment, such as gait speed. Gait speed can be easily added to the clinic visit as a fifth vital sign and has been demonstrated to predict outcomes in several studies of HCT patients. Regardless of what tool is implemented, with an aging population of HCT recipients, HCT physicians must adapt their practices to include geriatric assessments and meet this growing need.

2.13 Conclusions

We are now in the age of HCT where we can happily say that nearly every patient has a donor. HCT is safer and more accessible than ever before, which has extended HCT to older and older adults. The number of older adults receiving HCT is increasing every year due to heightened physician comfort and data showing that geriatric HCT is safe. This is true across multiple platforms and donor types. Now that HCT is more accessible to and safer for older adults, we must overcome the barriers to HCT referral in this population and identify those older adults who would benefit from this curative therapy. More detailed and widely used geriatric assessments, which also include measures of functional and frailty status, would help to increase referral to and receipt of HCT in older adults. If geriatric assessments are routinely used, more patients that are indeed eligible for HCT may be identified, rather than imprecisely ruled out based on age alone. Additionally, we may identify patients who may benefit from additional interventions that may make them HCT-eligible or improve their post-HCT survival.

As the transplant population ages, physicians should place a greater emphasis on optimizing our patient's frailty, nutrition, and functional status before HCT. This may come in the form of a prescribed physical therapist, personal trainer, supervised exercise programs, clinical nutrition consults, geriatric consultations, and medication adjustments. Moreover, we need to increase the enrollment of adults ≥65 years old in clinical trials and develop HCT platforms individualized to the older adults who, despite recent advances, still have outcomes that lag those of their younger counterparts. With better GVHD prevention strategies such as PTCy and abatacept, we are moving the needle forward to improve HCT outcomes for older adults. We are also fortunate that for the first time, we have new effective FDA-approved agents for GVHD treatment that will increase the safety of HCT for older adults and may move to the prophylaxis arena in future studies. Finally, understanding the biological underpinnings of aging may allow us to identify biomarkers of frailty to both guide treatment decisions and identify ways to minimize the influence of age on HCT outcomes and end HCT's association with accelerated aging.

References

Abboud R, Keller J, Slade M et al (2016) Severe cytokine-release syndrome after T cell–replete peripheral blood haploidentical donor transplantation is associated with poor survival and anti–IL-6 therapy is safe and well tolerated. Biol Blood Marrow Transplant 22(10):1851–1860

Abboud R, Gao F, Rettig MP et al (2021) A single-arm, open-label, pilot study of the JAK1 selective inhibitor itacitinib for the prophylaxis of graft-versus-host disease and cytokine release syndrome in T-cell replete haploidentical peripheral blood hematopoietic cell transplantation. Blood 138:100. https://doi.org/10.1182/blood-2021-144591

Abid MB, Estrada-Merly N, Zhang M-J et al (2023) Impact of donor age on allogeneic hematopoietic cell transplantation outcomes in older adults with acute myeloid leukemia. Transplant Cell Ther 29(9):578.e1–578.e9. https://doi.org/10.1016/j.jtct.2023.06.020

Abramson MH, Gutgarts V, Zheng J et al (2021) Acute kidney injury in the modern era of allogeneic hematopoietic stem cell transplantation. Clin J Am Soc Nephrol 16(9):1318–1327. https://doi.org/10.2215/cjn.19801220

Ackula H, Medawar G, Cerio C, Roberts TF, Meleveedu K (2021) De-escalation of Ptcy dosing in matched allogenic transplants: a single institution retrospective study. Blood 138(Supplement 1):4872–4872. https://doi.org/10.1182/blood-2021-153455

Al Malki MM, Yang D, Labopin M et al (2020) Comparing transplant outcomes in ALL patients after haploidentical with PTCy or matched unrelated donor transplantation. Blood Adv 4(9):2073–2083. https://doi.org/10.1182/bloodadvances.2020001499

Al-Ali HK, Jaekel N, Junghanss C et al (2012) Azacitidine in patients with acute myeloid leukemia medically unfit for or resistant to chemotherapy: a multicenter phase I/II study. Leuk Lymphoma 53(1):110–117

Alblooshi R, Kanfar S, Lord B et al (2021) Clinical prevalence and outcome of cardiovascular events in the first 100 days postallogeneic hematopoietic stem cell transplant. Eur J Haematol 106(1):32–39. https://doi.org/10.1111/ejh.13482

Al-Homsi AS, Cirrone F, Wo S et al (2023) PTCy, abatacept, and a short course of tacrolimus for GVHD prevention after haploidentical transplantation. Blood Adv 7(14):3604–3611. https://doi.org/10.1182/bloodadvances.2023010545

Ali H, Tsai NC, Synold T et al (2022) Peritransplantation ruxolitinib administration is safe and effective in patients with myelofibrosis: a pilot open-label study. Blood Adv. 6(5):1444–1453. https://doi.org/10.1182/bloodadvances.2021005035

Alousi AM, Le-Rademacher J, Saliba RM et al (2013) Who is the better donor for older hematopoietic transplant recipients: an older-aged sibling or a young, matched unrelated volunteer? Blood 121(13):2567–2573. https://doi.org/10.1182/blood-2012-08-453860

Ando M, Ando Y, Hasegawa Y et al (2001) Prognostic value of performance status assessed by patients themselves, nurses, and oncologists in advanced non-small cell lung cancer. Br J Cancer 85(11):1634–1639. https://doi.org/10.1054/bjoc.2001.2162

Andronesi A, Sorohan B, Burcea A et al (2022) Incidence and risk factors for acute kidney injury after allogeneic stem cell transplantation: a prospective study. Biomedicines 10(2). https://doi.org/10.3390/biomedicines10020262

Armenian SH, Horak D, Scott JM et al (2017) Cardiovascular function in long-term hematopoietic cell transplantation survivors. Biol Blood Marrow Transplant 23(4):700–705. https://doi.org/10.1016/j.bbmt.2017.01.006

Artz AS (2016) Biologic vs physiologic age in the transplant candidate. Hematology Am Soc Hematol Educ Program 2016(1):99–105. https://doi.org/10.1182/asheducation-2016.1.99

Artz A, Logan BR, Saber W et al (2023) The Composite Health Risk Assessment Model (CHARM) to predict 1-year Non-Relapse Mortality (NRM) among older recipients of allogeneic transplantation: a prospective BMT-CTN Study 1704. Blood 142(Supplement 1):109–109. https://doi.org/10.1182/blood-2023-173855

Auletta JJ, Kou J, Chen M et al (2023) Real-world data showing trends and outcomes by race and ethnicity in allogeneic hematopoietic cell transplantation: a report from the center for international blood and marrow transplant research. Transplant Cell Ther 29(6):346.e1–346.e10. https://doi.org/10.1016/j.jtct.2023.03.007

Backhaus D, Brauer D, Pointner R et al (2023) A high hematopoietic cell transplantation comorbidity Index (HCT-CI) does not impair outcomes after non-myeloablative allogeneic stem cell transplantation in acute myeloid leukemia patients 60 years or older. Bone Marrow Transplant 58(1):30–38. https://doi.org/10.1038/s41409-022-01833-0

Baron F, Labopin M, Peniket A et al (2015) Reduced-intensity conditioning with fludarabine and busulfan versus fludarabine and melphalan for patients with acute myeloid leukemia: A report from the A cute L eukemia W orking P arty of the E uropean G roup for B lood and M arrow T ransplantation. Cancer 121(7):1048–1055

Bazarbachi A, Labopin M, Moukalled N et al (2023) How much have post-transplant outcomes improved since 2000 for older patients with acute myeloid leukemia? A study from the EBMT acute leukemia working party. Blood 142:896. https://doi.org/10.1182/blood-2023-179489

Bejar R, Stevenson KE, Caughey B et al (2014) Somatic mutations predict poor outcome in patients with myelodysplastic syndrome after hematopoietic stem-cell transplantation. J Clin Oncol 32(25):2691

Bhatt VR, Wang T, Chen K et al (2022) Chronic graft-versus-host disease, nonrelapse mortality, and disease relapse in older versus younger adults undergoing matched allogeneic peripheral blood hematopoietic cell transplantation: a center for international blood and marrow transplant research analysis. Transplant Cell Ther. 28(1):34–42. https://doi.org/10.1016/j.jtct.2021.10.002

Blaise DP, Michel Boiron J, Faucher C et al (2005) Reduced intensity conditioning prior to allogeneic stem cell transplantation for patients with acute myeloblastic leukemia as a first-line treatment. Cancer 104(9):1931–1938. https://doi.org/10.1002/cncr.21418

Boddu PC, Kadia TM, Garcia-Manero G et al (2019) Validation of the 2017 European LeukemiaNet classification for acute myeloid leukemia with NPM1 and FLT3-internal tandem duplication genotypes. Cancer 125(7):1091–1100. https://doi.org/10.1002/cncr.31885

Bodelon C, Savage SA, Gadalla SM (2014) Telomeres in molecular epidemiology studies. Prog Mol Biol Transl Sci 125:113–131

Boissel N, Huguet F, Graux C et al (2021) Frontline consolidation with blinatumomab for high-risk philadelphia-negative acute lymphoblastic adult patients. Early results from the Graall-2014-QUEST Phase 2. Blood 138:1232

Bolaños-Meade J, Hamadani M, Wu J et al (2023) Post-transplantation cyclophosphamide-based graft-versus-host disease prophylaxis. N Engl J Med. 388(25):2338–2348. https://doi.org/10.1056/NEJMoa2215943

Brissot E, Labopin M, Moiseev I et al (2020) Post-transplant cyclophosphamide versus antithymocyte globulin in patients with acute myeloid leukemia in first complete remission undergoing allogeneic stem cell transplantation from 10/10 HLA-matched unrelated donors. J Hematol Oncol 13:1–10

Broglie L, Friend BD, Chhabra S et al (2023) Expanded HCT-CI definitions capture comorbidity better for younger patients of allogeneic HCT for nonmalignant diseases. Transplant Cell Ther 29(2):125.e1–125.e9

Buckstein R, Wells RA, Zhu N et al (2016) Patient-related factors independently impact overall survival in patients with myelodysplastic syndromes: an MDS-CAN prospective study. Br J Haematol 174(1):88–101. https://doi.org/10.1111/bjh.14033

Cao XY, Chen JQ, Wang H et al (2023) Addition of venetoclax to myeloablative conditioning regimens for allogeneic hematopoietic stem cell transplantation in high-risk AML. Ann Med 55(1):388–400. https://doi.org/10.1080/07853890.2022.2164610

Cashen AF, Schiller GJ, O'Donnell MR, DiPersio JF (2010) Multicenter, phase II study of decitabine for the first-line treatment of older patients with acute myeloid leukemia. J Clin Oncol 28(4):556–561

Church S, Rogers E, Rockwood K, Theou O (2020) A scoping review of the clinical frailty scale. BMC Geriatr 20(1):393. https://doi.org/10.1186/s12877-020-01801-7

Ciurea SO, Zhang MJ, Bacigalupo AA et al (2015) Haploidentical transplant with posttransplant cyclophosphamide vs matched unrelated donor transplant for acute myeloid leukemia. Blood. 126(8):1033–1040. https://doi.org/10.1182/blood-2015-04-639831

Ciurea SO, Kongtim P, Varma A et al (2020) Is there an optimal conditioning for older patients with AML receiving allogeneic hematopoietic cell transplantation? Blood 135(6):449–452. https://doi.org/10.1182/blood.2019003662

Cortes JE, Heidel FH, Hellmann A et al (2019) Randomized comparison of low dose cytarabine with or without glasdegib in patients with newly diagnosed acute myeloid leukemia or high-risk myelodysplastic syndrome. Leukemia 33(2):379–389. https://doi.org/10.1038/s41375-018-0312-9

Coulson AB, Royle K-L, Pawlyn C et al (2022) Frailty-adjusted therapy in Transplant Non-Eligible patients with newly diagnosed Multiple Myeloma (FiTNEss (UK-MRA Myeloma XIV Trial)): a study protocol for a randomised phase III trial. BMJ Open 12(6):e056147. https://doi.org/10.1136/bmjopen-2021-056147

Cui K, Zhang S, Li J (2023) Risk factors for acute kidney injury after hematopoietic stem cell transplantation: a systematic review and meta-analysis. Blood Purif 52(11–12):917–928. https://doi.org/10.1159/000533945

Cusatis R, Litovich C, Feng Z et al (2024) Current trends and outcomes in cellular therapy activity in the United States, including prospective patient-reported outcomes data collection in the center for international blood and marrow transplant research registry. Transplant Cell Ther. 30:917.e1. https://doi.org/10.1016/j.jtct.2024.06.021

Cutler C, Lee SJ, Arai S et al (2021) Belumosudil for chronic graft-versus-host disease after 2 or more prior lines of therapy: the ROCKstar Study. Blood 138(22):2278–2289. https://doi.org/10.1182/blood.2021012021

D'Mello MJJ, Ross SA, Briel M, Anand SS, Gerstein H, Paré G (2015) Association between shortened leukocyte telomere length and cardiometabolic outcomes. Circ Cardiovasc Genetics 8(1):82–90. https://doi.org/10.1161/CIRCGENETICS.113.000485

Dale W (2022) Why is geriatric assessment so infrequently used in oncology practices? The ongoing issue of nonadherence to this standard of care for older adults with cancer. JCO Oncol Pract (Wolters Kluwer Health) 18:475–477

Dale W, Williams GR, MacKenzie AR et al (2021) How is geriatric assessment used in clinical practice for older adults with cancer? A survey of cancer providers by the American Society of Clinical Oncology. JCO Oncol Pract 17(6):336–344

Damlaj M, Alkhateeb HB, Hefazi M et al (2016) Fludarabine-busulfan reduced-intensity conditioning in comparison with fludarabine-melphalan is associated with increased relapse risk in spite of pharmacokinetic dosing. Biol Blood Marrow Transplant 22(8):1431–1439

de Haan G, Lazare SS (2018) Aging of hematopoietic stem cells. Blood. 131(5):479–487

de Lima M, Anagnostopoulos A, Munsell M et al (2004) Nonablative versus reduced-intensity conditioning regimens in the treatment of acute myeloid leukemia and high-risk myelodysplastic syndrome: dose is relevant for long-term disease control after allogeneic hematopoietic stem cell transplantation. Blood. 104(3):865–872. https://doi.org/10.1182/blood-2003-11-3750

DeAngelo D, Stevenson K, Dahlberg S et al (2015) Long-term outcome of a pediatric-inspired regimen used for adults aged 18–50 years with newly diagnosed acute lymphoblastic leukemia. Leukemia 29(3):526–534

DeFilipp Z, Ciurea SO, Cutler C et al (2023a) Hematopoietic cell transplantation in the management of myelodysplastic syndrome: an evidence-based review from the American Society for Transplantation and Cellular Therapy Committee on Practice Guidelines. Transplant Cell Ther 29(2):71–81. https://doi.org/10.1016/j.jtct.2022.11.014

Defilipp Z, Kim HT, Knight L et al (2023b) Maintenance ruxolitinib is associated with low rates of chronic Gvhd requiring systemic treatment in older adults with AML/MDS undergoing allogeneic HCT: results of a phase II multicenter trial. Blood 142(Supplement 1):653–653. https://doi.org/10.1182/blood-2023-180423

Devine SM, Owzar K, Blum W et al (2015) Phase II study of allogeneic transplantation for older patients with acute myeloid leukemia in first complete remission using a reduced-intensity conditioning regimen: results from Cancer and Leukemia Group B 100103 (Alliance for Clinical Trials in Oncology)/Blood and Marrow Transplant Clinical Trial Network 0502. J Clin Oncol 33(35):4167

Devlin R, Gupta V (2016) Myelofibrosis: to transplant or not to transplant? Hematology Am Soc Hematol Educ Program 2016(1):543–551. https://doi.org/10.1182/asheducation-2016.1.543

DeZern AE, Gondek LP (2020) Stem cell donors should be screened for CHIP. Blood Adv. 4(4):784–788. https://doi.org/10.1182/bloodadvances.2019000394

Diamond HR, Ornellas MH, Orfao A et al (2011) Acute myeloid leukemia of donor origin after allogeneic stem cell transplantation from a sibling who harbors germline XPD and XRCC3 homozygous polymorphisms. J Hematol Oncol 4:1–8

DiNardo CD, Pratz KW, Letai A et al (2018) Safety and preliminary efficacy of venetoclax with decitabine or azacitidine in elderly patients with previously untreated acute myeloid leukaemia: a non-randomised, open-label, phase 1b study. Lancet Oncol 19(2):216–228

DiNardo CD, Jonas BA, Pullarkat V et al (2020) Azacitidine and venetoclax in previously untreated acute myeloid leukemia. N Engl J Med 383(7):617–629. https://doi.org/10.1056/NEJMoa2012971

Ding H, Chen C, Shaffer JR et al (2012) Telomere length and risk of stroke in Chinese. Stroke 43(3):658–663

Dlouha D, Maluskova J, Lesna IK, Lanska V, Hubacek J (2014) Comparison of the relative telomere length measured in leukocytes and eleven different human tissues. Physiol Res 63:S343

Döhner H, Estey E, Grimwade D et al (2017) Diagnosis and management of AML in adults: 2017 ELN recommendations from an international expert panel. Blood 129(4):424–447. https://doi.org/10.1182/blood-2016-08-733196

Dombret H, Seymour JF, Butrym A et al (2015) International phase 3 study of azacitidine vs conventional care regimens in older patients with newly diagnosed AML with> 30% blasts. Blood 126(3):291–299

Dombret H, Topp MS, Schuh AC et al (2019) Blinatumomab versus chemotherapy in first salvage or in later salvage for B-cell precursor acute lymphoblastic leukemia. Leuk Lymphoma 60(9):2214–2222

Dwabe S, Martinez H, Rodman J et al (2021) Evaluating outcomes using post-transplant cyclophosphamide/tacrolimus/mycophenolate mofetil as gvhd prophylaxis compared with methotrexate/tacrolimus in matched-related and matched-unrelated hematopoietic stem cell transplant in relation to the haploidentical setting in acute lymphoblastic leukemia: a single center experience. Blood 138:4874. https://doi.org/10.1182/blood-2021-149165

Elias HK, Bryder D, Park CY (2017) Molecular mechanisms underlying lineage bias in aging hematopoiesis. Semin Hematol (Elsevier) 54:4–11

Flannelly C, Tan BEX, Tan JL et al (2020) Barriers to hematopoietic cell transplantation for adults in the United States: a systematic review with a focus on age. Biol Blood Marrow Transplant 26(12):2335–2345. https://doi.org/10.1016/j.bbmt.2020.09.013

Flowers ME, Inamoto Y, Carpenter PA et al (2011) Comparative analysis of risk factors for acute graft-versus-host disease and for chronic graft-versus-host disease according to National Institutes of Health consensus criteria. Blood 117(11):3214–3219

Frey NV, Shaw PA, Hexner EO et al (2020) Optimizing chimeric antigen receptor T-cell therapy for adults with acute lymphoblastic leukemia. J Clin Oncol. 38(5):415–422. https://doi.org/10.1200/JCO.19.01892

Freyer CW, Babushok DV, Frey NV et al (2022) Low-dose total body irradiation added to fludarabine and busulfan reduced-intensity conditioning reduces graft

failure in patients with myelofibrosis. Transplant Cell Ther 28(9):590–596. https://doi.org/10.1016/j.jtct.2022.06.018
Frick M, Chan W, Arends CM et al (2019) Role of donor clonal hematopoiesis in allogeneic hematopoietic stem-cell transplantation. J Clin Oncol 37(5):375–385
Fried LP, Tangen CM, Walston J et al (2001) Frailty in older adults: evidence for a phenotype. J Gerontol A Biol Sci Med Sci 56(3):M146–M156. https://doi.org/10.1093/gerona/56.3.m146
Fuchs EJ, McCurdy SR, Solomon SR et al (2022) HLA informs risk predictions after haploidentical stem cell transplantation with posttransplantation cyclophosphamide. Blood 139(10):1452–1468. https://doi.org/10.1182/blood.2021013443
Gajra A, Jeune-Smith Y, Fortier S et al (2022) The use and knowledge of validated geriatric assessment instruments among US community oncologists. JCO Oncol Pract 18(7):e1081–e1090
Gale RP, Bortin MM, van Bekkum DW et al (1987) Risk factors for acute graft-versus-host disease. Br J Haematol 67(4):397–406. https://doi.org/10.1111/j.1365-2141.1987.tb06160.x
Gambella M, Bregante S, Raiola AM et al (2023) Haploidentical hematopoietic cell transplantation for myelofibrosis in the ruxolitinib era. Transplant Cell Therapy 29(1):49.e1–49.e7. https://doi.org/10.1016/j.jtct.2022.10.015
Gangat N, Ilyas R, Johnson IM et al (2023) Outcome of patients with acute myeloid leukemia following failure of frontline venetoclax plus hypomethylating agent therapy. Haematologica 108(11):3170–3174. https://doi.org/10.3324/haematol.2022.282677
Garcia JS, Kim HT, Murdock HM et al (2021) Adding venetoclax to fludarabine/busulfan RIC transplant for high-risk MDS and AML is feasible, safe, and active. Blood Adv 5(24):5536–5545. https://doi.org/10.1182/bloodadvances.2021005566
García-Cadenas I, Redondo S, Esquirol A et al (2023) Successful outcome in patients with myelofibrosis undergoing allogeneic donor hematopoietic cell transplantation using reduced doses of post-transplantation cyclophosphamide: challenges and review of the literature. Transplant Cell Ther 29(7):473.e1–473.e6. https://doi.org/10.1016/j.jtct.2023.04.008
Genovese G, Kähler AK, Handsaker RE et al (2014) Clonal hematopoiesis and blood-cancer risk inferred from blood DNA sequence. N Engl J Med 371(26):2477–2487
Ghobadi A, Slade M, Kantarjian H et al (2022) The role of allogeneic transplant for adult Ph+ ALL in CR1 with complete molecular remission: a retrospective analysis. Blood 140(20):2101–2112. https://doi.org/10.1182/blood.2022016194
Gibson CJ, Kennedy JA, Nikiforow S et al (2017) Donor-engrafted CHIP is common among stem cell transplant recipients with unexplained cytopenias. Blood 130(1):91–94. https://doi.org/10.1182/blood-2017-01-764951
Gibson CJ, Kim HT, Zhao L et al (2022) Donor clonal hematopoiesis and recipient outcomes after transplantation. J Clin Oncol 40(2):189–201. https://doi.org/10.1200/JCO.21.02286
Goekbuget N, Viardot A, Steffen B et al (2022) Outcome of 841 older patients (> 55 yrs) with newly diagnosed Ph/BCR-ABL negative ALL prospectively treated according to pediatric-based, age-adapted GMALL protocols. Blood 140(Supplement 1):121–123
Gökbuget N, Kneba M, Raff T et al (2012) Adult patients with acute lymphoblastic leukemia and molecular failure display a poor prognosis and are candidates for stem cell transplantation and targeted therapies. Blood 120(9):1868–1876. https://doi.org/10.1182/blood-2011-09-377713
Gökbuget N, Dombret H, Bonifacio M et al (2018) Blinatumomab for minimal residual disease in adults with B-cell precursor acute lymphoblastic leukemia. Blood 131(14):1522–1531. https://doi.org/10.1182/blood-2017-08-798322
Gökbuget N, Zugmaier G, Dombret H et al (2020) Curative outcomes following blinatumomab in adults with minimal residual disease B-cell precursor acute lymphoblastic leukemia. Leuk Lymphoma 61(11):2665–2673
Goldstone AH, Richards SM, Lazarus HM et al (2008) In adults with standard-risk acute lymphoblastic leukemia, the greatest benefit is achieved from a matched sibling allogeneic transplantation in first complete remission, and an autologous transplantation is less effective than conventional consolidation/maintenance chemotherapy in all patients: final results of the International ALL Trial (MRC UKALL XII/ECOG E2993). Blood. 111(4):1827–1833. https://doi.org/10.1182/blood-2007-10-116582
Gondek LP, Zheng G, Ghiaur G et al (2016) Donor cell leukemia arising from clonal hematopoiesis after bone marrow transplantation. Leukemia 30(9):1916–1920
Gowin K, Ballen K, Ahn KW et al (2020) Survival following allogeneic transplant in patients with myelofibrosis. Blood Adv 4(9):1965–1973. https://doi.org/10.1182/bloodadvances.2019001084
Gupta V, Tallman MS, He W et al (2010) Comparable survival after HLA-well-matched unrelated or matched sibling donor transplantation for acute myeloid leukemia in first remission with unfavorable cytogenetics at diagnosis. Blood 116(11):1839–1848
Gutiérrez-García G, Villarreal J, Garrote M et al (2020) Impact of severe acute kidney injury and chronic kidney disease on allogeneic hematopoietic cell transplant recipients: a retrospective single center analysis. Bone Marrow Transplant 55(7):1264–1271
Gyurkocza B, Storb R, Storer BE et al (2010) Nonmyeloablative allogeneic hematopoietic cell transplantation in patients with acute myeloid leukemia. J Clin Oncol. 28(17):2859–2867. https://doi.org/10.1200/jco.2009.27.1460
Hadjis AD, White GW, Catania C et al (2023) Frail and pre-frail older adults with acute myeloid leukemia have inferior survival due to higher rates of

disease-related deaths. Blood 142:376. https://doi.org/10.1182/blood-2023-188978

Hadjis AD, White GW, Gill SI et al (2024) Post-transplant frailty phenotype predicts survival. Transplant Cell Ther 30(2):S407–S408. https://doi.org/10.1016/j.jtct.2023.12.575

Hahn T, McCarthy PL Jr, Zhang M-J et al (2008) Risk factors for acute graft-versus-host disease after human leukocyte antigen–identical sibling transplants for adults with leukemia. J Clin Oncol 26(35):5728–5734

Hamaker M, Lund C, Te Molder M et al (2022) Geriatric assessment in the management of older patients with cancer–a systematic review (update). J Geriatr Oncol 13(6):761–777

Hamilton BK, Rybicki L, Abounader D et al (2017) Allogeneic hematopoietic cell transplantation for adult T cell acute lymphoblastic leukemia. Biol Blood Marrow Transplant 23(7):1117–1121. https://doi.org/10.1016/j.bbmt.2017.04.003

Hay KA, Gauthier J, Hirayama AV et al (2019) Factors associated with durable EFS in adult B-cell ALL patients achieving MRD-negative CR after CD19 CAR T-cell therapy. Blood. 133(15):1652–1663. https://doi.org/10.1182/blood-2018-11-883710

Hernández-Boluda JC, Pereira A, Kröger N et al (2021) Allogeneic hematopoietic cell transplantation in older myelofibrosis patients: a study of the chronic malignancies working party of EBMT and the Spanish Myelofibrosis Registry. Am J Hematol 96(10):1186–1194. https://doi.org/10.1002/ajh.26279

Herold S, Kuhn M, Stange T et al (2017) Donor cell leukemia: evidence for multiple preleukemic clones and parallel long term clonal evolution in donor and recipient. Leukemia 31(7):1637–1640

Hirsch P, Mamez A, Belhocine R et al (2016) Clonal history of a cord blood donor cell leukemia with prenatal somatic JAK2 V617F mutation. Leukemia 30(8):1756–1759

Hirschbühl K, Labopin M, Polge E et al (2023) Total body irradiation versus busulfan based intermediate intensity conditioning for stem cell transplantation in ALL patients >45 years—a registry-based study by the Acute Leukemia Working Party of the EBMT. Bone Marrow Transplant 58(8):874–880. https://doi.org/10.1038/s41409-023-01966-w

Hodder A, Mishra AK, Enshaei A et al (2024) Blinatumomab for first-line treatment of children and young persons with B-ALL. J Clin Oncol. 42(8):907–914. https://doi.org/10.1200/jco.23.01392

Holmes HM, Des Bordes JK, Kebriaei P et al (2014) Optimal screening for geriatric assessment in older allogeneic hematopoietic cell transplantation candidates. J Geriatr Oncol 5(4):422–430. https://doi.org/10.1016/j.jgo.2014.04.004

Holtan SG, Martens MJ, Al Malki MM et al (2023) Patient-reported outcomes of BMT CTN 1703: a randomized phase III study for Gvhd Prophylaxis – a quality of life evaluation. Blood 142(Supplement 1):897–897. https://doi.org/10.1182/blood-2023-187859

Hourigan CS, Dillon LW, Gui G et al (2020) Impact of conditioning intensity of allogeneic transplantation for acute myeloid leukemia with genomic evidence of residual disease. J Clin Oncol. 38(12):1273–1283. https://doi.org/10.1200/jco.19.03011

Huang LW, Sheng Y, Andreadis C et al (2020) Functional status as measured by geriatric assessment predicts inferior survival in older allogeneic hematopoietic cell transplantation recipients. Biol Blood Marrow Transplant 26(1):189–196. https://doi.org/10.1016/j.bbmt.2019.08.022

Huguet F, Leguay T, Raffoux E et al (2009) Pediatric-inspired therapy in adults with Philadelphia chromosome–negative acute lymphoblastic leukemia: the GRAALL-2003 study. J Clin Oncol 27(6):911–918

Imus PH, Blackford AL, Bettinotti M et al (2019) Severe cytokine release syndrome after haploidentical peripheral blood stem cell transplantation. Biol Blood Marrow Transplant 25(12):2431–2437. https://doi.org/10.1016/j.bbmt.2019.07.027

Imus PH, Pasca S, Tsai HL et al (2024) Recipient clonal hematopoiesis in allogeneic bone marrow transplantation for lymphoid malignancies. Blood Adv 8(14):3849–3858. https://doi.org/10.1182/bloodadvances.2023011761

Jaiswal S, Fontanillas P, Flannick J et al (2014) Age-related clonal hematopoiesis associated with adverse outcomes. N Engl J Med 371(26):2488–2498

Jaiswal S, Natarajan P, Silver AJ et al (2017) Clonal hematopoiesis and risk of atherosclerotic cardiovascular disease. N Engl J Med 377(2):111–121

Johnson IM, Bezerra ED, Farrukh F et al (2022) Cardiac events in patients with acute myeloid leukemia treated with venetoclax combined with hypomethylating agents. Blood Adv 6(17):5227–5231. https://doi.org/10.1182/bloodadvances.2022007333

Jones LW, Watson D, Herndon JE et al (2010) Peak oxygen consumption and long-term all-cause mortality in nonsmall cell lung cancer. Cancer 116(20):4825–4832

Jones LW, Courneya KS, Mackey JR et al (2012) Cardiopulmonary function and age-related decline across the breast cancer survivorship continuum. J Clin Oncol 30(20):2530–2537

Jung HW, Kim S, Jang IY, Shin DW, Lee JE, Won CW (2020) Screening value of timed up and go test for frailty and low physical performance in Korean older population: the Korean Frailty and Aging Cohort Study (KFACS). Ann Geriatr Med Res 24(4):259–266. https://doi.org/10.4235/agmr.20.0072

Kanate AS, Majhail NS, Savani BN et al (2020) Indications for hematopoietic cell transplantation and immune effector cell therapy: guidelines from the American Society for Transplantation and Cellular Therapy. Biol Blood Marrow Transplant 26(7):1247–1256. https://doi.org/10.1016/j.bbmt.2020.03.002

Kanduri SR, Cheungpasitporn W, Thongprayoon C et al (2020) Incidence and mortality of acute kidney injury in patients undergoing hematopoietic stem cell transplantation: a systematic review and meta-analysis. QJM 113(9):621–632. https://doi.org/10.1093/qjmed/hcaa072

Kantarjian HM, Thomas XG, Dmoszynska A et al (2012) Multicenter, randomized, open-label, phase III trial of decitabine versus patient choice, with physician advice, of either supportive care or low-dose cytarabine for the treatment of older patients with newly diagnosed acute myeloid leukemia. J Clin Oncol 30(21):2670–2677

Kantarjian H, Thomas D, Jorgensen J et al (2013) Results of inotuzumab ozogamicin, a CD22 monoclonal antibody, in refractory and relapsed acute lymphocytic leukemia. Cancer 119(15):2728–2736

Kantarjian HM, DeAngelo DJ, Stelljes M et al (2016) Inotuzumab ozogamicin versus standard therapy for acute lymphoblastic leukemia. N Engl J Med 375(8):740–753. https://doi.org/10.1056/NEJMoa1509277

Kantarjian H, Stein A, Gökbuget N et al (2017) Blinatumomab versus chemotherapy for advanced acute lymphoblastic leukemia. N Engl J Med 376(9):836–847. https://doi.org/10.1056/NEJMoa1609783

Karnofsky DA, Abelmann WH, Craver LF, Burchenal JH (1948) The use of the nitrogen mustards in the palliative treatment of carcinoma: with particular reference to bronchogenic carcinoma. Cancer 1:634

Kasamon YL, Bolaños-Meade J, Prince GT et al (2015) Outcomes of nonmyeloablative HLA-haploidentical blood or marrow transplantation with high-dose post-transplantation cyclophosphamide in older adults. J Clin Oncol 33(28):3152–3161. https://doi.org/10.1200/jco.2014.60.4777

Kashyap R, Khalid F, Srivastav S, Malik S, Lee PC-H, Khan Z (2022) Abatacept as salvage therapy for graft-versus-host disease in allogenic hematopoietic stem cell transplantation : a systematic review. Blood 140(Supplement 1):12882–12883. https://doi.org/10.1182/blood-2022-168445

Kawamura K, Kako S, Mizuta S et al (2017) Comparison of conditioning with fludarabine/busulfan and fludarabine/melphalan in allogeneic transplantation recipients 50 years or older. Biol Blood Marrow Transplant 23(12):2079–2087

Kelly CM, Shahrokni A (2016) Moving beyond Karnofsky and ECOG performance status assessments with new technologies. J Oncol 2016:6186543. https://doi.org/10.1155/2016/6186543

Kitko CL, Arora M, DeFilipp Z et al (2023) Axatilimab for chronic graft-versus-host disease after failure of at least two prior systemic therapies: results of a phase I/II study. J Clin Oncol 41(10):1864–1875. https://doi.org/10.1200/jco.22.00958

Klepin HD (2019) Ready for prime time: role for geriatric assessment to improve quality of care in hematology practice. Hematology Am Soc Hematol Educ Program 2019(1):53–58. https://doi.org/10.1182/hematology.2019001299

Kneis S, Straub E, Walz ID et al (2020) Gait analysis of patients after allogeneic hematopoietic cell transplantation reveals impairments of functional performance. Integr Cancer Ther 19:1534735420915782. https://doi.org/10.1177/1534735420915782

Kobayashi S, Kobayashi A, Osawa Y et al (2017) Donor cell leukemia arising from preleukemic clones with a novel germline DDX41 mutation after allogenic hematopoietic stem cell transplantation. Leukemia 31(4):1020–1022

Kogon A, Hingorani S (2010) Acute kidney injury in hematopoietic cell transplantation. Semin Nephrol (Elsevier) 30:615–626

Kohrt HE, Turnbull BB, Heydari K et al (2009) TLI and ATG conditioning with low risk of graft-versus-host disease retains antitumor reactions after allogeneic hematopoietic cell transplantation from related and unrelated donors. Blood 114(5):1099–1109. https://doi.org/10.1182/blood-2009-03-211441

Kollman C, Howe CW, Anasetti C et al (2001) Donor characteristics as risk factors in recipients after transplantation of bone marrow from unrelated donors: the effect of donor age. Blood 98(7):2043–2051

Koshy AG, Kim HT, Liegel J et al (2023) Phase 2 clinical trial evaluating abatacept in patients with steroid-refractory chronic graft-versus-host disease. Blood. 141(24):2932–2943. https://doi.org/10.1182/blood.2022019107

Kröger N, Iacobelli S, Franke G-N et al (2017) Dose-reduced versus standard conditioning followed by allogeneic stem-cell transplantation for patients with myelodysplastic syndrome: a prospective randomized phase III study of the EBMT (RICMAC Trial). J Clin Oncol 35(19):2157–2164. https://doi.org/10.1200/jco.2016.70.7349

Kröger N, Shahnaz Syed Abd Kadir S, Zabelina T et al (2018) Peritransplantation ruxolitinib prevents acute graft-versus-host disease in patients with myelofibrosis undergoing allogenic stem cell transplantation. Biol Blood Marrow Transplant 24(10):2152–2156. https://doi.org/10.1016/j.bbmt.2018.05.023

Kunte S, Rybicki L, Viswabandya A et al (2022) Allogeneic blood or marrow transplantation with haploidentical donor and post-transplantation cyclophosphamide in patients with myelofibrosis: a multicenter study. Leukemia 36(3):856–864. https://doi.org/10.1038/s41375-021-01449-1

Leonard JT, Hayes-Lattin B (2018) Reduced intensity conditioning allogeneic hematopoietic stem cell transplantation for acute lymphoblastic leukemia; current evidence, and improving outcomes going forward. Curr Hematol Malig Rep 13(4):329–340. https://doi.org/10.1007/s11899-018-0462-x

Li A, Wu QV, Davis C et al (2018) The Association between transplant-associated thrombotic microangiopathy and calcineurin inhibitor and sirolimus levels. Blood 132(Supplement 1):814–814. https://doi.org/10.1182/blood-2018-99-110400

Li D, Sun C-L, Kim H et al (2021) Geriatric assessment–driven intervention (GAIN) on chemotherapy-related toxic effects in older adults with cancer: a randomized clinical trial. JAMA Oncol 7(11):e214158

Liang EC, Craig J, Torelli S et al (2022) Allogeneic hematopoietic cell transplantation for adult acute lymphoblastic leukemia in the modern era. Transplant Cell Ther. 28(8):490–495. https://doi.org/10.1016/j.jtct.2022.05.010

Lin RJ, Shahrokni A, Dahi PB et al (2018) Pretransplant comprehensive geriatric assessment in hematopoietic cell transplantation: a single center experience. Bone Marrow Transplant 53(9):1184–1187. https://doi.org/10.1038/s41409-018-0151-3

Lin RJ, Elias HK, van den Brink MRM (2021) Immune reconstitution in the aging host: opportunities for mechanism-based therapy in allogeneic hematopoietic cell transplantation. Front Immunol 12:674093. https://doi.org/10.3389/fimmu.2021.674093

Liu MA, DuMontier C, Murillo A et al (2019) Gait speed, grip strength, and clinical outcomes in older patients with hematologic malignancies. Blood 134(4):374–382. https://doi.org/10.1182/blood.2019000758

Lopes J, Jorge S (2011) Acute kidney injury following HCT: incidence, risk factors and outcome. Bone Marrow Transplant 46(11):1399–1408

Lopes J, Jorge S, Neves M (2016) Acute kidney injury in HCT: an update. Bone Marrow Transplant 51(6):755–762

Lopez-Otin C, Blasco MA, Partridge L, Serrano M, Kroemer G (2013) The hallmarks of aging. Cell. 153(5):1194–1217. https://doi.org/10.1016/j.cell.2013.05.039

Lori SM, Masha K, Wendy S et al (2014) Geriatric assessment to predict survival in older allogeneic hematopoietic cell transplantation recipients. *Haematologica*. 99(8):1373–1379. https://doi.org/10.3324/haematol.2014.103655

Luznik L, Bolaños-Meade J, Zahurak M et al (2010) High-dose cyclophosphamide as single-agent, short-course prophylaxis of graft-versus-host disease. Blood 115(16):3224–3230. https://doi.org/10.1182/blood-2009-11-251595

Maiti A, Rausch CR, Cortes JE et al (2021) Outcomes of relapsed or refractory acute myeloid leukemia after frontline hypomethylating agent and venetoclax regimens. Haematologica. 106(3):894–898. https://doi.org/10.3324/haematol.2020.252569

Marks DI, Wang T, Pérez WS et al (2010) The outcome of full-intensity and reduced-intensity conditioning matched sibling or unrelated donor transplantation in adults with Philadelphia chromosome–negative acute lymphoblastic leukemia in first and second complete remission. Blood 116(3):366–374

Martinelli G, Boissel N, Chevallier P et al (2017) Complete hematologic and molecular response in adult patients with relapsed/refractory Philadelphia chromosome–positive B-precursor acute lymphoblastic leukemia following treatment with blinatumomab: results from a phase II, single-arm, multicenter study. J Clin Oncol 35(16):1795–1802

Mathisen MS, O'Brien S, Thomas D, Cortes J, Kantarjian H, Ravandi F (2011) Role of tyrosine kinase inhibitors in the management of Philadelphia chromosome-positive acute lymphoblastic leukemia. Curr Hematol Malig Rep 6(3):187–194. https://doi.org/10.1007/s11899-011-0093-y

McAdams MJ, Hyder M, Dimitrova D et al (2021) Phase I/II study of reduced dosing of Post-Transplantation Cyclophosphamide (PTCy) after HLA-haploidentical bone marrow transplantation. Blood 138(Supplement 1):101–101. https://doi.org/10.1182/blood-2021-146997

McClune BL, Weisdorf DJ, Pedersen TL et al (2010) Effect of age on outcome of reduced-intensity hematopoietic cell transplantation for older patients with acute myeloid leukemia in first complete remission or with myelodysplastic syndrome. J Clin Oncol. 28(11):1878–1887. https://doi.org/10.1200/JCO.2009.25.4821

McCurdy SR, Muth ST, Tsai HL et al (2018) Early fever after haploidentical bone marrow transplantation correlates with class II HLA-mismatching and myeloablation but not outcomes. Biol Blood Marrow Transplant 24(10):2056–2064. https://doi.org/10.1016/j.bbmt.2018.06.004

McCurdy SR, Hyder MA, Sabina R et al (2024) Interim results of a phase I/II trial of intermediate-dose post-transplantation cyclophosphamide after reduced intensity conditioned hla-mismatched bone marrow transplantation for older or unfit patients. Transplant Cell Ther 30(2):S2–S3. https://doi.org/10.1016/j.jtct.2023.12.021

McLornan D, Szydlo R, Koster L et al (2019) Myeloablative and reduced-intensity conditioned allogeneic hematopoietic stem cell transplantation in myelofibrosis: a retrospective study by the chronic malignancies Working Party of the European Society for Blood and Marrow Transplantation. Biol Blood Marrow Transplant 25(11):2167–2171. https://doi.org/10.1016/j.bbmt.2019.06.034

McSweeney PA, Niederwieser D, Shizuru JA et al (2001) Hematopoietic cell transplantation in older patients with hematologic malignancies: replacing high-dose cytotoxic therapy with graft-versus-tumor effects. Blood 97(11):3390–3400

Mehta RS, Saliba RM, Rondon G et al (2022) Post-transplantation cyclophosphamide versus tacrolimus and methotrexate graft-versus-host disease prophylaxis for HLA-matched donor transplantation. Transplant Cell Therapy 28(10):695.e1–695.e10. https://doi.org/10.1016/j.jtct.2022.07.021

Mehta RS, Marin D, Alousi A et al (2023) Haploidentical vs matched unrelated donors for patients with ALL: donor age matters more than donor type. Blood Adv. 7(8):1594–1603. https://doi.org/10.1182/bloodadvances.2022009240

Mehta RS, Ramdial J, Kebriaei P et al (2024) Haploidentical versus HLA-matched sibling donor HCT with PTCy prophylaxis: HLA factors and donor age considerations. Blood Adv 8:5306. https://doi.org/10.1182/bloodadvances.2024013853

Menezes MM, Marques AI, Chuva T et al (2022) Acute kidney injury after allogeneic hematopoietic stem cell transplantation – predictors and survival impact: a single center retrospective study. Nefrología (English Edition) 42(6):656–663. https://doi.org/10.1016/j.nefroe.2022.11.006

Meyer C, Mau L-W, Murphy EA et al (2019) Addressing knowledge gaps in acute myeloid leukemia to improve referral for hematopoietic cell transplantation consultation. J Natl Compr Cancer Netw 17(12):1473–1481

Mishra A, Preussler JM, Bhatt VR et al (2021) Breaking the age barrier: physicians' perceptions of candidacy for allogeneic hematopoietic cell transplantation in older adults. Transplant Cell Therapy 27(7):617.e1–617.e7. https://doi.org/10.1016/j.jtct.2021.03.028

Miyata M, Ichikawa K, Matsuki E, Watanabe M, Peltier D, Toubai T (2021) Recent advances of acute kidney injury in hematopoietic cell transplantation. Front Immunol 12:779881. https://doi.org/10.3389/fimmu.2021.779881

Mohile SG, Epstein RM, Hurria A et al (2020) Communication with older patients with cancer using geriatric assessment: a cluster-randomized clinical trial from the National Cancer Institute Community Oncology Research Program. JAMA Oncol 6(2):196–204

Mohile SG, Mohamed MR, Xu H et al (2021) Evaluation of geriatric assessment and management on the toxic effects of cancer treatment (GAP70+): a cluster-randomised study. Lancet 398(10314):1894–1904

Mohty M, Labopin M, Volin L et al (2010) Reduced-intensity versus conventional myeloablative conditioning allogeneic stem cell transplantation for patients with acute lymphoblastic leukemia: a retrospective study from the European Group for Blood and Marrow Transplantation. Blood 116(22):4439–4443

Muffly L, Pasquini MC, Martens M et al (2017) Increasing use of allogeneic hematopoietic cell transplantation in patients aged 70 years and older in the United States. Blood 130(9):1156–1164. https://doi.org/10.1182/blood-2017-03-772368

Murillo A, Cronin AM, Laubach JP et al (2019) Performance of the International Myeloma Working Group myeloma frailty score among patients 75 and older. J Geriatr Oncol 10(3):486–489. https://doi.org/10.1016/j.jgo.2018.10.010

Mushtaq MU, Shahzad M, Amin MK et al (2024) Outcomes with HLA-matched unrelated donor versus haploidentical hematopoietic cell transplantation. Leuk Lymphoma 65(4):493–502. https://doi.org/10.1080/10428194.2023.2300708

Myllymäki M, Redd R, Reilly CR et al (2020) Short telomere length predicts nonrelapse mortality after stem cell transplantation for myelodysplastic syndrome. Blood 136(26):3070–3081. https://doi.org/10.1182/blood.2020005397

Nademanee A, Schmidt GM, Parker P et al (1995) The outcome of matched unrelated donor bone marrow transplantation in patients with hematologic malignancies using molecular typing for donor selection and graft-versus-host disease prophylaxis regimen of cyclosporine, methotrexate, and prednisone. Blood 86:1228

Nakamura R, Saber W, Martens MJ et al (2021) Biologic assignment trial of reduced-intensity hematopoietic cell transplantation based on donor availability in patients 50-75 years of age with advanced myelodysplastic syndrome. J Clin Oncol 39(30):3328–3339. https://doi.org/10.1200/jco.20.03380

National Cancer Institute (2024) Bethesda M. SEER Cancer Stat Facts: Acute Myeloid Leukemia. Accessed May 21, 2024. https://seer.cancer.gov/statfacts/html/amyl.html

Newell LF, Williams T, Liu J et al (2021a) Engrafted donor-derived clonal hematopoiesis after allogenic hematopoietic cell transplantation is associated with chronic graft-versus-host disease requiring immunosuppressive therapy, but no adverse impact on overall survival or relapse. Transplant Cell Therapy 27(8):662.e1–662.e9

Newell LF, Williams T, Liu J et al (2021b) Engrafted donor-derived clonal hematopoiesis after allogenic hematopoietic cell transplantation is associated with chronic graft-versus-host disease requiring immunosuppressive therapy, but no adverse impact on overall survival or relapse. Transplant Cell Ther. 27(8):662.e1–662.e9. https://doi.org/10.1016/j.jtct.2021.04.014

Niederwieser C, Iacobelli S, Franke G-N et al (2024) Reduced intensity versus myeloablative conditioning for MDS: long-term results of an EBMT phase III study (RICMAC). Bone Marrow Transplant 59(8):1084–1091. https://doi.org/10.1038/s41409-024-02282-7

Oken MM, Creech RH, Tormey DC et al (1982) Toxicity and response criteria of the Eastern Cooperative Oncology Group. Am J Clin Oncol 5(6):649–656

Olin RL, Fretham C, Pasquini MC et al (2020) Geriatric assessment in older alloHCT recipients: association of functional and cognitive impairment with outcomes. Blood Adv. 4(12):2810–2820. https://doi.org/10.1182/bloodadvances.2020001719

Oran B, Ahn KW, Fretham C et al (2021) Fludarabine and melphalan compared with reduced doses of busulfan and fludarabine improve transplantation outcomes in older patients with myelodysplastic syndromes. Transplant Cell Therapy 27(11):921.e1–921.e10. https://doi.org/10.1016/j.jtct.2021.08.007

Parikh C, Coca S (2006) Acute renal failure in hematopoietic cell transplantation. Kidney Int 69(3):430–435

Parikh CR, Schrier RW, Storer B et al (2005) Comparison of ARF after myeloablative and nonmyeloablative hematopoietic cell transplantation. Am J Kidney Dis 45(3):502–509

Patriarca F, Bacigalupo A, Sperotto A et al (2008) Allogeneic hematopoietic stem cell transplantation in myelofibrosis: the 20-year experience of the Gruppo Italiano Trapianto di Midollo Osseo (GITMO). Haematologica 93(10):1514–1522. https://doi.org/10.3324/haematol.12828

Pattipaka T, Sarp S, Nakhaei P, Güneş S (2024) Ruxolitinib in patients with graft versus host disease (GvHD): findings from a compassionate use program. Bone Marrow Transplant 59(5):637–646. https://doi.org/10.1038/s41409-024-02207-4

Pereira MP, Remberger M, Chen C et al (2023) Choosing between older matched sibling donor and younger matched unrelated donor in allogeneic hematopoietic cell transplantation: comparison of clinical outcomes in acute myeloid leukemia and myelodysplastic syndrome. Transplant Cell Ther. 29(11):697.e1–697.e10. https://doi.org/10.1016/j.jtct.2023.08.009

Podsiadlo D, Richardson S (1991) The timed "up & go": a test of basic functional mobility for frail elderly persons. J Am Geriatr Soc 39(2):142–148. https://doi.org/10.1111/j.1532-5415.1991.tb01616.x

Pollyea DA, Pratz KW, Jonas BA et al (2018) Venetoclax in combination with hypomethylating agents induces rapid, deep, and durable responses in patients with AML ineligible for intensive therapy. Blood 132:285

Popat UR, Pasvolsky O, Bassett R Jr et al (2023) Myeloablative fractionated busulfan for allogeneic stem cell transplant in older patients or patients with comorbidities. Blood Adv 7(20):6196–6205. https://doi.org/10.1182/bloodadvances.2023010850

Pratz K, Dinardo C, Arellano ML et al (2023) P525: long-term outcomes of stem cell transplant in older patients with acute myeloid leukemia treated with venetoclax + HMA therapies. Hemasphere. 7(Suppl):LID – e68978fe. (2572-9241 (Electronic))

Raghunandan S, Qayed M, Watkins BK et al (2023) Abatacept for graft versus host disease prophylaxis in patients 60 years and older receiving mismatched unrelated donor transplantation for hematologic malignancies. Bone Marrow Transplant 58(11):1264–1266. https://doi.org/10.1038/s41409-023-02043-y

Raj RV, Hamadani M, Szabo A et al (2018) Peripheral blood grafts for T cell–replete haploidentical transplantation increase the incidence and severity of cytokine release syndrome. Biol Blood Marrow Transplant 24(8):1664–1670. https://doi.org/10.1016/j.bbmt.2018.04.010

Remberger M, Kumlien G, Aschan J et al (2002) Risk factors for moderate-to-severe chronic graft-versus-host disease after allogeneic hematopoietic stem cell transplantation. Biol Blood Marrow Transplant 8(12):674–682. https://doi.org/10.1053/bbmt.2002.v8.abbmt080674

Richa E, Papari M, Allen J et al (2009) Older age but not donor health impairs allogeneic granulocyte colony-stimulating factor (G-CSF) peripheral blood stem cell mobilization. Biol Blood Marrow Transplant 15(11):1394–1399

Rockwood K, Song X, MacKnight C et al (2005) A global clinical measure of fitness and frailty in elderly people. CMAJ 173(5):489–495. https://doi.org/10.1503/cmaj.050051

Roila F, Lupattelli M, Sassi M et al (1991) Intra and interobserver variability in cancer patients' performance status assessed according to Karnofsky and ECOG scales. Ann Oncol 2(6):437–439

Roth-Guepin G, Canaani J, Ruggeri A et al (2017) Allogeneic stem cell transplantation in acute lymphoblastic leukemia patients older than 60 years: a survey from the acute leukemia working party of EBMT. Oncotarget 8(68):112972–112979. https://doi.org/10.18632/oncotarget.22934

Saliba RM, Arora M, Spellman SR et al (2020) Acute graft-versus-host disease is less severe and associated with lower non-relapse mortality after haploidentical transplantation with post-cyclophosphamide prophylaxis. Biol Blood Marrow Transplant 26(3):S26–S27

Santoro N, Labopin M, Giannotti F et al (2018) Unmanipulated haploidentical in comparison with matched unrelated donor stem cell transplantation in patients 60 years and older with acute myeloid leukemia: a comparative study on behalf of the ALWP of the EBMT. J Hematol Oncol. 11(1):55. https://doi.org/10.1186/s13045-018-0598-0

Saraceni F, Scortechini I, Fiorentini A et al (2021) Conditioning regimens for frail patients with acute leukemia undergoing allogeneic stem cell transplant: how to strike gently. Clin Hematol Int 3(4):153–160. https://doi.org/10.2991/chi.k.210731.001

Savva GM, Donoghue OA, Horgan F, O'Regan C, Cronin H, Kenny RA (2012) Using timed up-and-go to identify frail members of the older population. J Gerontol Ser A 68(4):441–446. https://doi.org/10.1093/gerona/gls190

Schetelig J, Baldauf H, Rave C et al (2023) Unrelated donors below 35 years of age should be preferred over HLA-matched sibling donors older than 50 years for patients with myeloid malignancies. Blood 142(Supplement 1):1046–1046. https://doi.org/10.1182/blood-2023-180531

Schlenk RF, Döhner K, Mack S et al (2010) Prospective evaluation of allogeneic hematopoietic stem-cell transplantation from matched related and matched unrelated donors in younger adults with high-risk acute myeloid leukemia: German-Austrian trial AMLHD98A. J Clin Oncol 28(30):4642–4648

Schrappe M, Valsecchi MG, Bartram CR et al (2011) Late MRD response determines relapse risk overall and in subsets of childhood T-cell ALL: results of the AIEOP-BFM-ALL 2000 study. Blood 118(8):2077–2084. https://doi.org/10.1182/blood-2011-03-338707

Scott BL, Pasquini MC, Logan BR et al (2017) Myeloablative versus reduced-intensity hematopoietic cell transplantation for acute myeloid leukemia and myelodysplastic syndromes. J Clin Oncol. 35(11):1154–1161. https://doi.org/10.1200/JCO.2016.70.7091

Scott BL, Pasquini MC, Fei M et al (2021) Myeloablative versus reduced-intensity conditioning for hematopoietic cell transplantation in acute myelogenous leukemia and myelodysplastic syndromes-long-term follow-up of the BMT CTN 0901 clinical trial. Transplant Cell Ther. 27(6):483.e1–483.e6. https://doi.org/10.1016/j.jtct.2021.02.031

Sekeres MA, Taylor J (2022) Diagnosis and treatment of myelodysplastic syndromes: a review. JAMA 328(9):872–880. https://doi.org/10.1001/jama.2022.14578

Shaw BE, Jimenez-Jimenez AM, Burns LJ et al (2021) National marrow donor program-sponsored multicenter, phase II trial of HLA-mismatched unrelated donor bone marrow transplantation using post-transplant cyclophosphamide. J Clin Oncol. 39(18):1971–1982. https://doi.org/10.1200/JCO.20.03502

Shimoni A, Hardan I, Shem-Tov N et al (2006) Allogeneic hematopoietic stem-cell transplantation in AML and MDS using myeloablative versus reduced-intensity conditioning: the role of dose intensity. Leukemia 20(2):322–328. https://doi.org/10.1038/sj.leu.2404037

Shingai N, Morito T, Najima Y et al (2015) Early-onset acute kidney injury is a poor prognostic sign for allogeneic SCT recipients. Bone Marrow Transplant 50(12):1557–1562

Short NJ, Jabbour E, Sasaki K et al (2016) Impact of complete molecular response on survival in patients with Philadelphia chromosome-positive acute lymphoblastic leukemia. Blood 128(4):504–507. https://doi.org/10.1182/blood-2016-03-707562

Soo WK, King MT, Pope A, Parente P, Dārziņš P, Davis ID (2022) Integrated Geriatric Assessment and Treatment Effectiveness (INTEGERATE) in older people with cancer starting systemic anticancer treatment in Australia: a multicentre, open-label, randomised controlled trial. Lancet Healthy Longevity 3(9):e617–e627

Sørensen J, Klee M, Palshof T, Hansen H (1993) Performance status assessment in cancer patients. An inter-observer variability study. Br J Cancer 67(4):773–775

Sorror ML, Maris MB, Storb R et al (2005) Hematopoietic cell transplantation (HCT)-specific comorbidity index: a new tool for risk assessment before allogeneic HCT. Blood 106(8):2912–2919. https://doi.org/10.1182/blood-2005-05-2004

Sorror ML, Storer B, Storb RF (2009) Validation of the hematopoietic cell transplantation-specific comorbidity index (HCT-CI) in single and multiple institutions: limitations and inferences. Biol Blood Marrow Transplant 15(6):757–758

Sorror ML, Martin PJ, Storb RF et al (2014) Pretransplant comorbidities predict severity of acute graft-versus-host disease and subsequent mortality. Blood 124(2):287–295. https://doi.org/10.1182/blood-2014-01-550566

Sperling AS, Gibson CJ, Ebert BL (2017) The genetics of myelodysplastic syndrome: from clonal haematopoiesis to secondary leukaemia. Nat Rev Cancer 17(1):5–19. https://doi.org/10.1038/nrc.2016.112

Srour SA, Saliba RM, Bittencourt MCB et al (2019) Haploidentical transplantation for acute myeloid leukemia patients with minimal/measurable residual disease at transplantation. Am J Hematol 94(12):1382–1387. https://doi.org/10.1002/ajh.25647

Srour SA, Kongtim P, Rondon G et al (2020) Haploidentical transplants for patients with relapse after the first allograft. Am J Hematol 95(10):1187–1192. https://doi.org/10.1002/ajh.25924

Stelljes M, Raffel S, Alakel N et al (2024) Inotuzumab ozogamicin as induction therapy for patients older than 55 years with Philadelphia Chromosome-Negative B-Precursor ALL. J Clin Oncol 42(3):273–282. https://doi.org/10.1200/jco.23.00546

Stock W, Luger SM, Advani AS et al (2014) Favorable outcomes for older Adolescents and Young Adults (AYA) with Acute Lymphoblastic Leukemia (ALL): early results of U.S. Intergroup Trial C10403. Blood 124(21):796–796. https://doi.org/10.1182/blood.V124.21.796.796

Suárez-González J, Martinez-Laperche C, Martínez N et al (2018) Whole-exome sequencing reveals acquisition of mutations leading to the onset of donor cell leukemia after hematopoietic transplantation: a model of leukemogenesis. Leukemia 32(8):1822–1826

Sung AD, Koll T, Gier SH et al (2024) Preconditioning frailty phenotype influences survival and relapse for older allogeneic transplantation recipients. Transplant Cell Ther 30(4):415.e1–415.e16. https://doi.org/10.1016/j.jtct.2024.01.062

Tanaka J, Kanamori H, Nishiwaki S et al (2013) Reduced-intensity vs myeloablative conditioning allogeneic hematopoietic SCT for patients aged over 45 years with ALL in remission: a study from the Adult ALL Working Group of the Japan Society for Hematopoietic Cell Transplantation (JSHCT). Bone Marrow Transplant 48(11):1389–1394

Tauro S, Craddock C, Peggs K et al (2005) Allogeneic stem-cell transplantation using a reduced-intensity conditioning regimen has the capacity to produce durable remissions and long-term disease-free survival in patients with high-risk acute myeloid leukemia and myelodysplasia. J Clin Oncol 23(36):9387–9393. https://doi.org/10.1200/jco.2005.02.0057

Tefferi A, Guglielmelli P, Lasho TL et al (2018) MIPSS70+ Version 2.0: mutation and karyotype-enhanced international prognostic scoring system for primary myelofibrosis. J Clin Oncol. 36(17):1769–1770. https://doi.org/10.1200/JCO.2018.78.9867

Tilly H, Castaigne S, Bordessoule D et al (1990) Low-dose cytarabine versus intensive chemotherapy in the treatment of acute nonlymphocytic leukemia in the elderly. J Clin Oncol 8(2):272–279. https://doi.org/10.1200/jco.1990.8.2.272

Tonzi P, Hsu J, Cole K et al (2024) Clinical burden of GvHD in patients receiving CAST for prophylaxis following peripheral blood haploidentical hematopoietic stem cell transplant. Transplant Cell Ther 30(2, Supplement):S274. https://doi.org/10.1016/j.jtct.2023.12.367

Topp MS, Gökbuget N, Zugmaier G et al (2014) Phase II trial of the anti-CD19 bispecific T cell–engager blinatumomab shows hematologic and molecular remissions in patients with relapsed or refractory B-precursor acute lymphoblastic leukemia. J Clin Oncol 32(36):4134–4140

Topp MS, Gökbuget N, Stein AS et al (2015) Safety and activity of blinatumomab for adult patients with relapsed or refractory B-precursor acute lymphoblastic leukaemia: a multicentre, single-arm, phase 2 study. Lancet Oncol 16(1):57–66

Topp MS, Stein AS, Gökbuget N et al (2021) Blinatumomab as first salvage versus second or later salvage in adults with relapsed/refractory B-cell precursor acute lymphoblastic leukemia: results of a pooled analysis. Cancer Med 10(8):2601–2610

Ustun C, Le-Rademacher J, Wang HL et al (2019) Allogeneic hematopoietic cell transplantation compared to chemotherapy consolidation in older acute myeloid leukemia (AML) patients 60-75 years in first complete remission (CR1): an alliance (A151509), SWOG, ECOG-ACRIN, and CIBMTR study. Leukemia 33(11):2599–2609. https://doi.org/10.1038/s41375-019-0477-x

Uziel O, Lahav M, Shargian L et al (2020) Premature ageing following allogeneic hematopoietic stem cell transplantation. Bone Marrow Transplant 55(7):1438–1446. https://doi.org/10.1038/s41409-020-0839-z

Vaiserman A, Krasnienkov D (2020) Telomere length as a marker of biological age: state-of-the-art, open issues, and future perspectives. Front Genet 11:630186. https://doi.org/10.3389/fgene.2020.630186

Valcárcel D, Martino R, Caballero D et al (2008) Sustained remissions of high-risk acute myeloid leukemia and myelodysplastic syndrome after reduced-intensity conditioning allogeneic hematopoietic transplantation: chronic graft-versus-host disease is the strongest factor improving survival. J Clin Oncol. 26(4):577–584. https://doi.org/10.1200/jco.2007.11.1641

van Besien K, Artz A, Smith S et al (2005) Fludarabine, melphalan, and alemtuzumab conditioning in adults with standard-risk advanced acute myeloid leukemia and myelodysplastic syndrome. J Clin Oncol 23(24):5728–5738. https://doi.org/10.1200/jco.2005.15.602

Vasbinder A, Hoeger CW, Catalan T et al (2023) Cardiovascular events after hematopoietic stem cell transplant: incidence and risk factors. JACC Cardio Oncol 5(6):821–832. https://doi.org/10.1016/j.jaccao.2023.07.007

Vasu S, Kohlschmidt J, Mrózek K et al (2018) Ten-year outcome of patients with acute myeloid leukemia not treated with allogeneic transplantation in first complete remission. Blood Adv 2(13):1645–1650. https://doi.org/10.1182/bloodadvances.2017015222

Verger E, Salamero M, Conill C (1992) Can Karnofsky performance status be transformed to the Eastern Cooperative Oncology Group scoring scale and vice versa? Eur J Cancer 28(8–9):1328–1330

Wachsmuth LP, Patterson MT, Eckhaus MA, Venzon DJ, Gress RE, Kanakry CG (2019) Post-transplantation cyclophosphamide prevents graft-versus-host disease by inducing alloreactive T cell dysfunction and suppression. J Clin Invest. 129(6):2357–2373. https://doi.org/10.1172/jci124218

Wachsmuth LP, Patterson MT, Eckhaus MA, Venzon DJ, Kanakry CG (2020) Optimized timing of post-transplantation cyclophosphamide in mhc-haploidentical murine hematopoietic cell transplantation. Biol Blood Marrow Transplant 26(2):230–241. https://doi.org/10.1016/j.bbmt.2019.09.030

Walter RB, Pagel JM, Gooley TA et al (2010) Comparison of matched unrelated and matched related donor myeloablative hematopoietic cell transplantation for adults with acute myeloid leukemia in first remission. Leukemia 24(7):1276–1282

Wang Q, Zhan Y, Pedersen NL, Fang F, Hägg S (2018) Telomere length and all-cause mortality: a meta-analysis. Ageing Res Rev 48:11–20

Watkins B, Qayed M, McCracken C et al (2021) Phase II trial of costimulation blockade with abatacept for prevention of acute GVHD. J Clin Oncol 39(17):1865–1877. https://doi.org/10.1200/jco.20.01086

Webster JA, Reed M, Tsai HL et al (2023) Allogeneic blood or marrow transplantation with high-dose post-transplantation cyclophosphamide for acute lymphoblastic leukemia in patients age ≥55 years. Transplant Cell Ther. 29(3):182.e1–182.e8. https://doi.org/10.1016/j.jtct.2022.12.018

Weisdorf D, Hakke R, Blazar B et al (1991) Risk factors for acute graft-versus-host disease in histocompatible donor bone marrow transplantation. Transplantation 51(6):1197–1202

Welford J, Rafferty R, Hunt K et al (2022) The Clinical Frailty Scale can indicate prognosis and care requirements on discharge in oncology and haemato-oncology inpatients: a cohort study. Eur J Cancer Care 31(6):e13752. https://doi.org/10.1111/ecc.13752

Wertheimer T, Dohse M, Afram G et al (2021) Abatacept as salvage therapy in chronic graft-versus-host disease—a retrospective analysis. Ann Hematol 100(3):779–787. https://doi.org/10.1007/s00277-021-04434-x

Williams GR, Hopkins JO, Klepin HD et al (2023) Practical assessment and management of vulnerabilities in older patients receiving systemic cancer therapy: ASCO guideline questions and answers. JCO Oncol Pract 19(9):718–723. https://doi.org/10.1200/op.23.00263

Winters AC, Bosma G, Abbott D et al (2022) Outcomes are similar after allogeneic hematopoietic stem cell transplant for newly diagnosed acute myeloid leukemia patients who received venetoclax + azacitidine versus intensive chemotherapy. Transplant Cell Ther 28(10):694.e1–694.e9. https://doi.org/10.1016/j.jtct.2022.07.022

Wood W, Deal A, Reeve B et al (2013) Cardiopulmonary fitness in patients undergoing hematopoietic SCT: a pilot study. Bone Marrow Transplant 48(10):1342–1349

Xiao H, Shi J, Luo Y et al (2011) First report of multiple CEBPA mutations contributing to donor origin of leukemia relapse after allogeneic hematopoietic stem cell transplantation. Blood 117(19):5257–5260

Xie M, Lu C, Wang J et al (2014) Age-related mutations associated with clonal hematopoietic expansion and malignancies. Nat Med 20(12):1472–1478. https://doi.org/10.1038/nm.3733

Xu X, Hu H, Lin Y et al (2019) Differences in leukocyte telomere length between coronary heart disease and normal population: a multipopulation meta-analysis. Biomed Res Int 2019(1):5046867

Yanada M, Matsuo K, Suzuki T, Naoe T (2006) Allogeneic hematopoietic stem cell transplantation as part of postremission therapy improves survival for adult patients with high-risk acute lymphoblastic leukemia. Cancer 106(12):2657–2663. https://doi.org/10.1002/cncr.21932

Zeiser R, Bubnoff NV, Butler J et al (2020a) Ruxolitinib for glucocorticoid-refractory acute graft-versus-host disease. N Engl J Med 382(19):1800–1810. https://doi.org/10.1056/NEJMoa1917635

Zeiser R, Polverelli N, Ram R et al (2020b) Ruxolitinib (RUX) vs best available therapy (BAT) in patients with steroid-refractory/steroid-dependent chronic graft-vs-host disease (cGVHD): primary findings from the phase 3, randomized REACH3 study. Blood 136:22–24

Zeiser R, Polverelli N, Ram R et al (2021) Ruxolitinib for glucocorticoid-refractory chronic graft-versus-host disease. N Engl J Med 385(3):228–238

Zhang H, Dong T, Liu Z et al (2023) Phase I/II study of reduced dosing of Post-Transplantation Cyclophosphamide (PTCy) after HLA-identical peripheral blood hematopoietic stem cell transplantation. Blood 142(Supplement 1):650–650. https://doi.org/10.1182/blood-2023-189430

Zhang X, Zhao X, Chen S et al (2024) Addition of ruxolitinib to standard graft-versus-host disease prophylaxis for allogeneic stem cell transplantation in aplastic anemia patients. Bone Marrow Transplant 59(7):997–1005. https://doi.org/10.1038/s41409-024-02266-7

Zheng X, Gao H, Lu N et al (2023) Efficacy of venetoclax combined with decitabine conditioning regimen for allogeneic hematopoietic stem cell transplantation in high-risk and elderly patients with myeloid neoplasms. Ann Hematol 102(12):3603–3611. https://doi.org/10.1007/s00277-023-05500-2

Conditioning Regimen Options

3

Zheng-Li Xu and Xiao-Jun Huang

Abstract

In allogeneic hematopoietic stem cell transplantation (allo-HSCT), the conditioning regimen plays a crucial role in the transplantation process, aiming to eliminate any leukemia cells and suppress the recipient's immune system to promote engraftment. Conditioning regimens are categorized into myeloablative (MAC), non-myeloablative (NMA), and reduced-intensity conditioning (RIC) based on their intensity. The choice of regimen depends on disease type, patient age, comorbidities, and donor source. The conditioning regimens vary between matched sibling donors and alternative donors. In haploidentical transplantation, the mainstream non-T cell depletion protocols included the granulocyte colony-stimulating factor (G-CSF)/antithymocyte globulin (ATG)-based and post-transplantation cyclophosphamide (PT-Cy)-based strategies. In hematologic nonmalignancies such as aplastic anemia, Fanconi anemia, and thalassemia, the tailored conditioning regimens are crucial for improving outcomes. The diversified and personalized conditioning regimens are the future direction in the field of transplantation. Future directions may include integrating new targeted therapies and immunotherapies into conditioning regimens to further optimize outcomes.

Keywords

Conditioning regimen · Intensity · G-CSF · ATG · PTCy

3.1 Introduction

Allogeneic hematopoietic stem cell transplantation (allo-HSCT) is a cornerstone treatment for various hematologic malignancies and disorders. The conditioning regimen is a critical component of the transplantation process, designed to eradicate residual malignant cells and suppress the recipient's immune system to prevent graft rejection, thereby facilitating donor cell engraftment. The selection of a conditioning regimen is tailored to the patient's specific condition, disease status, and other factors such as age, comorbidities, and the donor's HLA compatibility. The regimens can be myeloablative, using high doses

Z.-L. Xu · X.-J. Huang (✉)
Peking University People's Hospital, Peking University Institute of Hematology, National Clinical Research Center for Hematologic Disease, Beijing, China
e-mail: xjhrm@medmail.com.cn

© The Author(s), under exclusive license to Springer Nature Switzerland AG 2025

K. Ballen (ed.), *Clinical Management of Challenges in Allogeneic Stem Cell Transplant*, Advances in Experimental Medicine and Biology 1475, https://doi.org/10.1007/978-3-031-84988-6_3

of chemotherapy or radiation therapy, or reduced intensity, which are less toxic and often used in older patients or those with compromised health.

It is recommended to emphasize the importance of individualized conditioning regimens to maximize the therapeutic effect while minimizing toxicity and adverse effects, such as graft-versus-host disease (GVHD) and infections. The choice of conditioning regimen is a balance between achieving the best possible outcome and managing the risks associated with transplantation. Ongoing research aims to refine these regimens to improve patient outcomes and quality of life post-transplantation. In this chapter, we will introduce the intensity category of conditioning regimens and discuss the choice and optimization of conditioning regimens in malignant and non-malignant hematologic diseases.

3.2 The Category Based on the Intensity of Conditioning Regimens

Conditioning regimens are categorized based on their intensity, serving different purposes in the allo-HSCT. The myeloablative conditioning regimens (MAC) cause irreversible cytopenia, and stem cell support is mandatory. Nonmyeloablative conditioning (NMA) regimens cause minimal cytopenia and can be given also without stem cell support. Reduced-intensity conditioning (RIC) regimens do not fit the criteria for MAC or NMA regimens: They cause cytopenia of variable duration and should be given with stem cell support, as the cytopenia may not be reversible (Bacigalupo et al. 2009).

3.2.1 The Myeloablative Conditioning Regimens (MAC)

The classic myeloablative conditioning regimen has a dosage that lies between the maximum tolerable dose for normal organs and the maximum dose that can kill leukemia cells, creating a so-called "therapeutic window" that is relatively safe but very narrow. Patients who are elderly or have preexisting organ dysfunction often cannot tolerate it. The advantage of this conditioning regimen is that it can quickly and intensely eliminate leukemia cells. However, due to the high-intensity conditioning's strong bone marrow toxicity, it often leads to severe mucosal barrier damage, significantly increasing the chances of patients developing severe infections and acute graft-versus-host disease (GVHD). The treatment-related mortality rate due to complications such as infections, GVHD, and hepatic vein occlusion disease (HVOD) may exceed 15–30% in adult patients. Early allogeneic hematopoietic stem cell transplantation (allo-HSCT) was limited to patients under 50 years old and in good general condition, but due to the advancement of HLA typing technology, supportive treatment methods, and conditioning regimens, the transplantation-related mortality rate of myeloablative HSCT has significantly decreased, and most experts believe that the age limit for patients can be extended to 55 years old (Zhang et al. 2021).

MAC regimens included (i) ≥100 mg/kg or 3.6 g/m^2 IV cyclophosphamide, (ii) ≥12 Gy TBI, (iii) ≥16 mg/kg PO busulfan or 9.6 mg/kg IV busulfan (Xu et al. 2018).

3.2.1.1 TBI-Based Conditioning

The total body irradiation (TBI)-based conditioning regimen, first used by the father of bone marrow transplantation, Thomas, is still the mainstream protocol for acute lymphoblastic leukemia (ALL). Compared to chemotherapy-based conditioning regimens, TBI has unique advantages. It has good immunosuppressive and tumoricidal effects and also has the advantage of overcoming tumor drug resistance and drug barriers in the central nervous system and testes. The intensity of TBI-based conditioning regimens depends on the total dose, dose rate, and whether it is fractionated. Several studies have shown that there are several randomized controlled or multicenter large sample retrospective studies indicating that for ALL patients, compared with chemotherapy-based conditioning regimens, TBI-based myeloablative conditioning regimens may better eliminate leukemia cells, significantly reduce the relapse rate, and improve disease-free survival (DFS) rates (Peters et al. 2021; Kebriaei et al. 2018). Among them, the combination of TBI and etoposide is widely used in the conditioning regimen for acute leukemia, especially ALL.

However, due to the lack of precision and organ targeting of TBI, excessive radiation doses can affect normal tissues or organs (lungs, heart, liver, and kidneys), leading to radiation-related toxicity and an increase in transplant-related mortality (TRM). The emergence of helical tomotherapy has made it possible to achieve total marrow irradiation (TMI) and total marrow and lymphatic irradiation (TMLI). TMI/TMLI is gradually being used for transplantation conditioning, which can control the radiation dose in areas with high disease burden or high-risk parts, eliminate leukemia cells while reducing the toxicity of radiation therapy to important organs, and appropriately increase the radiation therapy index.

3.2.1.2 Non-TBI-Based Conditioning

To avoid the many toxic reactions of TBI and to expand the indications for allo-HSCT, the use of conditioning regimens combining busulfan (Bu) with cyclophosphamide (Cy) has gradually increased since the 1980s. Bu, as a component of the conditioning regimen, has many advantages. It has a broad-spectrum antitumor effect, no cross-resistance with other alkylating agents, and can penetrate the blood-brain barrier, providing both preventive and therapeutic effects against central nervous system leukemia. The systemic exposure of Bu is closely related to efficacy and side effects; if the dose is too low, the risks of graft rejection and post-transplant relapse will significantly increase; if the dose is too high, it increases the risk of seizures and HVOD. Precise dosing of Bu through pharmacokinetic monitoring can minimize side effects while achieving the optimal therapeutic window. In addition, adjusting the timing and sequence of administration of Bu and other conditioning drugs can significantly enhance the cytotoxic effect of the combination regimen with other drugs, significantly reducing the relapse rate without increasing transplant-related mortality. Treosulfan, a bifunctional alkylating drug, showed strong myelotoxic, immunosuppressive, and antileukemic properties. An open-label, randomized, noninferiority, phase 3 trial from Europe has compared the efficacy of treosulfan plus fludarabine and busulfan plus flufarabine among patients with myeloid malignancy. The treosulfan dose was 14 g/m^2/day or 10 g/m^2/day for 3 days, and the busulfan dose was 3.2 mg/kg/day for 2 days. The 2-year overall survival, transplantation-related mortality, and nonrelapse mortality were all improved in the treosulfan group compared with the busulfan group (Beelen et al. 2020). Several studies have explored the effects of conditioning regimens containing thiotepa, among which the TBF (thiotepa, Bu, Flu) regimen is currently a more commonly used conditioning regimen (Saraceni et al. 2018, 2019). A study followed up for a long term on 253 patients with acute myeloid leukemia (AML) who underwent allo-HSCT, and compared with the conditioning regimen of Bu combined with Flu, the TBF conditioning regimen significantly reduced the 5-year cumulative relapse rate and significantly improved the 5-year overall survival (OS) (Sora et al. 2020). Furthermore, thiotepa has good penetration into the central nervous system and milder adverse reactions such as bone marrow suppression, and it has been widely used in pretransplant conditioning for ALL. A retrospective study from the European Blood and Marrow Transplantation Association (EBMT) showed that for adult ALL, the efficacy of the thiotepa-based conditioning regimen is similar to the traditional Cy combined with TBI regimen in terms of 2-year overall survival and disease-free survival (Swoboda et al. 2022).

There have been several randomized controlled or multicenter large sample retrospective studies comparing the efficacy and safety of TBI/Cy and Bu/Cy regimens in allo-HSCT. Early studies showed that TBI/Cy seemed to have a better antitumor effect, with relatively lower patient relapse rates, and the difference was particularly evident in ALL patients. However, whether it can improve overall survival remains unclear, as the results of the studies are not consistent. Relatively, the Bu/

Cy conditioning regimen may have a higher relapse rate and VOD risk, but it may be related to the unstable pharmacokinetics of oral Bu in early studies. A study by the EBMT including 1479 patients with CR AML showed that the 2-year TRM for the TBI/Cy group and the Bu/Cy group was 18% versus 16%, the 2-year relapse rate was 21% versus 25%, and the 2-year DFS was 61% versus 59%, with no statistical difference (Nagler et al. 2013). A more recent open-label, randomized phase III trial has indicated that busulfan-based regimen has noninferior efficiency and safety as a TBI-based regimen in allo-HSCT for adult standard-risk B-ALL in first complete remission (CR1) (Zhang et al. 2023). Most scholars currently believe that the efficacy of TBI/Cy and Bu/Cy regimens is basically the same, and the specific choice can be made according to the patient's condition (Table 3.1).

Another commonly used MAC regimen is the myeloablative dose busulfan plus fludarabine (Bu/Flu), of which the regimen-related toxicity is considered to be lower than Bu/Cy. In a network meta-analysis, a group of MAC regimens has been investigated in AML patients in CR undergoing allo-HSCT from HLA-matched related or unrelated donors, including 4-day busulfan (for a median total dose of 16 mg/kg orally, or 12.8 mg/kg intravenously) plus cyclophosphamide (Bu/Cy), Bu/Cy plus etoposide (Bu/Cy/VP-16), 3-day busulfan (for a median total dose of 12 mg/kg orally, or 9.6 mg/kg intravenously) plus fludarabine (Bu3/Flu), Bu3/Flu plus cytarabine (Bu3/Flu/Ara-C) or thiotepa (Bu3/Flu/TT), 4-day busulfan plus fludarabine (Bu4/Flu), cyclophosphamide plus single-dose (≥5.5 Gy) or fractionated (for a median total dose of 12 Gy) TBI (Cy/TBI), cyclophosphamide plus higher dose TBI (total dose 15.75 Gy) (Cy/higher TBI), and 8-Gy fractionated TBI plus fludarabine (TBI 8 Gy/Flu). The analysis only included 3-day busulfan (total 9.6 mg/kg) plus fludarabine and thiotepa (Bu3/Flu/TT) for the final analysis, since 4-day busulfan (total 12.8 mg/kg) plus fludarabine and thiotepa (Bu4/Flu/TT) is excessively toxic and highly not recommended. The pooled results indicated that Bu3/Flu/TT, Bu3/Flu/Ara-C, and TBI/ Flu are associated with lower nonrelapse mortality (NRM) and cumulative incidence of relapse when compared with Bu/Cy (Luo et al. 2023).

Table 3.1 Conditioning regimens in allo-HSCT for severe aplastic anemia

Conditioning regimen	Drug	Dose	Days	Donor type
Cy-ATG	Cy r-ATG	200 mg/kg 10 mg/kg	−5 to −2 −5 to −2	MSD/URD
FluCy-ATG	Flu Cy r-ATG	150 mg/m^2 100 mg/kg 10 mg/kg	−6 to −2 −5 to −2 −5 to −2	MSD/URD
BuCy-ATG	Bu Cy r-ATG	6.4 mg/kg 200 mg/kg 10 mg/kg	−7 to −6 −5 to −2 −5 to −2	HID
BuFluCy-ATG	Bu Flu Cy r-ATG	6.4 mg/kg 150 mg/m^2 100 mg/kg 10 mg/kg	−7 to −6 −6 to −2 −5 to −2 −5 to −2	HID
BuFlu-PTCy	Bu Flu Cy PTCy	6.4 mg/kg 200 mg/m^2 29 mg/kg 120 mg/kg	−8 to −7 −8 to −4 −3, −2 +3, +4	HID
TBIFluATG-PTCy	r-ATG Flu Cy TBI PTCy	4.5 mg/kg 150 mg/m^2 29 mg/kg 200–400 cGy 100 mg/kg	−9 to −7 −6 to −2 −6 to −5 −1 +3, +4	HID

ATG antithymocyte globulin, *Bu* busulfan, *Cy* cyclophosphamide, *Flu* fludarabine, *HID* haploidentical donor, *MSD* matched sibling donor, *URD* unrelated donor, *PTCy* post-transplantation cyclophosphamide

3.2.2 The Nonmyeloablative Conditioning (NMA) and Reduced-Intensity Conditioning Regimens (RIC)

Due to the important role of the graft versus leukemia (GVL) effect of donor cells in allo-HSCT, the clinical application of nonmyeloablative and reduced-intensity conditioning regimens is increasing, especially suitable for elderly and patients with higher Hematopoietic Cell Transplantation Comorbidity Index (HCT-CI). Fludarabine (Flu) is often used instead of cyclophosphamide (Cy) to reduce the toxic side effects of Cy while taking advantage of the strong immunosuppressive effect of Flu to facilitate the engraftment of donor hematopoietic stem cells.

Unlike myeloablative conditioning regimens, nonmyeloablative conditioning regimens do not emphasize the direct killing effect on leukemia cells. They mainly rely on immunosuppression to induce the recipient's immune tolerance to the donor, allowing the donor cells to engraft stably and form a stable chimera. Subsequently, the GVL effect is exerted through immune-active cells in the graft or proliferated and differentiated from donor hematopoietic stem cells, as well as through donor lymphocyte infusion (DLI) after transplantation, thus achieving the purpose of cure. Therefore, this type of transplantation is more like a platform for adoptive cellular immunotherapy. Nonmyeloablative conditioning has the following characteristics: (i) minimal tumor clearance; (ii) minimal bone marrow toxicity; (iii) contains potent immunosuppressants; and (iv) mostly mixed chimera or gradually achieving complete engraftment. Examples of NMA regimens include fludarabine and cyclophosphamide (Flu-Cy), TBI 2 Gy, TBI 1 Gy, total lymphoid radiation (TLI), and antithymocyte globulin (ATG) (Bacigalupo et al. 2009). The most widely used nonmyeloablative conditioning regimen is 2Gy TBI-Flu, TLI + antithymocyte globulin (ATG), and Flu-Cy. A phase II multicenter randomized study comparing nonmyeloablative allo-HCT indicated that patients included in the TLI-ATG arm had lower incidence of chronic GVHD, higher incidence of relapse, and similar OS in comparison to patients included in the Flu-TBI arm (Baron et al. 2015).

Many clinical studies have shown that although the direct killing effect of nonmyeloablative conditioning regimens on tumor cells is relatively minor, compared with patients undergoing chemotherapy at the same time, transplantation can significantly improve patient prognosis, especially for advanced patients, indicating that nonmyeloablative allo-HSCT is feasible through "immune clearance" rather than "myeloablation" to achieve donor hematopoietic stem cell engraftment, and can effectively treat malignant hematologic diseases relying on subsequent GVL effects. In addition, although nonmyeloablative allo-HSCT significantly reduces TRM, especially early TRM, relapse, GVHD, and infection are still the main problems affecting the OS and quality of life of patients. For high-risk malignant hematologic patients, relapse is still the main reason for transplant failure. It is currently believed that nonmyeloablative allo-HSCT is mainly suitable for patients with minimal tumor burden, slow disease progression, sensitive to the GVL effect, and not suitable for conventional transplantation. In addition, nonmyeloablative allo-HSCT still has many issues, such as the timing of transplantation, when to start DLI, DLI dosage and course, and the prevention and treatment of post-DLI GVHD, all of which require further in-depth discussion.

As for the reduced-intensity conditioning regimen, the intensity is between myeloablative and nonmyeloablative conditioning regimens, and it is generally considered to meet the following standards: Bu <9mg/kg, TBI <8Gy, Mel <140 mg/m^2 (Xu et al. 2018). Although the bone marrow suppression of the reduced-intensity conditioning regimen can recover on its own, it usually requires the support of donor hematopoietic stem cells to avoid severe complications caused by long-term severe cytopenia. Theoretically, the organ damage of the reduced-intensity conditioning regimen is less than that of the myeloablative conditioning regimen, the patient's transfusion requirements are low, the risk of early infection is reduced, and tissue damage and inflammatory response are relatively mild, thus reducing the risk of GVHD; at the same time, the intensity of the reduced-intensity conditioning regimen is greater than that of the nonmyeloablative conditioning regimen, which is conducive to reducing the post-transplant relapse rate. There are many types of reduced-intensity regimens, usually with 90–180 mg/m^2 of Flu as the basic immunosuppressive means, on the basis of which reduced-dose TBI or alkylating agents (such as Bu, Mel) are combined to exert antitumor effects, and other regimens also include ATG or alemtuzumab to reduce the incidence and severity of GVHD. Like the nonmyeloablative conditioning regimen, the reduced-intensity regimen has greatly improved the patient's tolerance to the conditioning regimen, and for patients not suitable for traditional myeloablative allo-HSCT, early TRM can be significantly reduced. Previous large-scale randomized controlled trial (RCT) clinical studies compared the efficacy differences between myeloablative and reduced-intensity allo-HSCT in the treatment of myelodysplastic syndrome (MDS) and AML, showing that the nonrelapse mortality of reduced-intensity allo-HSCT was significantly reduced, the incidence of grade II–IV aGVHD, grade III–IV aGVHD, and cGVHD was significantly reduced, subgroup analysis showed that for AML patients, there was no statistical difference in OS and relapse rate, and there was a trend of reduced nonrelapse mortality and grade II–IV aGVHD in the reduced-intensity allo-HSCT group; for MDS patients, OS, DFS, and relapse rates of the two conditioning regimens were basically similar; for intermediate-risk patients, the efficacy of the reduced-intensity regimen was similar to that of the myeloablative regimen, but for high-risk patients, the prognosis of myeloablative allo-HSCT was better (Scott et al. 2017; Niederwieser et al. 2024; Kroger et al. 2017). With the arrival of an aging society, the use of reduced-intensity allo-HSCT in the clinic will increase day by day. Some studies have shown that adding new targeted drugs to the reduced-intensity conditioning regimen may further reduce the relapse rate without increasing nonrelapse mortality. Adding the BCL-2 inhibitor venetoclax to the FluBu2-reduced conditioning regimen does not affect hematopoietic engraftment and does not aggravate GVHD; it is a safe and effective conditioning regimen for high-risk MDS and AML patients (Garcia et al. 2021). Using more efficient, low-toxicity new targeted drugs in the conditioning regimen, while reducing the types and doses of chemotherapy drugs, is one of the future exploration directions in the field of transplantation.

3.2.3 Intensified Conditioning Regimens

For high-risk or relapsed AML/ALL, conventional MAC transplants still have a high relapse rate. Therefore, combining the conditioning with drugs that have strong antileukemic effects, such as daunorubicin or etoposide, can enhance the antileukemic action of the transplant treatment, but also increase the early toxic reactions and related nonrelapse mortality (NRM) after transplantation. It is recommended for relatively young patients without significant organ damage and those at high risk according to the HCT-CI.

For untreated nonremission or refractory AML, administering high-dose chemotherapy before transplantation to reduce the leukemia burden can achieve a higher rate of complete remission. After the end of chemotherapy, with an interval of 3–7 days or directly in sequence with transplant conditioning of different intensities, the 2-year overall survival (OS) and disease-free survival (DFS) can reach 30–50% (Schmid et al. 2006; Wang et al. 2019). The FLAMSA chemotherapy regimen (Flu 30 mg/(m^2•d) for 5 days + Ara-C 2 g/(m^2•d) for 4 days + AMSA 100 mg/(m^2•d) for 4 days) followed by the RIC (reduced-intensity conditioning) Cy/TBI (Cy 80–120 mg/kg + TBI 4Gy) regimen after a 3-day interval is expected to have an overall survival rate of about 60% for untreated nonremission AML (Schmid et al. 2006). For high-risk adult AML patients aged 45–65 years in CR1/CR2, the FLAMSA followed by RIC conditioning also has good efficacy, and compared with Flu-Treo-based conditioning, it significantly reduces post-transplant relapse and improves relapse-free survival (RFS). The TCE (Thiotepa 5 mg/(kg•d) for 2 days + Cy 400 mg/(m^2•d) for 4 days + VP-16100 mg/(m^2•d) for 4 days) or CA (clofarabine 30 mg/(m^2•d) for 5 days + Ara-C 1 g/(m^2•d) for 5 days) regimens followed by reduced-intensity Flu/Bu or Bu/Cy conditioning for the treatment of refractory malignant hematological diseases have certain efficacy. The FLAG-IDA or CLAGM regimens are re-induction regimens for relapsed AML, and when followed by iv-Bu-based conditioning for the treatment of refractory AML, they both achieve a high remission rate, with 2-year OS/DFS reaching 40–50% (Wang et al. 2019, 2020).

For untreated or refractory ALL, strong chemotherapy followed by conditioning regimens has not achieved satisfactory efficacy and is not recommended for clinical use (Bazarbachi et al. 2020). In the MAC conditioning regimen for Cy-TBI, the addition of intermediate to high doses of etoposide (VP-16 at 30–40 mg/kg) for the treatment of high-risk acute lymphoblastic leukemia (ALL), such as positive measurable residual disease (MRD) in the first complete remission (CR1) bone marrow or accompanied by high-risk cytogenetic abnormalities, or in the second complete remission (CR2), significantly reduces the risk of post-transplant relapse without increasing post-transplant nonrelapse mortality (NRM) (Arai et al. 2018). In recent years, for patients with relapsed/refractory leukemia, an increasing number of new treatment technologies and drugs, such as immunotoxins, bispecific antibodies, and chimeric antigen receptor T-cells combined with hematopoietic stem cell transplantation, aim to achieve the best remission before transplantation, thereby improving the efficacy of the allo-HSCT.

3.3 The Conditioning Regimens for Hematologic Malignancy

3.3.1 Conditioning Regimens of Allo-HSCT from Matched Sibling Donors

The conditioning protocols were mature in allo-HSCT from matched sibling donors (MSD), and the two most popular ones were the Cy/TBI (TBI 12Gy followed by IV Cy 60 mg/kg × 2 days) and the Bu/Cy protocol (Bu 4 mg/kg × 4 days and Cy 60 mg/kg × 2 days) from the European and American area (Nagler and Shimoni 2019). In China, the modified TBI/Cy protocol (TBI 770 cGy followed by IV 1.8 mg/m^2/day × 2 days) and Bu/Cy protocol (Bu 3.2 mg/kg/day × 3 days and Cy 1.8 mg/m^2/day × 2 days) were commonly used (Zhang et al. 2021). Recently, the optimization of the conditioning regimen in MSD-HSCT has been explored. A multicenter, prospective study has suggested that ATG should be included in the conditioning regimens to further improve the prognosis of patients with hematologic malignancies after MSD-HSCT (Chang et al. 2020). This prospective randomized controlled trial (RCT) study demonstrated that low-dose ATG (4.5 mg/kg) can effectively decrease the incidence of aGvHD without affecting relapse or nonrelapse mortality after MSD-HSCT.

3.3.2 Conditioning Regimens of Allo-HSCT from Alternative Donors

In recent decades, haploidentical donor stem cell transplantation (haplo-SCT), especially non-T-cell depletion in vitro strategies, including the granulocyte colony-stimulating factor (G-CSF)/antithymocyte globulin (ATG)-based Beijing Protocol and the post-transplantation cyclophosphamide (PT-Cy)-based Baltimore protocol, have made great advances. The optimization of different protocols has been continuously evolving.

3.3.2.1 The G-CSF/ATG-Based Protocol

The key elements of the G-CSF/ATG-based protocol (Beijing Protocol) consist of G-CSF mobilization in donors and ATG application in recipients for GVHD prophylaxis (Xu and Huang 2023). In the earlier exploration, Huang's team found that in vivo G-CSF administration induced T-cell hyporesponsiveness in bone marrow (BM) or peripheral blood (PB) grafts from healthy donors (Jun et al. 2004). Huang's team also firstly demonstrated the cellular and molecular mechanism of G-CSF inducing T-cell tolerance, indicating that G-CSF activates STAT3 via the receptor CSF3R; then, phosphorylated STAT3 enters the nucleus and upregulates suppressor of cytokine signaling 1 (SOCS1) expression levels, and high expression levels of SOCS1 inhibited T-cell proliferation and inflammatory cytokine secretion by inhibiting CSF3R/JAK2/STAT3 signaling and NLRP3 inflammasome signaling activation (Guo et al. 2022). Based on the theory that G-CSF induces T-cell tolerance, Prof. Huang led the team and established a novel protocol with the combination of G-CSF in donors and antithymocyte globulin (ATG) in recipients. Subsequently, a series of key techniques including biomolecular markers guided GVHD prophylaxis, disease prevention and treatment, as well as donor selection in haploidentical transplantation without in vitro T-cell depletion, which gradually developed into the world's first haplo-HSCT system, known as "Beijing Protocol" (Apperley et al. 2016).

Haplo-SCT with a G-CSF/ATG-based protocol has achieved comparable outcomes with those transplanted from MSDs in the treatment of leukemia or MDS (Wang et al. 2015, 2016). The commonly used conditioning regimens included modified Bu/Cy/ATG or TBI/Cy/ATG (Zhang et al. 2021). The modified Bu/Cy/ATG were described as intravenous Bu 9.6 mg/kg, Cy 1.8 g/m^2/day for 2 days plus ATG. The modified TBI/Cy/ATG were 770 cGy for 1 day, Cy 1.8 g/m^2/day for 2 days plus ATG. The ATG (rabbit antithymocyte globulin, Sangstat, Lyon, France) is used in a dose range of 6–10 mg/kg for HID-HSCT.

The optimal dosage of ATG in the G-CSF/ATG-based modality has been continuously under investigation. An open-label, prospective, randomized trial was initiated, aiming to compare two different total doses of rabbit ATG (6 mg/kg vs. 10 mg/kg) in patients receiving haplo-SCT. The incidences of grade II–IV and III–IV aGvHD were higher in the ATG-6 group but the dose of 6 mg/kg ATG decreased the risk of Epstein–Barr Virus (EBV) reactivation. With the extension of the follow-up period, the 5-year probability of GVHD-free relapse-free survival (GRFS) was higher in the ATG-10 group (Chang et al. 2020). Another multicenter, randomized study was initiated to compare 10.0 mg/kg ATG with 7.5 mg/kg ATG following G-CSF/ATG-based haplo-SCT. The 100-day cumulative incidences of aGvHD and cGvHD were comparable, but 7.5 mg/kg ATG for GvHD prophylaxis was associated with reduced EBV and cytomegalovirus (CMV) infections compared with 10.0 mg/kg ATG. Extended follow-up indicated similar 5-year probabilities of GRFS between the two groups (Wang et al. 2021). The personalized optimal dose of ATG should be further investigated based on the balance of GVHD and infection.

For older adults (55 years of age and older) or individuals with significant comorbidities, reduced-intensity conditioning that substitutes (or partially) cyclophosphamide with fludarabine has been recommended. A prospective study focused on an RIC regimen for HID-HSCT. The study involved 50 individuals aged 55 and above, preconditioned with busulfan (at a dosage of 3.2 mg/kg per day, intravenously for a duration of 3 days), fludarabine (30 mg/m^2 per day, intravenously for 5 days), cyclophosphamide (1.0 g/m^2 per day, intravenously for 2 days), and antithymocyte globulin (ATG) (2.5 mg/kg per day, intravenously for 4 days). The reported outcomes for the 1-year post-transplantation were a NRM rate of 23.3%, DFS of 60.2%, and OS of 63.5% (NCT03412409) (Sun et al. 2021). Thus, this novel RIC conditioning regimen followed by HID-HSCT might be a promising option for older patients or those with comorbidity pre-HSCT.

3.3.2.2 The PT-Cy Based Protocol

Initially, PTCy-based allo-HSCT from haploidentical donors was performed utilizing a non-myeloablative conditioning approach that included fludarabine (150 mg/m^2), cyclophosphamide (14.5 mg/kg for 2 days), and a low dosage of TBI (2 cGy) (Luznik et al. 2008). This regimen produced favorable results in terms of engraftment, GVHD, and NRM, but the relapse rate reached 51% at 1 year. Thus, the high incidence of relapses underscored the necessity for a more intensified conditioning strategy.

Accumulating data have confirmed the efficacy and safety of TBI-based and Bu-based MAC regimens for selected patients undergoing PTCy-based haplo-HCT. A retrospective study analyzed 150 patients with AML who underwent PTCy-based haplo-BMT with MAC (Chiusolo et al. 2018). Younger patients were eligible for Flu (120 mg/m^2)/TBI (9–12 Gy), and older patients received TBF. The TBF regimen consisted of thiotepa (10 mg/kg), Bu (9.6 mg/kg), and Flu (150 mg/m^2), and the Bu dose was adjusted to 6.4 mg/kg for patients aged over 60 years. The 4-year OS was 72% for patients in complete remission (CR) and 26% for those with advanced disease. Data from EBMT have compared two MAC regimens (Flu/TBI, $n = 117$; TBF, $n = 119$), in patients with ALL who underwent PTCy-based haplo-HCT. When patients transplanted in CR were considered, Flu/TBI presented a lower risk of NRM, but a higher risk of relapse, resulting in no significant difference in OS between Flu/TBI and TBF (Swoboda et al. 2022). Another prospective study from Johns Hopkins reported the results of PTCy-based haplo-BMT with TBI (12 Gy)/Cy (100 mg/kg) or Bu (12.8 mg/kg)/Cy (100 mg/kg) in 96 adult and pediatric patients with hematologic malignancy. The 3-year NRM, relapse, and OS were 11%, 43%, and 54%, respectively (Symons et al. 2020).

In the PTCy-based haplo-HCT, Flu/Bu and Flu/MEL are among the most frequently used as RIC regimens. Several Bu- and MEL-based RIC regimens have been explored. A study from the ALWP of the EBMT compared the reduced-intensity TBF regimen with Flu (150 mg/m^2)/Cy (29 mg/kg)/TBI (2 Gy) in 490 patients with AML in CR. The TBF RIC regimen consisted of a maximal total busulfan dose of 6.4 mg/kg and a total thiotepa dose ranging from 5 mg/kg to 12 mg/kg. The TBF RIC group had lower relapse, comparable NRM, better DFS, and better OS than the Flu/Cy/TBI NMAC group in patients aged <60 years (Devillier et al. 2022). The Flu/MEL-based conditioning regimen has been investigated among older patients. Ciurea et al. reported on 43 AML patients with a median age of 61 years who received Flu (160 mg/m^2)/MEL (100–140 mg/m^2) in combination with thiotepa (5 mg/kg) or TBI (2 Gy). With the median follow-up of 19 months, 2-year overall survival, progression-free survival (PFS), and relapse incidence were 42%, 42%, and 24% (Ciurea et al. 2018).

At the moment, there is no clear consensus on the preference between MAC or RIC in PTCy-based haplo-HCT. Although MAC could potentially lower disease relapse, this advantageous outcome does not appear to result in improved overall survival, primarily due to the higher risk of nonrelapse mortality. It is imperative to conduct future randomized studies that concentrate on PTCy-based haplo-HCT to derive significant and conclusive findings.

Several published studies have attempted to compare the clinical outcomes of different haploidentical protocols. Recently, the clinical outcomes of G-CSF/ATG, PTCy, and PTCy combined with low-dose ATG (PTCy+ATG^{low}) have been demonstrated based on data from the Chinese Bone Marrow Transplantation Registry Group (CBMTRG). This study is the first to compare the clinical outcomes of patients via the propensity score matching method (PSM) to reduce confounding bias. Compared with those in the PTCy group, the incidences of 28-day neutrophil engraftment and 100-day platelet engraftment were significantly greater in the G-CSF/ATG group. Patients in the G-CSF/ATG group had lower nonrelapse mortality, 3-year overall survival, and leukemia-free survival rates than those in the PTCy group (Xu et al. 2024). Data from a larger number of patients and prospective randomized controlled trials are necessary to clarify the clinical outcomes of different haplo-HSCT protocols.

3.3.3 The Option of Different Intensity Conditioning Regimens

In the allo-HSCT treating leukemia or MDS, the MAC regimens include traditional TBICy, BuCy, and its modified regimens. The commonly used RIC regimens are fludarabine-containing or ATG-included regimens. The intensified regimens generally include the addition of a drug, such as Ara-C, VP16, Melphalan, TBI,

Fludarabine, or Tespamin, to standard condition regimens. It is primarily used for refractory and relapsed malignant patients.

The optimal conditioning regimen for a patient should be selected based on the type and status of the disease, comorbidities, underlying conditions, and donor source.

For example, a standard-intensity conditioning regimen is used for younger patients (younger than 55 years old), and RIC regimens are used for patients older than 55 years old and patients with poor organ function or HSCT-CI ≥3, regardless of age. Intensified regimens are used for young patients with refractory and relapsed diseases. Intensified regimens may reduce the incidence of relapse to some extent, but they often increase TRM. Therefore, these regimens may not significantly increase overall survival. RIC regimens are often better tolerated but require immunosuppression agents and cell therapy to reduce the relapse risk. Therefore, RIC regimens are combined with other regimens, such as fludarabine. HSCT with standard intensification conditioning regimen followed by immunosuppression adjustment or cell therapy to enhance the graft versus leukemia (GVL) effect is also feasible for refractory and relapsed patients.

3.4 The Conditioning Regimens for Hematologic Nonmalignancy

3.4.1 Aplastic Anemia

3.4.1.1 HSCT from HLA-Matched Sibling or Unrelated Donor

HLA-matched HSCT is the preferred treatment for young patients with severe aplastic anemia (SAA). The conditioning regimen of 200 mg/kg cyclophosphamide (Cy) combined with antithymocyte globulin (ATG) for allo-HSCT is currently the most commonly used conditioning regimen for matched related allogeneic transplantation, with a high success rate of up to 95% and an overall survival rate close to 90%. However, for older patients, the toxicity of Cy at a dose of 200 mg/kg is significant and poorly tolerated. Therefore, in recent years, researchers have attempted to use a conditioning regimen of Flu combined with a lower dose of Cy to reduce the toxic side effects of high-dose Cy. A large-scale retrospective study indicated that for recipients of HLA-matched sibling donor transplantation, Flu/Cy/ATG or Cy/ATG led to the best 5-year survival of 91% in the two groups (Bejanyan et al. 2019). For unrelated allo-HSCT, current recommendations include Flu-Cy-ATG (FCA) with low-dose TBI or Flu-Cy-CAMPATH (FCC): Cy at a dose between 50 and 100 mg/kg (total dose), TBI between 2 and 4 Gy, and Flu from 100 to 150 mg/m^2 (Bacigalupo 2017).

However, with conditioning consisting of Cy 200 mg/kg plus r-ATG 10 mg/kg, the incidence rates of mixed chimerism have been reported to be 20.3% and 35.7% in matched sibling (MSD) or unrelated donor (MUD) transplantation (Xu et al. 2021). Furthermore, mixed chimerism has significantly increased secondary graft rejection rates from 0.4% to 14.8% and decreased OS from 89.6% to 72.7% compared with donor chimerism (Xu et al. 2021). A previous pilot study has indicated that the addition of low-dose busulfan (3.2 mg/kg) decreased the incidence of mixed chimerism and improved event-free survival without increasing side effects (Xu et al. 2022a). As the intensity of the conditioning regimen was associated with the incidence of MC, a more intensive conditioning regimen should be further explored to facilitate donor chimerism in HLA-matched donor transplantation.

3.4.1.2 HSCT from Haploidentical Donor

Nowadays, haplo-HSCT using different regimens has overcome the difficulty of graft failure and severe graft-versus-host disease (GVHD) and achieved inspiring survival outcomes in SAA. The regimens consist mainly of G-CSF/ATG-based and PT-Cy-based non-T-cell depletion strategies. In particular, the G-CSF and ATG-based regimen includes the largest sample size, and the successful wide use of the G-CSF and ATG-based regimen has promoted haplo-HSCT, a higher priority in SAA patients without matched related or unrelated donors in China (Zhang et al. 2021).

- *The G-CSF/ATG-Based Protocol*

The G-CSF/ATG-based protocol consisted of Bu/Cy/ATG or Bu/Flu/Cy/ATG in the treatment of SAA. In 2012, Huang's group first reported the successful use of G-CSF/ATG-based haplo-SCT for treating severe aplastic anemia (SAA). The regimen added busulfan (Bu; 3.2 mg/kg/day administered intravenously in divided doses on days −7 and −6) to Cy + ATG (50 mg/kg/day Cy and 2.5 mg/kg/day rabbit ATG for 4 days). A prospective multicenter study has been launched to further verify the safety and efficacy of G-CSF/ATG-based haplo-SCT in SAA. The initial outcomes showed that all patients surviving for more than 28 days achieved donor myeloid engraftment, and the long-term outcomes indicated that the estimated 9-year OS and failure free survival (FFS) for the whole cohort were 85.4% and 84.0%, respectively (Xu et al. 2016, 2022b). Another multicenter study was conducted based on a registry database, indicating that upfront haplo-SCT with the G-CSF/ATG-based protocol achieved similar 9-year OS with MSD-SCT (87.1% vs. 89.3%) in SAA (Xu et al. 2022c).

Regimen-related toxicity (RRT) is also a concern because severe cardiotoxicity was reported in 5.6% of SAA patients who received conditioning with Bu/Cy/ATG. The modified regimen Bu/Flu/Cy/ATG is emerging to improve results among SAA patients at risk of severe cardiotoxicity. This Bu/Flu/Cy/ATG conditioning utilized busulfan (Bu, 3.2 mg/kg for 2 days), low-dose Cy (100 mg/kg), fludarabine (150 mg/m^2), and 10 mg/kg r-ATG. Compared to Bu/Cy/ATG conditioning using high-dose Cy of 200 mg/kg, the incidence of severe cardiotoxicity of Bu/Flu/Cy/ATG conditioning was significantly decreased. The 1-year OS was similar, with 88.8% and 84.7% in the Bu/Flu/Cy/ATG and Bu/Cy/ATG groups (Lin et al. 2022).

- *The PT-Cy-Based Protocol*

The initially attempted conditioning regimen consisted of 14.5 mg/kg Cy (days −6 and−5), 30 mg/m^2 fludarabine (Flu; days −6 to −2), and low-dose TBI (200 cGy). All patients received high-dose Cy (50 mg/kg) on days +3 and + 4. The reported rate of GF was 25–33%, and the reported OS was 67–78% (Clay et al. 2014; Prata et al. 2020). Subsequently, DeZern modified the PT-Cy protocol by adding r-ATG (0.5 mg/kg on day −9 and 2 mg/kg on days −8 and −7) on the basis of the traditional PT-Cy regimen, with the day-28 cumulative incidence of neutrophil recovery 94%, and the 1-year overall survival of 81% (DeZern et al. 2022). Although 200 cGy was sufficient for engraftment in the above study, 400 cGy appeared to result in better engraftment, without increased toxicity, in other series and in the treatment-naive setting (DeZern et al. 2020).

In recent decades, the modification of conditioning has been evolving. TBI or Bu is the basic component in current conditioning regimens, but these have potentially adverse effects on fertility. Long-term follow-up of young patients who underwent haploidentical transplantation, including fertility, is necessary.

3.4.2 Fanconi Anemia

Fanconi anemia (FA) is a rare autosomal recessive disease characterized by congenital malformations, progressive BM dysfunction, and a high risk of developing malignancies. In particular, genetic susceptibility predisposes FA patients to regimen-related toxicity (RRT), but insufficient regimen intensity leads to an increased risk of graft failure.

In the last decade, a series of modifications to allo-HSCT regimens, specifically dose reductions of alkylating agents and irradiation, as well as the addition of Flu to conditioning, have dramatically improved outcomes for patients transplanted for FA (Ebens et al. 2017). Data from the Center for International Blood and Marrow Transplant Research (CIBMTR) have compared outcomes of irradiation-containing (irradiation+ Cy±ATG) to non-irradiation-containing (Cy alone, Cy+ATG, Busulfan+ATG, or Flu+Cy) conditioning regimens before MSD-HCT for FA and indicated 5-year overall survival of 78% and 81% in the two groups (Pasquini et al. 2008). The recent results from the CIBMTR have found that patients with Flu and Cy had significantly better OS than those conditioned with Flu and Cy plus Bu or TBI in HSCT from matched siblings and unrelated donors (Xu et al. 2023). Currently, most studies in haplo-HCT for FA have included radiation in the conditioning regimens. Although the dose of irradiation is low (100–200 cGy), secondary tumor rates have been reported ranging from 3% to 7%. Another study focused on the unmanipulated haplo-HCT in patients with FA with radiation-free conditioning, with the majority of Flu-Cy regimen (FC, 48.2%) and Bu-Flu-Cy regimen (BFC, 41.1%), reported myeloid engraftment of 96% and survival of 81% (Xu et al. 2022d). Generally, the FluCy-ATG regimen (Flu 150 mg/m^2, Cy 5–20 mg/kg/day × 4 days, and rabbit ATG 10 mg/kg) with or without low-dose TBI may be used for alternative donor transplantation.

3.4.3 Thalassemia

Thalassemia is a hemoglobinopathy that, in its more severe forms, has an extremely poor prognosis. Alloimmunization from frequent transfusions in thalassemia increases the risk of GF in HSCT. Therefore, intensified conditioning regimens instead of standard conditioning regimens should generally be used in patients with thalassemia.

In the allo-HSCT from unrelated or matched sibling donors, NF-08-TM HSCT protocol based on intravenous busulfan (Bu), cyclophosphamide (Cy), fludarabine (Flu), and thiotepa (TT) has led to 3-year OS of 92% and 90% in the two groups (Li et al. 2012). In haplo-HSCT treating thalassemia, both the PT-Cy-based and G-CSF/ATG-based protocols have been continuously tried. A study with PT-Cy-based protocol included the conditioning regimen of Bu (520 mg/m^2), Flu (210 mg/m^2) plus ATG (4.5 mg/kg), and showed a (PGF) primary graft failure rate of 6.5% and 2-year OS of 95% (Anurathapan et al. 2016). The FBCA regimen has been attempted in G-CSF/ATG-based protocol, including Flu (25 mg/m^2/day for 6 days), Bu (3.2 mg/kg/day for 4 days), Cy (60 mg/kg/day for 2 days), and r-ATG (2.5 mg/kg/day for 5 days). All eight patients survived and achieved independence from blood transfusion (Sun et al. 2018).

3.5 Summary

In general, the conditioning regimen is influenced by various factors such as the type of disease, the state of the disease, age, comorbidities, the source of the transplant, the degree of HLA compatibility between donor and recipient, and the GVHD prevention strategy. The selection of an appropriate conditioning regimen varies among different transplant centers, necessitating multicenter prospective studies for comparative analysis of efficacy.

With the development and maturation of new targeted therapies, CAR-T, and other immunotherapies in the field of hematologic disease treatment, integrating the aforementioned strategies into the conditioning regimen has become a research hotspot and has preliminarily shown a promising application prospect, which may be a new way to optimize the conditioning regimen and improve the efficacy for patients with hematologic diseases. With the advances in transplant technology and the expansion of transplant indications, how to maximize the GVL effect while controlling GVHD, reduce late complications, and improve the quality of life remains a clinical challenge. Diversified and personalized conditioning regimens are the development direction in the field of allo-HSCT.

References

Anurathapan U et al (2016) Hematopoietic stem cell transplantation for homozygous beta-thalassemia and beta-thalassemia/hemoglobin E patients from haploidentical donors. Bone Marrow Transplant 51(6):813–818

Apperley J et al (2016) Haploidentical hematopoietic stem cell transplantation: a global overview comparing Asia, the European Union, and the United States. Biol Blood Marrow Transplant 22(1):23–26

Arai Y et al (2018) Improved prognosis with additional medium-dose VP16 to CY/TBI in allogeneic transplantation for high risk ALL in adults. Am J Hematol 93(1):47–57

Bacigalupo A (2017) How I treat acquired aplastic anemia. Blood 129(11):1428–1436

Bacigalupo A et al (2009) Defining the intensity of conditioning regimens: working definitions. Biol Blood Marrow Transplant 15(12):1628–1633

Baron F et al (2015) Non-myeloablative allogeneic hematopoietic cell transplantation following fludarabine plus 2 Gy TBI or ATG plus 8 Gy TLI: a phase II randomized study from the Belgian Hematological Society. J Hematol Oncol 8:4

Bazarbachi AH et al (2020) Allogeneic stem-cell transplantation with sequential conditioning in adult patients with refractory or relapsed acute lymphoblastic leukemia: a report from the EBMT Acute Leukemia Working Party. Bone Marrow Transplant 55(3):595–602

Beelen DW et al (2020) Treosulfan or busulfan plus fludarabine as conditioning treatment before allogeneic haemopoietic stem cell transplantation for older patients with acute myeloid leukaemia or myelodysplastic syndrome (MC-FludT.14/L): a randomised, non-inferiority, phase 3 trial. Lancet Haematol 7(1):e28–e39

Bejanyan N et al (2019) Choice of conditioning regimens for bone marrow transplantation in severe aplastic anemia. Blood Adv 3(20):3123–3131

Chang YJ et al (2020) Antithymocyte globulin for matched sibling donor transplantation in patients with hematologic malignancies: a multicenter, open-label, randomized controlled study. J Clin Oncol 38(29):3367–3376

Chiusolo P et al (2018) A modified post-transplant cyclophosphamide regimen, for unmanipulated haploidentical marrow transplantation, in acute myeloid leukemia: a multicenter study. Biol Blood Marrow Transplant 24(6):1243–1249

Ciurea SO et al (2018) Haploidentical transplantation for older patients with acute myeloid leukemia and myelodysplastic syndrome. Biol Blood Marrow Transplant 24(6):1232–1236

Clay J et al (2014) Nonmyeloablative peripheral blood haploidentical stem cell transplantation for refractory severe aplastic anemia. Biol Blood Marrow Transplant 20(11):1711–1716

Devillier R et al (2022) Reduced intensity versus non-myeloablative conditioning regimen for haploidentical transplantation and post-transplantation cyclophosphamide in complete remission acute myeloid leukemia: a study from the ALWP of the EBMT. Bone Marrow Transplant 57(9):1421–1427

DeZern AE et al (2020) Haploidentical BMT for severe aplastic anemia with intensive GVHD prophylaxis including posttransplant cyclophosphamide. Blood Adv 4(8):1770–1779

DeZern AE et al (2022) Haploidentical bone marrow transplantation in patients with relapsed or refractory severe aplastic anaemia in the USA (BMT CTN 1502): a multicentre, single-arm, phase 2 trial. Lancet Haematol 9(9):e660–e669

Ebens CL, MacMillan ML, Wagner JE (2017) Hematopoietic cell transplantation in Fanconi anemia: current evidence, challenges and recommendations. Expert Rev Hematol 10(1):81–97

Garcia JS et al (2021) Adding venetoclax to fludarabine/busulfan RIC transplant for high-risk MDS and AML is feasible, safe, and active. Blood Adv 5(24):5536–5545

Guo H et al (2022) Multiomics analysis identifies SOCS1 as restraining T cell activation and preventing graft-versus-host disease. Adv Sci 9(21):e2200978

Jun HX, Jun CY, Yu ZX (2004) In vivo induction of T-cell hyporesponsiveness and alteration of immunological cells of bone marrow grafts using granulocyte colony-stimulating factor. Haematologica 89(12):1517–1524

Kebriaei P et al (2018) Intravenous Busulfan compared with Total body irradiation pretransplant conditioning for adults with acute lymphoblastic leukemia. Biol Blood Marrow Transplant 24(4):726–733

Kroger N et al (2017) Dose-reduced versus standard conditioning followed by allogeneic stem-cell transplantation for patients with myelodysplastic syndrome: a prospective randomized phase III study of the EBMT (RICMAC trial). J Clin Oncol 35(19):2157–2164

Li C et al (2012) A novel conditioning regimen improves outcomes in beta-thalassemia major patients using unrelated donor peripheral blood stem cell transplantation. Blood 120(19):3875–3881

Lin F et al (2022) A modified conditioning regimen based on low-dose cyclophosphamide and fludarabine for haploidentical hematopoietic stem cell transplant in severe aplastic anemia patients at risk of severe cardiotoxicity. Clin Transpl 36(1):e14514

Luo C et al (2023) Myeloablative conditioning regimens in adult patients with acute myeloid leukemia undergoing allogeneic hematopoietic stem cell transplantation in complete remission: a systematic review and network meta-analysis. Bone Marrow Transplant 58(2):175–185

Luznik L et al (2008) HLA-haploidentical bone marrow transplantation for hematologic malignancies using nonmyeloablative conditioning and high-dose, posttransplantation cyclophosphamide. Biol Blood Marrow Transplant 14(6):641–650

Nagler A, Shimoni A (2019) In: Carreras E et al (eds) Conditioning, in the EBMT handbook: hematopoietic stem cell transplantation and cellular therapies. Springer, Cham, pp 99–107

Nagler A et al (2013) Allogeneic hematopoietic stem-cell transplantation for acute myeloid leukemia in remission: comparison of intravenous busulfan plus cyclophosphamide (Cy) versus total-body irradiation plus Cy as conditioning regimen–a report from the acute leukemia working party of the European group for blood and marrow transplantation. J Clin Oncol 31(28):3549–3556

Niederwieser C et al (2024) Reduced intensity versus myeloablative conditioning for MDS: long-term results of an EBMT phase III study (RICMAC). Bone Marrow Transplant 59(8):1084–1091

Pasquini R et al (2008) HLA-matched sibling hematopoietic stem cell transplantation for fanconi anemia: comparison of irradiation and nonirradiation containing conditioning regimens. Biol Blood Marrow Transplant 14(10):1141–1147

Peters C et al (2021) Total body irradiation or chemotherapy conditioning in childhood ALL: a multinational, randomized, noninferiority phase III study. J Clin Oncol 39(4):295–307

Prata PH et al (2020) Haploidentical transplantation and posttransplant cyclophosphamide for treating aplastic anemia patients: a report from the EBMT Severe Aplastic Anemia Working Party. Bone Marrow Transplant 55(6):1050–1058

Saraceni F et al (2018) Thiotepa, busulfan and fludarabine compared to busulfan and cyclophosphamide as conditioning regimen for allogeneic stem cell transplant from matched siblings and unrelated donors for acute myeloid leukemia. Am J Hematol 93(10):1211–1219

Saraceni F et al (2019) Fludarabine-treosulfan compared to thiotepa-busulfan-fludarabine or FLAMSA as conditioning regimen for patients with primary refractory or relapsed acute myeloid leukemia: a study from the Acute Leukemia Working Party of the European Society for Blood and Marrow Transplantation (EBMT). J Hematol Oncol 12(1):44

Schmid C et al (2006) Long-term survival in refractory acute myeloid leukemia after sequential treatment with chemotherapy and reduced-intensity conditioning for allogeneic stem cell transplantation. Blood 108(3):1092–1099

Scott BL et al (2017) Myeloablative versus reduced-intensity hematopoietic cell transplantation for acute myeloid leukemia and myelodysplastic syndromes. J Clin Oncol 35(11):1154–1161

Sora F et al (2020) Allogeneic hemopoietic stem cell transplants in patients with acute myeloid leukemia (AML) prepared with Busulfan and Fludarabine (BUFLU) or Thiotepa, Busulfan, and Fludarabine (TBF): a retrospective study. Biol Blood Marrow Transplant 26(4):698–703

Sun Q et al (2018) Haploidentical haematopoietic stem cell transplantation for thalassaemia major based on an FBCA conditioning regimen. Br J Haematol 182(4):554–558

Sun YQ et al (2021) Haploidentical stem cell transplantation with a novel conditioning regimen in older patients: a prospective single-arm phase 2 study. Front Oncol 11:639502

Swoboda R et al (2022) Total body irradiation plus fludarabine versus thiotepa, busulfan plus fludarabine as a myeloablative conditioning for adults with acute lymphoblastic leukemia treated with haploidentical hematopoietic cell transplantation. A study by the Acute Leukemia Working Party of the EBMT. Bone Marrow Transplant 57(3):399–406

Symons HJ et al (2020) Myeloablative haploidentical BMT with posttransplant cyclophosphamide for hematologic malignancies in children and adults. Blood Adv 4(16):3913–3925

Wang Y et al (2015) Haploidentical vs identical-sibling transplant for AML in remission: a multicenter, prospective study. Blood 125(25):3956–3962

Wang Y et al (2016) Haploidentical transplant for myelodysplastic syndrome: registry-based comparison with identical sibling transplant. Leukemia 30(10):2055–2063

Wang L et al (2019) Clinical outcome of FLAG-IDA chemotherapy sequential with Flu-Bu3 conditioning regimen in patients with refractory AML: a parallel study from Shanghai Institute of Hematology and Institut Paoli-Calmettes. Bone Marrow Transplant 54(3):458–464

Wang Y et al (2020) Impact of prophylactic/preemptive donor lymphocyte infusion and intensified conditioning for relapsed/refractory leukemia: a real-world study. Sci China Life Sci 63(10):1552–1564

Wang Y et al (2021) Optimizing antithymocyte globulin dosing in haploidentical hematopoietic cell transplantation: long-term follow-up of a multicenter, randomized controlled trial. Sci Bull 66(24):2498–2505

Xu ZL, Huang XJ (2023) Haploidentical transplants with a G-CSF/ATG-based protocol: experience from China. Blood Rev 62:101035

Xu LP et al (2016) Haplo-identical transplantation for acquired severe aplastic anaemia in a multicentre prospective study. Br J Haematol 175(2):265–274

Xu L et al (2018) The consensus on indications, conditioning regimen, and donor selection of allogeneic hematopoietic cell transplantation for hematological diseases in China-recommendations from the Chinese Society of Hematology. J Hematol Oncol 11(1):33

Xu ZL et al (2021) The incidence, clinical outcome, and protective factors of mixed chimerism following hematopoietic stem cell transplantation for severe aplastic anemia. Clin Transpl 35(2):e14160

Xu LP et al (2022a) Bulsufan decreases the incidence of mixed chimaerism in HLA-matched donor transplantation for severe aplastic anaemia. Bone Marrow Transplant 57(7):1204–1206

Xu LP et al (2022b) Long-term follow-up of haploidentical transplantation in relapsed/refractory severe aplastic anemia: a multicenter prospective study. Sci Bull 67(9):963–970

Xu ZL et al (2022c) Comparable long-term outcomes between upfront haploidentical and identical sibling donor transplant in aplastic anemia: a national registry-based study. Haematologica 107(12):2918–2927

Xu L et al (2022d) Unmanipulated haploidentical haematopoietic cell transplantation with radiation-free conditioning in Fanconi anaemia: a retrospective analysis from the Chinese Blood and Marrow Transplantation Registry Group. Br J Haematol 199(3):401–410

Xu L et al (2023) Fludarabine- and low-dose cyclophosphamide-based conditioning regimens provided favorable survival and engraftment for unmanipulated hematopoietic cell transplantation from unrelated donors and matched siblings in patients with Fanconi anemia: results from the CBMTR. Bone Marrow Transplant 58(1):106–108

Xu ZL et al (2024) Clinical outcomes of three haploidentical transplantation protocols for hematologic malignancies based on data from the Chinese Bone Marrow Transplantation Registry Group. Haematologica. https://doi.org/10.3324/haematol.2024.286040

Zhang XH et al (2021) The consensus from the Chinese Society of Hematology on indications, conditioning regimens and donor selection for allogeneic hematopoietic stem cell transplantation: 2021 update. J Hematol Oncol 14(1):145

Zhang H et al (2023) Busulfan plus cyclophosphamide versus total body irradiation plus cyclophosphamide for adults acute B lymphoblastic leukemia: an open-label, multicenter, phase III trial. J Clin Oncol 41(2):343–353

Graft Source Choice

4

Sanghee Hong and Mitchell E. Horwitz

Abstract

Bone marrow grafts offer a lower risk of graft-versus-host disease (GVHD) but require a surgical procedure and are associated with slower engraftment rates. Peripheral blood stem cells (PBSCs) enable faster engraftment and easier collection but have historically been associated with a higher risk of GVHD. Umbilical cord blood (UCB) grafts are readily available and carry no donor risk, with a higher tolerance for human leukocyte antigen mismatches and a lower risk of GVHD, but they are constrained by limited cell doses and slower engraftment. Recent advancements in transplant techniques—such as post-transplant cyclophosphamide to reduce GVHD rates in PBSC grafts, and cord blood expansion methods to increase the effectiveness of UCB grafts—have significantly improved transplant outcomes.

Each graft source offers unique benefits and drawbacks that impact the selection process, factoring in patient age, disease type, donor availability, engraftment speed, immune recovery, GVHD risk, and logistical and financial considerations. These complexities highlight the need for a comprehensive and tailored approach to graft source selection, alongside adherence to guidelines and best practices, to optimize outcomes in allogeneic hematopoietic cell transplantation.

Keywords

Graft source · Peripheral blood · Bone marrow · Umbilical cord

4.1 Background

Hematopoietic cells (HCs) can be sourced from bone marrow, peripheral blood, placenta, and umbilical cord blood (UCB). Besides their primary role as progenitor cells, HCs are also employed for other therapeutic purposes, including post-transplant donor lymphocyte infusion (Appelbaum Frederick 2007).

Within the broader category of hematopoietic cells, hematopoietic stem cells (HSCs) represent a small but vital population of pluripotent cells located in the bone marrow. These cells possess the remarkable ability to self-renew without differentiation. Self-renewal refers to the capacity of HSCs to replicate themselves without undergoing differentiation (Seita and Weissman 2010). Since mature blood cells have a relatively short lifespan, such as an average of 110–120 days of the lifespan of red blood cells, HSCs consistently

S. Hong (✉) · M. E. Horwitz
Division of Hematologic Malignancies and Cellular Therapy, Department of Medicine, Duke University School of Medicine, Durham, NC, USA
e-mail: sanghee.hong@duke.edu

© The Author(s), under exclusive license to Springer Nature Switzerland AG 2025

K. Ballen (ed.), *Clinical Management of Challenges in Allogeneic Stem Cell Transplant*, Advances in Experimental Medicine and Biology 1475, https://doi.org/10.1007/978-3-031-84988-6_4

Table 4.1 Characteristics of different graft sources

	Bone marrow	Peripheral blood stem cells	Umbilical cord blood
Advantage	Lower risk of chronic GVHD	Higher cell yield, leading to faster engraftment	Readily available, collected without risk to mother or baby
	Suitable for both related and unrelated donors	No need for general anesthesia	Lower risk of GVHD
	No need for stem cell priming	Convenient collection in outpatient setting	Greater tolerance for HLA mismatch, increasing the donor pool
		Suitable for both related and unrelated donors	
Disadvantage	Requires general anesthesia for the donor	Higher risk of chronic GVHD compared to bone marrow	Limited cell dose
	Higher donor risk (surgical adverse risk)	Donor may experience side effects from mobilizing agents	Longer time to engraftment, which increases risks of infection and other complications
	Longer time to engraftment	Mobilization process may be uncomfortable for donors	Unable to proceed with donor lymphocyte infusion or stem cell boost if necessary
	Limited cell yield		

produce more differentiated progenitors while ensuring the HSC pool remains stable over a lifetime. This balance between self-renewal and differentiation is crucial for maintaining the proper size of the HSC population.

Traditionally, stem cell grafts have been obtained from three primary sources: peripheral blood, bone marrow, and UCB. When allogeneic hematopoietic cell transplantation (allo-HCT) was first introduced, HSCs were collected exclusively from bone marrow (Thomas et al. 1957), but now peripheral blood and UCB became important sources of HSC as well. Each of these sources offers distinct advantages and challenges that can impact both the short-term and long-term outcomes of transplantation (Table 4.1). For instance, peripheral blood stem cells (PBSCs) have gained popularity due to their ease of collection and rapid engraftment (Fig. 4.1); however, they may also be associated with a higher incidence of graft-versus-host disease (GVHD). Conversely, bone marrow-derived grafts are associated with a more gradual engraftment process but may offer a lower risk of GVHD. Graft source and transplant technique may heavily influence the outcome of allo-HCT.

The choice of graft source is not merely a technical decision; it involves a comprehensive assessment of the patient's clinical condition, disease characteristics, donor availability, and institutional protocols. As the landscape of stem cell transplantation continues to evolve with technological advancements and changes in treatment algorithms, including the use of haploidentical donors and the exploration of novel cell sources, understanding the implications of graft source selection has become increasingly vital.

4.2 Bone Marrow Stem Cell Graft

Bone marrow, the spongy tissue residing within bones, harbors HSCs essential for the reconstitution of the blood and immune system. The main structures of bone marrow include the bone, the vasculature, and a network of reticular stromal cells, with hematopoiesis occurring in the spaces between the vessels, bone, and reticular cells. This graft source has been extensively used since the advent of hematopoietic stem cell transplantation (HSCT) in the 1950s and continues to be a cornerstone in contemporary trans-

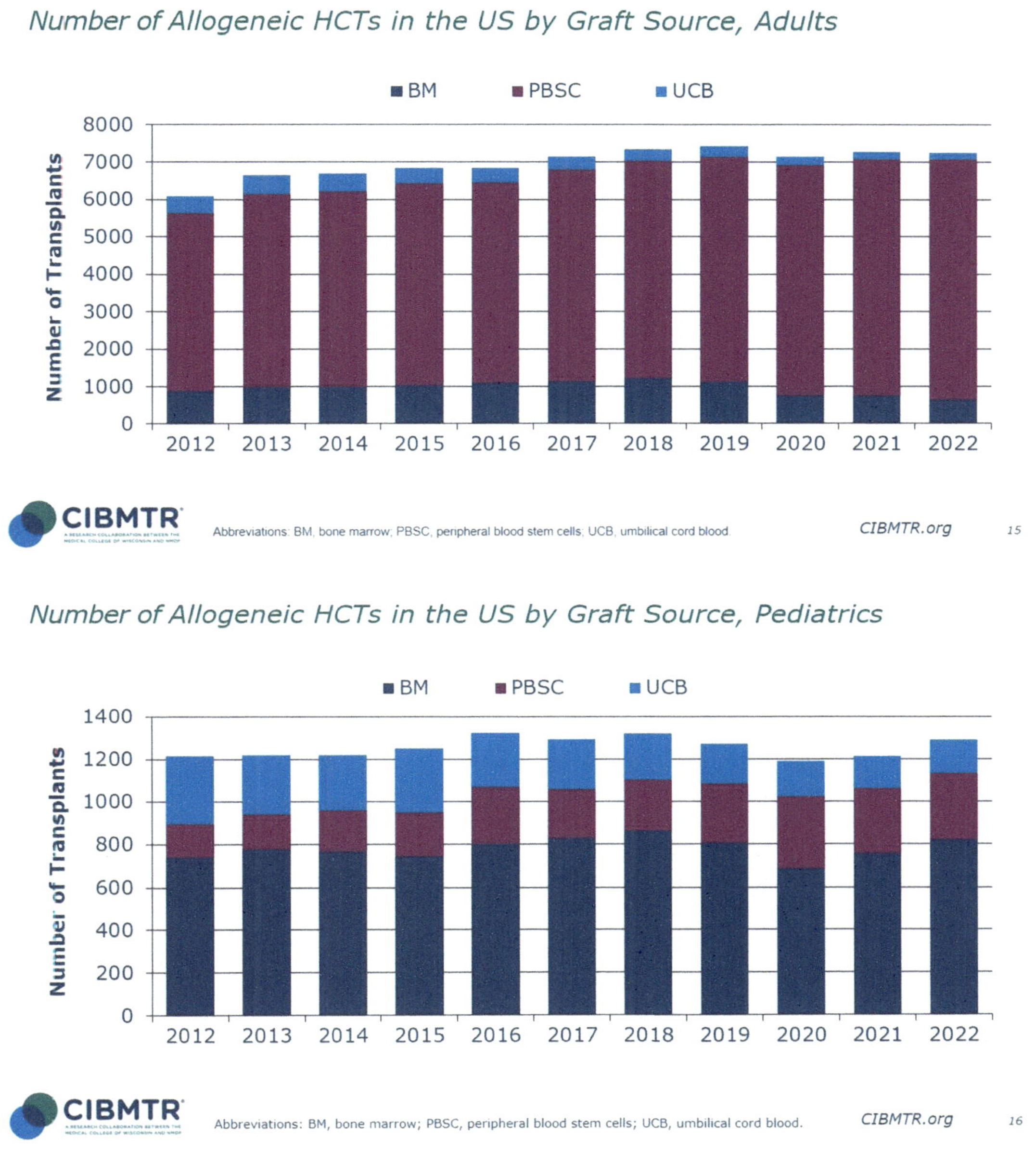

Fig. 4.1 Trend of graft source use in the United States. (Source: Courtesy of Center for International Blood and Marrow Transplant Research Summary Slides, 2023 (Cusatis et al. 2023))

plant medicine (Thomas et al. 1957, 1973). Bone marrow graft usage has decreased since the advent of granulocyte-colony stimulating factor (G-CSF) as a stem cell mobilizing agent, which has allowed for the collection of stem cells from the peripheral blood. However, bone marrow grafts still account for 11% of total products provided by the National Marrow Donor Program (Russell et al. 1996; Lane 1996). One of the challenges in bone marrow graft is the logistical complexity in comparison with PBSCs, as described below.

4.2.1 Logistics of Bone Marrow Collection

The logistics of bone marrow collection necessitate a coordinated effort between the donor, the collection team, and the transplant center, as detailed in the recently published white paper (Murakami et al. 2024). The process begins with donor identification and screening, ensuring compatibility and minimizing the risk of complications. The donor undergoes a surgical procedure under general or regional anesthesia, where multiple aspirations are performed from the posterior iliac crests to harvest the marrow. The aspirated marrow is then filtered, processed, and transported to the recipient in a time-sensitive manner to preserve cell viability. This logistical pathway, while complex, is well established and supported by standardized protocols to ensure the safety and efficacy of the graft.

At donor consent, reviewing possible complications of the harvest procedure in detail helps the donor and their caregiver to understand and prepare for the upcoming procedure. The harvest site pain, mechanical back pain, fatigue, nausea, vomiting (related to anesthesia), sore throat (related to intubation), and IV site pain are the most common complications occurring in ≥20% of cases (Rowley et al. 2001; Heldal et al. 2002; Pulsipher et al. 2013). Less frequently (<20%), donors may experience fever, bleeding or hematoma at harvest site, minor wound infection, hypotension, urinary retention, syncope, and chipped teeth (Rowley et al. 2001; Heldal et al. 2002; Pulsipher et al. 2013).

4.2.2 Optimal Cell Dose

The optimal cell dose is a critical determinant of transplant success, influencing engraftment kinetics and long-term outcomes (Gorin et al. 2003; McCurdy et al. 2019; Kollman et al. 2016). In bone marrow transplantation, the target dose is typically 4×10^8 total nucleated cells (TNCs) per kg of the recipient's ideal body weight with a recommended minimum yield of 2.5×10^8 TNC/kg. In a multicenter trial of haploidentical bone marrow transplantation for severe aplastic anemia (BMT CTN 1502), $>2.5 \times 10^8$ TNC/kg was associated with better 1-year GVHD-free survival at 85% (64–94%) (DeZern et al. 2020). Achieving this cell dose by ensuring bone marrow quality is imperative as it directly correlates with the speed of hematopoietic recovery and overall survival rates. Over the past few decades, the prevalence of obesity in the United States has been steadily increasing. Recently, in 2012–2016, mean TNC concentration has fallen to 18.7×10^6 cells/mL from 21.8×10^6 cells/mL (1994–1996), possibly due to an increase in lower-volume collection centers leading to poorer-quality harvests (Prokopishyn et al. 2019). Insufficient cell doses are linked with delay or failure in engraftment and fatal infections. Consequently, meticulous attention is paid to the volume and quality of the marrow harvested to meet the required cell dose thresholds. Recently, a white paper for best practices for bone marrow harvest guides details of ideal communication with donors, anesthesia, procedure and troubleshooting, postprocedural donor assessments, and follow-up (Murakami et al. 2024).

4.2.3 Disease Indications for Bone Marrow Grafts

Bone marrow as a graft source is particularly advantageous in specific disease contexts. In severe aplastic anemia, bone marrow grafts are associated with less risk of GVHD, leading to improved overall survival when compared with PBSC grafts in all age groups in related and unrelated donor settings (Eapen et al. 2011; Bacigalupo et al. 2012; Chu et al. 2011; Schrezenmeier et al. 2007). Conversely, bone marrow grafts in Fanconi Anemia are associated with less secondary cancers without significant benefits in GVHD or overall survival at a median follow-up of 6 years (Peffault de Latour et al. 2013). The choice of bone marrow is often guided by its unique biological properties and the specific needs of the underlying disease.

4.3 Advantages of Using Bone Marrow Grafts

4.3.1 Lower Risk of Graft-Versus-Host Disease (GVHD)

GVHD remains a significant challenge in allogeneic transplantation, characterized by the donor immune cells attacking the recipient's tissues. Bone marrow grafts were associated with a lower incidence and severity of GVHD compared to peripheral blood stem cell (PBSC) grafts (Gorin et al. 2003; Mehta et al. 2021). From a retrospective study, the 1-year rate of chronic GVHD was 8% when compared to 21% from PBSC graft (Mehta et al. 2021). In another registry-based study of allogeneic transplants for acute myeloid leukemia (AML), also 2-year chronic GVHD incidence was 41% vs. 53% in PB grafts from a randomized trial ($p = 0.01$) (Anasetti et al. 2012). However, these studies were performed before the wide use of post-transplant cyclophosphamide (PT-Cy) as a GVHD prophylaxis (Bolaños-Meade et al. 2023). However, recent haploidentical transplantation using PT-Cy prophylaxis for patients with AML was associated with a lower risk of acute GVHD, but no significant difference was observed in chronic GVHD or nonrelapse mortality (Devillier et al. 2024).

This reduced risk is attributed to the lower content of mature T cells in bone marrow, which are the primary mediators of GVHD. Chronic GVHD begins with naive T cells differentiating into inflammatory subsets, leading to thymic damage, impaired antigen presentation, and abnormal T- and B-cell activation (MacDonald et al. 2017). This process may generate alloantibodies and autoantibodies, contributing to tissue damage and conditions like scleroderma, with new therapies targeting T-cell depletion, cytokine inhibition, and regulatory T-cell restoration. Consequently, patients receiving bone marrow grafts experience fewer GVHD-related complications, improving their overall quality of life and long-term survival outcomes (Lee et al. 2016). This advantage makes bone marrow an attractive graft source, particularly for patients at high risk of GVHD or those with preexisting comorbidities that could exacerbate GVHD symptoms. As a result, some centers prefer bone marrow, particularly for haploidentical HSCT.

4.4 Disadvantages of Bone Marrow Grafts (Table 4.2)

4.4.1 Procedural Risk

The collection of bone marrow necessitates a surgical procedure performed under general or regional anesthesia. This requirement introduces small risks inherent to surgical interventions, such as anesthesia-related complications, infection, and bleeding. While varing by center practice, >99% of volunteer unrelated donors were performed under general anesthesia (Murakami et al. 2024). The reported rate of severe adverse incidence is 4.32 per 10,000 donations, with one reported incidence of death due to postoperative pulmonary embolism (Halter et al. 2009). For both related and unrelated donors, 36% and 31% experienced grade 2–4 pain requiring perioperative management, and also 24% and 17% required management for grade 2–4 toxicities from dizziness, site reactions, nausea, and syncope, respectively (Pulsipher et al. 2019). The invasive nature of the procedure also limits its applicability in certain donor populations, such as those with contraindications to anesthesia or significant medical comorbidities. These factors can pose logistical and ethical challenges, particularly in volunteer donors, necessitating careful donor selection and thorough informed consent processes (Murakami et al. 2024).

4.4.2 Delayed Hematopoietic Recovery

Compared to PBSC grafts, bone marrow grafts are associated with delayed hematopoietic recovery (Anasetti et al. 2012). The engraftment kinetics are slower, demonstrated by slower neutrophil and platelet recovery by an average of 3–4 days

Table 4.2 Known complications of bone marrow harvests[a]

Common (20–85%)	Less common (<20%)	Uncommon (about 1%)	Rare (<1%)
Bone marrow site and mechanical back pain	Hematoma, bleeding	Deep infection, bacteremia, abscess	Severe pain
Fatigue	Superficial wound infection	Prolonged pain/sacroiliitis	Prolonged paralysis after anesthesia
Nausea, vomiting	Urinary retention	Neuropathy	Anaphylaxis
Sore throat	Syncope, hypotension	Prolonged hospitalization	Myocardial infarction
Headache	Chipped teeth	Seizure	Arrhythmia
	Fever		Pulmonary embolism
			Fat embolism
			Transfusion complication
			Malignant hyperthermia
			Pulmonary edema
			Stroke
			Death

[a]Reformatted from Murakami et al. (2024)

when compared to PBSC graft (Pavletic et al. 1997). Slow engraftment exposes patients to prolonged periods of pancytopenia and increased vulnerability to infections and bleeding. This delayed recovery necessitates extended supportive care, including transfusions, antimicrobial prophylaxis, and vigilant monitoring for complications. The slower engraftment can be particularly challenging in patients with aggressive diseases or those who have undergone intensive pretransplant conditioning regimens, where rapid hematopoietic recovery is crucial for overall survival and disease control.

4.4.3 Limited Cell Dose

To highlight that the concentration of pluripotent hematopoietic stem cells (HSCs) is higher in bone marrow, it is important to note that the minimum cell dose required for engraftment is significantly lower in bone marrow grafts compared to peripheral blood stem cell (PBSC) grafts. While the generally accepted minimum dose for PBSCs is around 2×10^6 CD34+ cells per kilogram, the minimum dose for bone marrow is less clearly defined but is likely much lower than 2×10^6/kg. Bone marrow grafts typically provide a lower overall cell dose compared to PBSC grafts; data from the European registry shows a median bone marrow TNC dose of 2.7×10^8/kg (range: $0.17–29 \times 10^8$/kg) versus 9.3×10^8/kg (range: $0.02–77 \times 10^8$/kg) for PBSC grafts (Gorin et al. 2003). The limited cell dose in bone marrow grafts can be a significant drawback, particularly in adult recipients where higher cell doses are required to improve survival and prevent relapse (Gorin et al. 2003; Savani et al. 2016). Inadequate cell doses ($<5 \times 10^6$/kg of CD34+ dose) can compromise engraftment, leading to poor graft function, graft failure, or delayed hematopoietic recovery (Man et al. 2022; Zhao et al. 2019; Sun et al. 2019; Alchalby et al. 2016). To mitigate this risk, sometimes despite best practices (Murakami et al. 2024), multiple bone marrow harvest sessions may be required, increasing the burden on donors and complicating the logistical aspects of collection. This limitation underscores the importance of maximizing cell yield during the initial harvest and considering alternative strategies, such as combining bone marrow with PBSC grafts to achieve the desired cell dose.

4.4.4 Conclusion

Bone marrow grafts remain a vital and viable graft source in the landscape of hematopoietic stem cell transplantation. While the use of bone marrow graft is influenced by the logistics of collection, optimal cell dose considerations, and disease-specific indications, bone marrow graft provides diverse stem cell populations and lower GVHD. Depending on the indications, bone marrow grafts may be the preferred graft source for some allogeneic hematopoietic cell transplantation.

4.5 Peripheral Blood Stem Cells

Peripheral blood stem cells (PBSCs) have emerged as a prominent graft source for allogeneic hematopoietic cell transplantation (HSCT), offering several advantages and challenges compared to traditional bone marrow transplantation (BMT). The logistics of PBSC collection and the increasing trend in their usage have significantly impacted the field of HSCT. The Center for International Blood and Marrow Transplant Research (CIBMTR) reports a notable increase in PBSC utilization for allogeneic transplants, highlighting its importance in contemporary clinical practice, as in Fig. 4.1.

PBSCs are collected from the peripheral blood of donors via leukapheresis, following mobilization with growth factors such as G-CSF for 4–5 days. This nonsurgical procedure contrasts with the more invasive bone marrow harvest and provides a higher yield of CD34+ stem and progenitor cells, which is beneficial for engraftment and hematopoietic recovery.

4.5.1 Reduced GVHD Risks of PBSC Grafts with Post-transplant Cyclophosphamide

PT-Cy rapidly induces in vivo alloreactive T-cell depletion through active suppressive mechanisms, primarily driven by the early recovery of CD4 + CD25 + Foxp3+ regulatory T cells, leading to the prevention of GVHD (Wachsmuth et al. 2019). In 2023, a pivotal phase III randomized controlled trial (BMT CTN 1703) demonstrated superior GVHD-free, relapse-free survival at 1 year using PT-Cy, tacrolimus, mycophenolate mofetil prophylaxis in reduced-intensity conditioning PBSC grafts from human leukocyte antigen (HLA)-matched related or unrelated donor or 7/8 mismatched unrelated donor, compared to tacrolimus and methotrexate prophylaxis (Bolaños-Meade et al. 2023). Following the dissemination of this BMT CTN 1703 trial result, PT-Cy has been incorporated into standard GVHD prophylaxis not only for reduced-intensity PBSC grafts but also for myeloablative conditioning regimens. This shift in GVHD prophylaxis has led to a significant reduction in the risks of both acute and chronic GVHD for PBSC grafts.

This topic will be further discussed in Chap. 5, Optimal GVHD Prophylaxis.

4.5.2 Current Trend and Logistics

The shift toward PBSCs can be attributed to several factors, including the ease of collection, higher stem cell yield, and faster hematopoietic recovery post-transplant. There has been a steady increase in the number of PBSC transplants in both adult and pediatric recipients, and preference toward PBSC grafts was amplified during the COVID-19 pandemic given operating room availability leading to a 29% decrease in unrelated donor bone marrow harvests globally (Cusatis et al. 2023; Jöris et al. 2022; Parmar et al. 2023). This trend reflects the logistical advantages and clinical benefits associated with PBSC transplantation.

The logistics of PBSC collection involve mobilizing hematopoietic stem cells (HSCs) from the bone marrow into the peripheral blood using growth factors like G-CSF. Donors undergo several days of subcutaneous G-CSF injections, followed by leukapheresis to collect the mobilized stem cells. This process is less invasive compared to bone marrow harvest, which requires general anesthesia and multiple marrow aspirations.

4.6 Advantages of PBSCs

4.6.1 Faster Hematopoietic Recovery

As discussed earlier, a study from the European Blood and Marrow Transplant Registry found that the median TNC collected from PB grafts was more than three times of bone marrow grafts (Gorin et al. 2003). Also, most peripherally mobilized cells are more terminally differentiated cells, reducing the duration of neutropenia and associated risks of life-threatening infections (Anasetti et al. 2012; Savani et al. 2016). There are conflicting data on decreased or similar rates of acute GVHD in PBSC grafts when compared to bone marrow grafts (Anasetti et al. 2012; Cutler et al. 2001). However, overall survival and relapse between PBSC and bone marrow grafts were found not to be statistically different (Anasetti et al. 2012).

4.6.2 Easier Collection Procedure

The ease of PBSC collection is a significant factor contributing to the increased usage of this graft source. The nonsurgical nature of PBSC collection is another crucial benefit when compared to bone marrow harvests.

For related and unrelated donors, 44% and 37% of grade 2–4 skeletal pain were experienced on day 1 of collection, respectively (Pulsipher et al. 2019). Similarly, grade 2–4 of other toxicities were experienced in 18% of related donors and 9% of unrelated donors (Pulsipher et al. 2019). These rates of skeletal pain and other toxicities were significantly lower than those from bone marrow donation. Additionally, the recovery time for PBSC donors is shorter compared to bone marrow donors, allowing for a quicker return to normal activities. Many donors can resume normal activities typically a few days up to 1 month after the procedure, as the side effects of G-CSF are generally mild and transient (Pulsipher et al. 2019). This ease of collection makes PBSC donation more appealing to potential donors, enhancing the donor pool for allogeneic HSCT.

4.7 Disadvantages of PBSCs

4.7.1 Increased Risk of Chronic Graft-Versus-Host Disease

PBSC allo-HSCT was traditionally associated with a higher risk of chronic GVHD compared to bone marrow transplantation, as discussed earlier in the chapter (Mehta et al. 2021; Anasetti et al. 2012; Flowers et al. 2002). The higher T-cell content in PBSC grafts is believed to contribute to the increased incidence of chronic GVHD (Cutler et al. 2001). While T cells play a crucial role in the graft-versus-leukemia (GVL) effect, their presence also increases the risk of GVHD.

However, the comparisons were made when PBSC grafts were used with antiquated GVHD prophylaxis without PT-Cy. There are no randomized controlled trials comparing bone marrow and PBSC grafts with PT-Cy GVHD prophylaxis. However, given the degree of improvement in both acute (grade III–IV from 15% to 6%) and chronic GVHD (35%–22% at 1 year) (Bolaños-Meade et al. 2023), PBSC grafts may no longer be at a significant disadvantage at the risks of chronic GVHD.

4.7.2 Longer Days of Growth Factor Administration and Collection

G-CSF injections are typically administered for 4–5 days before leukapheresis, and 2–49% of donors require more than one leukapheresis session to achieve target cell doses (Anderlini et al. 1997; Richa et al. 2009; Wang et al. 2008; Ings et al. 2006; Schroeder et al. 2017). Advanced age of the donor is the most well-known factor associated with needing more than one leukapheresis session (Anderlini et al. 1997; Richa et al. 2009). Additional known factors associated with poor day 1 CD34+ cell yield include female donor sex, venous access, and a higher flow rate (Wang et al. 2008). This prolonged process can potentially affect donors' willingness to donate in the future. Additionally, many donors require central access for leukapheresis, which increases their risk of central-line-related infections and thrombosis.

Furthermore, while this study dates to the late 1990s, it estimated the cost of peripheral blood stem cell transplantation (PBSCT) following G-CSF mobilization at $6688, compared to $4262 for bone marrow harvesting (Bennett et al. 1999).

4.7.3 Complications of Growth Factors

The use of growth factors like G-CSF is generally safe, but it is not without complications (Anderlini 2009). Common side effects include skeletal pain, fatigue, insomnia, nausea, injection site redness, and headaches, which are usually mild and self-limiting (Newburger 2006; Pamphilon et al. 2008; Anderlini et al. 1996). However, rare but serious adverse events such as splenic rupture and severe allergic reactions can occur. It is essential to carefully monitor donors during the mobilization process and provide appropriate supportive care to manage any complications. No significant long-term side effects of G-CSF mobilization were noted, and donor experience was reported to be similar after 3 weeks of donation for both PBSC and bone marrow donors (Switzer et al. 2014).

4.7.4 Modern Techniques

4.7.4.1 Mobilization: Plerixafor and Motixafortide

Recent advancements in mobilization techniques have further improved the efficiency of PBSC collection. Plerixafor (Mozobil) and motixafortide (Aphexda) are two agents that have been approved by the US Food and Drug Administration (FDA) for use in autologous stem cell mobilization along with G-CSF in 2008 and 2023, respectively. A difference in their indications is that plerixafor was approved for both lymphoma and multiple myeloma (plerixafor) and multiple myeloma only (motixafortide). They are now being explored in the allogeneic donor setting.

Plerixafor is a low-affinity, short-acting CXCR4 antagonist that, when used in combination with G-CSF, significantly enhances the mobilization of HSCs into the peripheral blood. In a retrospective review, 4.1% of 1008 allo-HSCT required one dose of plerixafor due to poor collection after G-CSF yield that resulted in a median 2.9 (range 0.8–7.7) fold increase in CD34+ yield (Zhuang et al. 2022). When the first collection after G-CSF mobilization was <63% of the collection goal, the addition of plerixafor was found to be helpful in meeting the collection goal (Zhuang et al. 2022). Also, a phase II trial using a single-dose plerixafor 240 mcg/kg to mobilize HLA-identical sibling donors resulted in a median CD34+ dose of 4.7×10^6 /kg of recipient body weight (range $0.9–9.6 \times 10^6$), within 2 days of collection (Chen et al. 2019). None of the 64 donors from this trial experienced bone pain (Chen et al. 2019). Transplant outcomes of plerixafor mobilized allo-HCT were not significantly different from the controlled G-CSF mobilized PBSC graft cases (Chen et al. 2019; Devine et al. 2008).

Motixafortide, another CXCR4 antagonist with high affinity and slow receptor dissociation rate has demonstrated potent stem cell mobilizing effects as a single agent (Crees et al. 2023). A single-center phase 2 trial for sibling donor mobilization after a single-dose 1.0 and 1.25 mg/kg motixafortide subcutaneously resulted in 85% and 100% collection of CD34+ dose of $\geq 2 \times 10^6$/kg within two leukapheresis sessions, respectively (Crees et al. 2023). Interestingly, motixafortide preferentially mobilized naive and central memory T cells, which potentially leads to better engraftment and long-term hematopoietic reconstitution (Crees et al. 2023; Pasquini et al. 2012; Montoro et al. 2020; Bleakley et al. 2022; Oevermann et al. 2014; Handgretinger 2012; Pfeiffer et al. 2010).

4.7.5 Conclusion

The use of PBSCs as a graft source for allogeneic HSCT offers several advantages, including higher stem/progenitor cell yield, faster hematopoietic recovery, and a nonsurgical collection procedure. However, challenges such as increased risk of chronic GVHD and inconvenience and complications related to growth factor administration balanced with these advantages. Recently, PT-CY-based GVHD prophylaxis has substantially reduced GVHD rates, enhancing the safety and efficacy of PBSC transplantation. The technological advancements have contributed to the growing preference for PBSCs as the graft source in allogeneic HSCT.

4.8 Umbilical Cord Blood

UCB has for many years been a valuable source of HSC for allogeneic HSCT. The success of a UCB transplant in a child in 1986 marked the beginning of UCB being recognized as an alternative source in specific situations (Gluckman et al. 1997). Initially, the logistics of UCB collection, storage, and transplantation posed significant challenges, but advancements in these areas have made it a viable alternative to traditional bone marrow and peripheral blood sources. The UCB graft is placental blood that is collected from the umbilical cord postpartum. The process is noninvasive and poses no risk to the mother or newborn. The collected blood is then cryopreserved in public or private cord blood banks until needed for transplantation.

The logistics of UCB involve several critical steps. After a baby is born and the umbilical cord is clamped and cut, the remaining blood in the umbilical cord and placenta is collected using a sterile needle and bag. This process must be conducted quickly and efficiently to ensure the maximum yield and viability of the HSCs. The collected blood is then transported to a cord blood bank, where it undergoes testing for infectious diseases, HLA typing, and cell count analysis. After these assessments, the UCB is cryopreserved using controlled-rate freezing and stored in liquid nitrogen tanks at temperatures below −150 °C.

The use of UCB as a graft source offers several logistical advantages. Unlike bone marrow and peripheral blood, which require donor availability and a degree of urgency, UCB units are readily available off the shelf. This immediacy can be crucial for patients with aggressive diseases requiring rapid intervention. Moreover, the availability of UCB units from diverse ethnic backgrounds enhances the likelihood of finding suitable matches for patients from minority groups, addressing a significant gap in donor availability. The establishment of large public cord blood banks and international registries has further facilitated access to UCB, making it a globally accessible graft source.

4.9 Advantages of Umbilical Cord Blood as a Graft Source

4.9.1 Reduced Risk of GVHD

One of the most significant advantages of UCB transplantation is the reduced risk of GVHD. HLA-mismatched UCB transplant was shown to have comparable survival outcomes when compared with bone marrow grafts and also with lower rates of acute GVHD from mismatched UCB compared to mismatched bone marrow grafts (Laughlin et al. 2004). The immunologically naive nature of UCB cells contributes to a lower incidence and severity of GVHD compared to other graft sources (Wing et al. 2003). This is particularly beneficial for patients with high-risk hematologic malignancies, as it can improve overall survival and quality of life post-transplant.

The reduced risk of GVHD with UCB is attributed to several factors. First, UCB cells have lower levels of mature T cells, which are primarily responsible for initiating GVHD. Second, UCB contains a higher proportion of regulatory T cells (T_{reg}) and NK cells, which can modulate the immune response and reduce the likelihood of GVHD (Wing et al. 2003; Godfrey et al. 2005). Furthermore, the neonatal immune cells in UCB are more adaptable and less likely to mount a strong immune response against the recipient's tissues (Michaëlsson et al. 2006; Burt 2013).

4.9.2 Greater Tolerance and Broader Donor Availability

UCB is known for its greater tolerance of HLA mismatches, which broadens the pool of potential donors. While traditional HCT requires a high degree of HLA matching to reduce complications, UCB transplantation can be successful with lower levels of matching (4/6–6/6) (Laughlin et al. 2004; Gluckman et al. 2004). Consequently, UCB provides a viable option for patients who cannot find a fully matched donor within their family or registries.

The broader donor availability of UCB is particularly important for patients from ethnic minority groups who are underrepresented in bone marrow donor registries. UCB banks often have a diverse collection of units, increasing the chances of finding a suitable match for these patients. Additionally, the process of finding a match is faster with UCB, as the units are already collected, typed, and cryopreserved. This is a significant advantage for patients with rapidly progressing diseases who need immediate transplantation.

4.9.3 Disease Indication

For pediatric bone marrow failure syndromes, primary immunodeficiency disorders, and hemoglobinopathies such as sickle cell disease and thalassemia, the outcomes of UCB transplantation have continued to improve, and UCB is the preferred graft source (Spees et al. 2019; Ebens et al. 2018; Vander Lugt et al. 2020). In children with inherited metabolic disorders, transplant outcomes using noncarrier-matched sibling bone marrow (BM) are comparable to those using fully matched single UCB, and both sources are preferred in this indication (Boelens et al. 2013; Mallhi et al. 2017; Aldenhoven and Kurtzberg 2015).

4.10 Disadvantages of Umbilical Cord Blood as a Graft Source

4.10.1 Limited Cell Dose and Delayed Engraftment

A significant limitation of UCB is the limited number of HSCs in a single UCB unit. While UCB contains the highest concentration of primitive HSCs among the three graft sources, the overall low absolute number and low concentration of hematopoietic progenitor cells contribute to slow engraftment kinetics and an increased risk of graft failure. Single cord unit with TNC 2.5×10^7/kg or greater for myeloablative transplant for leukemia or myelodysplastic syndrome was associated with less transplant-related mortality (Barker et al. 2010). This is particularly challenging for adult recipients who require a higher cell dose for successful engraftment. The limited cell dose can lead to delayed hematopoietic recovery, increasing the risk of infections, bleeding, and other complications in the immediate post-transplant period. Hence, for adult patients, pioneered by Barker et al. (2005), double UCB transplantation is commonly used to extend the application of umbilical cord transplantation (Scaradavou et al. 2013). American Society of Transplantation and Cellular Therapy published a guideline for UCB transplantation with a minimum TNC dose for single-unit grafts as 2.5×10^7/kg and for double-unit grafts as 1.5×10^7/kg for each graft and with a minimum CD34+ dose for single-unit grafts as 1.5×10^7/kg and for double-unit grafts as 1.0×10^7/kg for each graft (Politikos et al. 2020).

4.10.2 Slower Immune Reconstitution

Compared to bone marrow and peripheral blood stem cell transplants, UCB transplantation is associated with slower immune reconstitution. The delayed recovery of immune function leaves patients vulnerable to opportunistic infections and relapse of the underlying disease. This necessitates prolonged prophylactic measures and close monitoring, which can impact the overall treatment course and patient outcomes.

The slower immune reconstitution with UCB transplantation is due to several factors. The limited number of HSCs in UCB leads to a slower production of new immune cells. Additionally, the neonatal immune cells in UCB take longer to mature and become fully functional (Kim and Broxmeyer 2011). This prolonged period of immune deficiency requires extensive use of antimicrobial prophylaxis and vigilant monitoring for infections, which can be resource-intensive and costly.

BMT CTN 1101 trial was a randomized controlled trial to compare double UCB transplantation with haploidentical bone marrow transplantation following reduced-intensity conditioning. This trial showed a significantly lower rate of neutrophil recovery after UCB transplant (95% vs. 99% in haploidentical transplantation, $p = 0.05$) and worse 5-year nonrelapse mortality (24% vs. 10%, $p = 0.002$) likely reflecting the downstream effect of the delayed engraftment and slower immune reconstitution (Fuchs et al. 2021; O'Donnell et al. 2022).

4.10.3 Potential Cost Issues

The cost of UCB transplantation includes several components. The initial collection and processing of UCB involve specialized equipment and trained personnel. The long-term storage of UCB units in liquid nitrogen tanks requires continuous maintenance and monitoring. The processes of collection, testing, cryopreservation, and storage are expensive, and these costs are often passed on to patients and healthcare systems. Additionally, the need for double UCB units or other strategies to augment cell dose further increases the financial burden.

Cost considerations are a crucial factor in the decision-making process for both healthcare providers and patients when choosing a graft source. Furthermore, the use of multiple UCB units or advanced techniques such as ex vivo expansion adds to the overall cost to the healthcare system. Additionally, due to the longer expected time to engraftment and immune reconstitution, double UCB transplantation is associated with a longer initial length of stay. Overall, the estimated total cost of double UCB transplant is $782,190, the highest among all allo-HSCT types (Kim et al. 2024). The cost of maintaining cord blood banks is indirectly linked to the expenses associated with umbilical cord blood transplants. These financial considerations must be weighed against the clinical benefits of UCB transplantation, especially in healthcare systems with limited resources.

However, when considering the accumulating medical cost post-transplant, UCB transplant for young (<65 years) patients was projected to be less costly by approximately $119,000 than haploidentical bone marrow transplantation using PT-CY GVHD prophylaxis, based on a phase III randomized controlled trial (Ramsey et al. 2023).

4.10.4 Modern Techniques to Enhance Umbilical Cord Blood Transplantation

4.10.4.1 Cord Blood Expansion Techniques

To address the limitation of the cell dose in UCB, several cord blood expansion techniques, such as ex vivo expansion, where UCB cells are cultured with mesenchymal stem cells (MSCs) and agents expanding early progenitor cells (EPC), have been developed.

The only FDA-approved cord blood expansion product is omidubicel, which is an ex vivo expanded CD133+ fraction and a nonexpanded CD133- fraction of a single cord blood using nicotinamide riboside. Nicotinamide is a form of vitamin B3 leading to HSC differentiation and bone marrow homing (Peled et al. 2012). A phase III randomized controlled trial of patient-specific expanded cord blood (omidubicel) and standard double cord transplantation demonstrated superiority in faster neutrophil (median 12, vs. 22 days) and platelet recovery (median 37, vs. 50 days) as well as lower incidence of bacterial and fungal infections (Horwitz et al. 2021). A recent pooled analysis of long-term outcomes of omidubicel use in five clinical trials combined showed 3-year overall survival at 62.7% and disease-free survival at 56.4% (Lin et al. 2023).

Recently, UM171 enhanced the activity of the CRL3 complex, which uses the KBTBD4 protein to identify its targets. This leads to the breakdown of parts of the LSD1/RCOR1 corepressor complex, restoring H3K4me2 and H3K27ac epigenetic marks that decrease quickly during the ex vivo culture of HSCs. A phase I/II trial revealed UM171 expanded cord blood transplant with median neutrophil recovery at 18 days and median platelet recovery at 42 days (Cohen et al. 2020). Reviewing long-term data, UM171 expanded cord blood transplants were associated with lower nonrelapse mortality and higher graft-versus-host disease-free, relapse-free survival compared to a historical cohort of standard double UBC transplantation and matched unrelated donor transplant (Cohen et al. 2023). This technique is still in development and requires further validation and research to fully understand its potential and applications.

Otherwise, efforts of the following ex vivo UCB expansion techniques have stalled due to persistent challenges. Co-transplanting MSCs with UCB has been investigated due to their ability to create a supportive microenvironment for HSCs, modulate the immune response, and promote hematopoiesis. While MSC co-culture with UCB demonstrated significant increases in TNCs and CD34+ cells, translating to shorter neutrophil and platelet engraftment times in some cases, consistent clinical benefits have not been conclusively established (de Lima et al. 2012). Other techniques, such as expansion of EPCs to increase HSC numbers and *Notch*-mediated ex vivo expansion of CD34+ cells, have shown promise in preclinical models but require further validation (Peled et al. 2012; Nam et al. 2014; Delaney et al. 2010). Similarly, StemRegenin-1, an aryl hydrocarbon receptor antagonist, demonstrated potential by shortening engraftment times in early-phase trials, but broader clinical applicability remains uncertain (Wagner Jr. et al. 2016; Boitano et al. 2010; Angelos et al. 2017).

4.10.4.2 Haploidentical Cord Blood Transplantation

Combining UCB with haploidentical transplantation (using a partially matched family donor) has been employed at selected transplant centers that report improved time to engraftment and decreased long-term relapse and nonrelapse mortality (Huang et al. 2022; van Besien and Childs 2016). However, another report stated similar transplant outcomes between double UCB and haploidentical cord blood HSCT (Jones et al. 2023). This strategy was thought to leverage the benefits of both graft sources: the immediate availability and lower GVHD risk of UCB, along with the higher cell dose and faster engraftment of haploidentical grafts.

4.10.5 Conclusion

Umbilical cord blood is a valuable graft source for allogeneic hematopoietic cell transplantation, offering unique advantages and facing specific challenges. Its reduced risk of GVHD and broader donor availability make it an attractive option, particularly for patients without fully matched donors. However, limitations such as the limited cell dose, delayed engraftment, slower immune reconstitution, and potential cost issues must be addressed. Modern techniques such as cord blood expansion, haploidentical and UCB combination, and co-transplantation with MSCs offer promising solutions to these challenges and are likely to enhance the role of UCB in the future of allo-HSCT.

4.11 Factors Influencing Choice of Graft Source

In the realm of allogeneic HSCT, the selection of the graft source is a critical decision that can significantly impact patient outcomes. Each source has unique advantages and limitations, and the choice of graft source is influenced by a multitude of factors, including patient-specific characteristics, disease type, donor availability, and logistical considerations.

4.11.1 Donor Availability and HLA Matching

The degree of HLA matching between the donor and recipient is a critical determinant of transplant success and complication rates. HLA-matched sibling donors are traditionally the ideal choice, but the chances of finding the HLA-matched sibling donors are expected to decrease with (1) increasing aging of the general population, (2) increasing age limit of allo-HSCT candidacy, and (3) decreased birth rates in the last several decades (Zuckerman 2017; Gragert et al. 2014). When such donors are unavailable, unrelated donor registries are utilized but racial/ethnic minorities are at a higher risk of inability to find the donor. UCB, on the other hand, offers greater tolerance for HLA mismatches, making it a viable option for patients who cannot find a fully matched donor.

BM and PBSC donors must undergo a rigorous selection process and be available at the time of transplant, which can be challenging in urgent situations. In contrast, UCB units are cryopreserved and readily available off the shelf, providing a critical advantage for patients requiring immediate transplantation. The establishment of large public cord blood banks and international registries has enhanced the accessibility of UCB for diverse patient populations.

4.11.2 Age and Size of the Recipient

Patient age and size are pivotal factors in graft source selection. Pediatric patients often receive UCB or BM transplants due to the lower incidence of GVHD associated with these sources. The smaller total blood volume and cell dose requirements in children make UCB a feasible option despite its limited cell dose. In contrast, adult patients typically require a higher cell dose for successful engraftment, which makes PBSCs a more suitable choice due to their higher HSC yield.

4.11.3 Disease Type and Status

The underlying disease and its status also play a crucial role in determining the graft source. For patients with acute leukemias or myelodysplastic syndromes, rapid engraftment and immune reconstitution are essential to prevent disease relapse and infections. PBSCs, with their higher cell dose and faster engraftment kinetics, are often preferred in these cases. Conversely, for patients with chronic leukemias or nonmalignant conditions, where avoidance of GVHD is critical, BM or UCB may be chosen for their lower GVHD risk.

4.11.4 Engraftment Speed

The speed of engraftment is a crucial consideration in graft source selection. PBSCs are associated with the fastest engraftment due to the higher number of HSCs and progenitor cells they contain. This rapid engraftment reduces the duration of neutropenia and the associated risks of infection and bleeding. BM transplants have intermediate engraftment kinetics, while UCB transplants generally have the slowest engraftment due to the limited cell dose and the immaturity of the HSCs.

4.11.5 Immune Reconstitution

The rate of immune reconstitution post-transplant is another important factor. PBSCs provide the fastest immune recovery, which is vital for protecting patients against opportunistic infections. BM offers a balance between immune reconstitution and GVHD risk, while UCB is associated with slower immune recovery, necessitating prolonged prophylactic measures and close monitoring. The slower immune reconstitution with UCB is a trade-off for its lower GVHD risk and greater HLA mismatch tolerance.

4.11.6 Risk of Graft-Versus-Host Disease

GVHD is a significant complication of allo-HSCT, where the donor's immune cells attack the recipient's tissues. PBSC transplants are associated with the highest risk of GVHD due to the higher content of mature T cells. Recently, the risks of GVHD following PBSC grafts have decreased substantially after the advent of post-transplant cyclophosphamide as a part of GVHD prophylaxis. BM transplants have a lower risk of GVHD compared to PBSCs but higher than UCB. The choice of graft source must balance

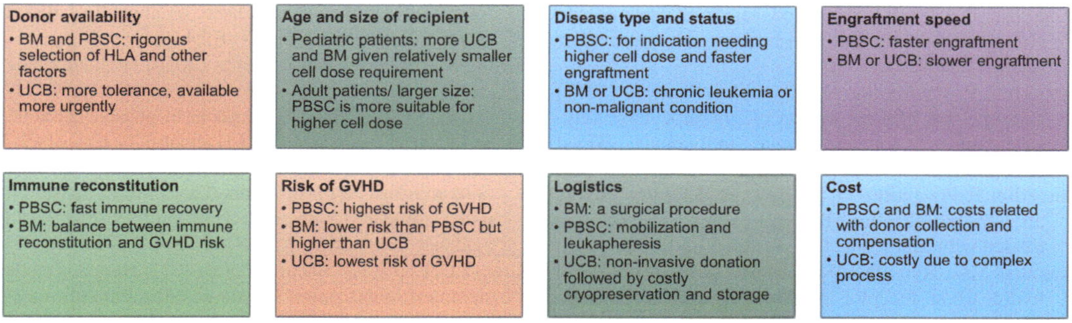

Fig. 4.2 Factors of graft source choice. Acronym: BM, bone marrow; GVHD, graft-versus-host disease; PBSC, peripheral blood stem cells; UCB, umbilical cord blood

the need for disease control through graft-versus-tumor effects with the risk of GVHD.

4.11.7 Logistics of Collection and Storage

The logistics of graft collection and storage are significant practical considerations. BM donation requires a surgical procedure under anesthesia, which can be a barrier for some donors. PBSC donation involves mobilizing HSCs into the peripheral blood with growth factors and collecting them through apheresis, which is less invasive but time-consuming. UCB donation is straightforward and noninvasive, performed at the time of birth without risk to the mother or newborn. However, the processes of cryopreservation and long-term storage of UCB units are complex and costly.

4.11.8 Cost

The financial aspects of allo-HSCT are non-negligible, with each graft source presenting different cost implications. PBSC and BM transplants involve costs related to donor mobilization, collection, and potential donor compensation. UCB transplants, while eliminating donor-related costs, involve significant expenses in the processing, testing, cryopreservation, and storage of the cord blood units. Additionally, strategies to augment the cell dose, such as double UCB transplants or ex vivo expansion, further increase the cost. The overall financial burden must be considered, particularly in resource-limited settings (Fig. 4.2).

4.12 Conclusion

The choice of graft source in allo-HSCT is a multifaceted decision influenced by patient-specific characteristics, disease type, donor availability, engraftment kinetics, immune reconstitution, GVHD risk, logistical considerations, and financial factors. Each graft source—BM, PBSCs, and UCB—offers unique advantages and faces specific limitations. The decision-making process must balance these factors to optimize patient outcomes. Technological advances and clinical innovations continue to emerge, enhancing the efficacy and feasibility of each graft source and expanding the options available for patients undergoing allo-HSCT.

References

Alchalby H, Yunus DR, Zabelina T et al (2016) Incidence and risk factors of poor graft function after allogeneic stem cell transplantation for myelofibrosis. Bone Marrow Transplant 51:1223–1227

Aldenhoven M, Kurtzberg J (2015) Cord blood is the optimal graft source for the treatment of pediatric patients with lysosomal storage diseases: clinical outcomes and future directions. Cytotherapy 17:765–774

Anasetti C, Logan BR, Lee SJ et al (2012) Peripheral-blood stem cells versus bone marrow from unrelated donors. N Engl J Med 367:1487–1496

Anderlini P (2009) Effects and safety of granulocyte colony-stimulating factor in healthy volunteers. Curr Opin Hematol 16:35–40

Anderlini P, Przepiorka D, Champlin R et al (1996) Biologic and clinical effects of granulocyte colony-stimulating factor in normal individuals. Blood 88:2819–2825

Anderlini P, Przepiorka D, Seong C et al (1997) Factors affecting mobilization of CD34+ cells in normal donors treated with filgrastim. Transfusion 37:507–512

Angelos MG, Ruh PN, Webber BR et al (2017) Aryl hydrocarbon receptor inhibition promotes hemato-lymphoid development from human pluripotent stem cells. Blood 129:3428–3439

Appelbaum Frederick R (2007) Hematopoietic-cell transplantation at 50. N Eng J Med 357:1472–1475

Bacigalupo A, Socié G, Schrezenmeier H et al (2012) Bone marrow versus peripheral blood as the stem cell source for sibling transplants in acquired aplastic anemia: survival advantage for bone marrow in all age groups. Haematologica 97:1142–1148

Barker JN, Weisdorf DJ, DeFor TE et al (2005) Transplantation of 2 partially HLA-matched umbilical cord blood units to enhance engraftment in adults with hematologic malignancy. Blood 105:1343–1347

Barker JN, Scaradavou A, Stevens CE (2010) Combined effect of total nucleated cell dose and HLA match on transplantation outcome in 1061 cord blood recipients with hematologic malignancies. Blood 115:1843–1849

Bennett CL, Waters TM, Stinson TJ et al (1999) Valuing clinical strategies early in development: a cost analysis of allogeneic peripheral blood stem cell transplantation. Bone Marrow Transplant 24:555–560

Bleakley M, Sehgal A, Seropian S et al (2022) Naive T-cell depletion to prevent chronic graft-versus-host disease. J Clin Oncol 40:1174–1185

Boelens JJ, Aldenhoven M, Purtill D et al (2013) Outcomes of transplantation using various hematopoietic cell sources in children with Hurler syndrome after myeloablative conditioning. Blood 121:3981–3987

Boitano AE, Wang J, Romeo R et al (2010) Aryl hydrocarbon receptor antagonists promote the expansion of human hematopoietic stem cells. Science 329:1345–1348

Bolaños-Meade J, Hamadani M, Wu J et al (2023) Post-transplantation cyclophosphamide-based graft-versus-host disease prophylaxis. N Engl J Med 388:2338–2348

Burt TD (2013) Fetal regulatory T cells and peripheral immune tolerance in utero: implications for development and disease. Am J Reprod Immunol 69:346–358

Chen YB, Le-Rademacher J, Brazauskas R et al (2019) Plerixafor alone for the mobilization and transplantation of HLA-matched sibling donor hematopoietic stem cells. Blood Adv 3:875–883

Chu R, Brazauskas R, Kan F et al (2011) Comparison of outcomes after transplantation of G-CSF-stimulated bone marrow grafts versus bone marrow or peripheral blood grafts from HLA-matched sibling donors for patients with severe aplastic anemia. Biol Blood Marrow Transplant 17:1018–1024

Cohen S, Roy J, Lachance S et al (2020) Hematopoietic stem cell transplantation using single UM171-expanded cord blood: a single-arm, phase 1-2 safety and feasibility study. Lancet Haematol 7:e134–e145

Cohen S, Bambace N, Ahmad I et al (2023) Improved outcomes of UM171–expanded cord blood transplantation compared with other graft sources: real-world evidence. Blood Adv 7:5717–5726

Crees ZD, Rettig MP, Bashey A et al (2023) Hematopoietic stem cell mobilization for allogeneic stem cell transplantation by motixafortide, a novel CXCR4 inhibitor. Blood Adv 7:5210–5214

Cusatis RLC, Feng Z, Allbee-Johnson M, Shaw BE (2023) Current uses and outcomes of cellular therapies: CIBMTR summary slides. CIBMTR

Cutler C, Giri S, Jeyapalan S et al (2001) Acute and chronic graft-versus-host disease after allogeneic peripheral-blood stem-cell and bone marrow transplantation: a meta-analysis. J Clin Oncol 19:3685–3691

de Lima M, McNiece I, Robinson SN et al (2012) Cord-blood engraftment with ex vivo mesenchymal-cell coculture. N Engl J Med 367:2305–2315

Delaney C, Heimfeld S, Brashem-Stein C et al (2010) Notch-mediated expansion of human cord blood progenitor cells capable of rapid myeloid reconstitution. Nat Med 16:232–236

Devillier R, Galimard JE, Blaise D et al (2024) Peripheral blood stem cell versus bone marrow graft for patients ≥60 years undergoing reduced intensity conditioning haploidentical transplantation for acute myeloid leukemia in complete remission: an analysis of the Acute Leukemia Working Party of the European Society for Blood and Marrow Transplantation. Am J Hematol 99:1250–1256

Devine SM, Vij R, Rettig M et al (2008) Rapid mobilization of functional donor hematopoietic cells without G-CSF using AMD3100, an antagonist of the CXCR4/SDF-1 interaction. Blood 112:990–998

DeZern AE, Zahurak ML, Symons HJ et al (2020) Haploidentical BMT for severe aplastic anemia with intensive GVHD prophylaxis including posttransplant cyclophosphamide. Blood Adv 4:1770–1779

Eapen M, Le Rademacher J, Antin JH et al (2011) Effect of stem cell source on outcomes after unrelated donor transplantation in severe aplastic anemia. Blood 118:2618–2621

Ebens CL, DeFor TE, Tryon R et al (2018) Comparable outcomes after HLA-matched sibling and alternative donor hematopoietic cell transplantation for children with Fanconi anemia and severe aplastic anemia. Biol Blood Marrow Transplant 24:765–771

Flowers MED, Parker PM, Johnston LJ et al (2002) Comparison of chronic graft-versus-host disease after transplantation of peripheral blood stem cells versus bone marrow in allogeneic recipients: long-term follow-up of a randomized trial. Blood 100:415–419

Fuchs EJ, O'Donnell PV, Eapen M et al (2021) Double unrelated umbilical cord blood vs HLA-haploidentical

bone marrow transplantation: the BMT CTN 1101 trial. Blood 137:420–428
Gluckman E, Rocha V, Boyer-Chammard A et al (1997) Outcome of cord-blood transplantation from related and unrelated donors. Eurocord Transplant Group and the European Blood and Marrow Transplantation Group. N Engl J Med 337:373–381
Gluckman E, Rocha V, Arcese W et al (2004) Factors associated with outcomes of unrelated cord blood transplant: guidelines for donor choice. Exp Hematol 32:397–407
Godfrey WR, Spoden DJ, Ge YG et al (2005) Cord blood CD4(+)CD25(+)-derived T regulatory cell lines express FoxP3 protein and manifest potent suppressor function. Blood 105:750–758
Gorin NC, Labopin M, Rocha V et al (2003) Marrow versus peripheral blood for geno-identical allogeneic stem cell transplantation in acute myelocytic leukemia: influence of dose and stem cell source shows better outcome with rich marrow. Blood 102:3043–3051
Gragert L, Eapen M, Williams E et al (2014) HLA match likelihoods for hematopoietic stem-cell grafts in the U.S. Registry. N Eng J Med 371:339–348
Halter J, Kodera Y, Ispizua AU et al (2009) Severe events in donors after allogeneic hematopoietic stem cell donation. Haematologica 94:94–101
Handgretinger R (2012) Negative depletion of CD3(+) and TcRαβ(+) T cells. Curr Opin Hematol 19:434–439
Heldal D, Brinch L, Tjønnfjord G et al (2002) Donation of stem cells from blood or bone marrow: results of a randomised study of safety and complaints. Bone Marrow Transplant 29:479–486
Horwitz ME, Stiff PJ, Cutler C et al (2021) Omidubicel vs standard myeloablative umbilical cord blood transplantation: results of a phase 3 randomized study. Blood 138:1429–1440
Huang F, Zeng X, Fan Z et al (2022) Haplo-peripheral blood stem cell plus cord blood grafts for hematologic malignancies might Lead to lower relapse compared with Haplo-peripheral blood stem cell plus bone marrow grafts. Transplant Cell Ther 28:849.e1–849.e8
Ings SJ, Balsa C, Leverett D et al (2006) Peripheral blood stem cell yield in 400 normal donors mobilised with granulocyte colony-stimulating factor (G-CSF): impact of age, sex, donor weight and type of G-CSF used. Br J Haematol 134:517–525
Jones MK, Kent A, Gil K et al (2023) Outcomes of haplo-cord and dual-cord transplants: a single-center retrospective analysis. Blood 142:7084–7084
Jöris MM, Schmidt AH, Bernas SN et al (2022) Impact of COVID-19 pandemic on global unrelated stem cell donations in 2020-report from world marrow donor association. Bone Marrow Transplant 57:1021–1024
Kim YJ, Broxmeyer HE (2011) Immune regulatory cells in umbilical cord blood and their potential roles in transplantation tolerance. Crit Rev Oncol Hematol 79:112–126
Kim NV, McErlean G, Yu S et al (2024) Healthcare resource utilization and cost associated with allogeneic hematopoietic stem cell transplantation: a scoping review. Transplant Cell Ther 30:542.e1–542.e29
Kollman C, Spellman SR, Zhang MJ et al (2016) The effect of donor characteristics on survival after unrelated donor transplantation for hematologic malignancy. Blood 127:260–267
Lane TA (1996) Allogeneic marrow reconstitution using peripheral blood stem cells: the dawn of a new era. Transfusion 36:585–589
Laughlin MJ, Eapen M, Rubinstein P et al (2004) Outcomes after transplantation of cord blood or bone marrow from unrelated donors in adults with leukemia. N Engl J Med 351:2265–2275
Lee SJ, Logan B, Westervelt P et al (2016) Comparison of patient-reported outcomes in 5-year survivors who received bone marrow vs peripheral blood unrelated donor transplantation: long-term follow-up of a randomized clinical trial. JAMA Oncol 2:1583–1589
Lin C, Schwarzbach A, Sanz J et al (2023) Multicenter long-term follow-up of allogeneic hematopoietic cell transplantation with omidubicel: a pooled analysis of five prospective clinical trials. Transplant Cell Ther 29:338.e1–338.e6
MacDonald KPA, Hill GR, Blazar BR (2017) Chronic graft-versus-host disease: biological insights from preclinical and clinical studies. Blood 129:13–21
Mallhi KK, Smith AR, DeFor TE et al (2017) Allele-level HLA matching impacts key outcomes following umbilical cord blood transplantation for inherited metabolic disorders. Biol Blood Marrow Transplant 23:119–125
Man Y, Lu Z, Yao X et al (2022) Recent advancements in poor graft function following hematopoietic stem cell transplantation. Front Immunol 13:911174
McCurdy SR, Kanakry CG, Tsai HL et al (2019) Development of grade II acute graft-versus-host disease is associated with improved survival after myeloablative HLA-matched bone marrow transplantation using single-agent post-transplant cyclophosphamide. Biol Blood Marrow Transplant 25:1128–1135
Mehta RS, Saliba RM, Alsfeld LC et al (2021) Bone marrow versus peripheral blood grafts for haploidentical hematopoietic cell transplantation with post-transplantation cyclophosphamide. Transplant Cell Ther 27:1003.e1–1003.e13
Michaëlsson J, Mold JE, McCune JM et al (2006) Regulation of T cell responses in the developing human fetus. J Immunol 176:5741–5748
Montoro J, Ceberio I, Hilden P et al (2020) Ex vivo T cell-depleted hematopoietic stem cell transplantation for adult patients with acute myelogenous leukemia in first and second remission: long-term disease-free survival with a significantly reduced risk of graft-versus-host disease. Biol Blood Marrow Transplant 26:323–332
Murakami MA, Connelly-Smith L, Spitzer T et al (2024) Bone marrow harvest: a white paper of best practices by the NMDP marrow alliance. Transplant Cell Ther 30:663–680

Nam TS, Park KH, Shawl AI et al (2014) Critical role for NAD glycohydrolase in regulation of erythropoiesis by hematopoietic stem cells through control of intracellular NAD content. J Biol Chem 289:16362–16373

Newburger PE (2006) Disorders of neutrophil number and function. Hematol Am Soc Hematol Educ Program:104–110

O'Donnell PV, Brunstein CG, Fuchs EJ et al (2022) Umbilical cord blood or HLA-haploidentical transplantation: real-world outcomes versus randomized trial outcomes. Transplant Cell Ther 28:109.e1–109.e8

Oevermann L, Michaelis SU, Mezger M et al (2014) KIR B haplotype donors confer a reduced risk for relapse after haploidentical transplantation in children with ALL. Blood 124:2744–2747

Pamphilon D, Nacheva E, Navarrete C et al (2008) The use of granulocyte-colony-stimulating factor in volunteer unrelated hemopoietic stem cell donors. Transfusion 48:1495–1501

Parmar G, Allan DS, Morris G et al (2023) The effect of the COVID-19 pandemic on unrelated allogeneic hematopoietic donor collections and safety. Curr Oncol 30:3549–3556

Pasquini MC, Devine S, Mendizabal A et al (2012) Comparative outcomes of donor graft CD34+ selection and immune suppressive therapy as graft-versus-host disease prophylaxis for patients with acute myeloid leukemia in complete remission undergoing HLA-matched sibling allogeneic hematopoietic cell transplantation. J Clin Oncol 30:3194–3201

Pavletic ZS, Bishop MR, Tarantolo SR et al (1997) Hematopoietic recovery after allogeneic blood stem-cell transplantation compared with bone marrow transplantation in patients with hematologic malignancies. J Clin Oncol 15:1608–1616

Peffault de Latour R, Porcher R, Dalle JH et al (2013) Allogeneic hematopoietic stem cell transplantation in Fanconi anemia: the European Group for Blood and Marrow Transplantation experience. Blood 122:4279–4286

Peled T, Shoham H, Aschengrau D et al (2012) Nicotinamide, a SIRT1 inhibitor, inhibits differentiation and facilitates expansion of hematopoietic progenitor cells with enhanced bone marrow homing and engraftment. Exp Hematol 40:342–55 e1

Pfeiffer MM, Feuchtinger T, Teltschik HM et al (2010) Reconstitution of natural killer cell receptors influences natural killer activity and relapse rate after haploidentical transplantation of T- and B-cell depleted grafts in children. Haematologica 95:1381–1388

Politikos I, Davis E, Nhaissi M et al (2020) Guidelines for cord blood unit selection. Biol Blood Marrow Transplant 26:2190–2196

Prokopishyn NL, Logan BR, Kiefer DM et al (2019) The concentration of total nucleated cells in harvested bone marrow for transplantation has decreased over time. Biol Blood Marrow Transplant 25:1325–1330

Pulsipher MA, Chitphakdithai P, Logan BR et al (2013) Acute toxicities of unrelated bone marrow versus peripheral blood stem cell donation: results of a prospective trial from the National Marrow Donor Program. Blood 121:197–206

Pulsipher MA, Logan BR, Kiefer DM et al (2019) Related peripheral blood stem cell donors experience more severe symptoms and less complete recovery at one year compared to unrelated donors. Haematologica 104:844–854

Ramsey SD, Bansal A, Li L et al (2023) Cost-effectiveness of unrelated umbilical cord blood transplantation versus HLA-haploidentical related bone marrow transplantation: evidence from BMT CTN 1101. Transplant Cell Ther 29:464.e1–464.e8

Richa E, Papari M, Allen J et al (2009) Older age but not donor health impairs allogeneic granulocyte colony-stimulating factor (G-CSF) peripheral blood stem cell mobilization. Biol Blood Marrow Transplant 15:1394–1399

Rowley SD, Donaldson G, Lilleby K et al (2001) Experiences of donors enrolled in a randomized study of allogeneic bone marrow or peripheral blood stem cell transplantation. Blood 97:2541–2548

Russell N, Gratwohl A, Schmitz N (1996) The place of blood stem cells in allogeneic transplantation. Br J Haematol 93:747–753

Savani BN, Labopin M, Blaise D et al (2016) Peripheral blood stem cell graft compared to bone marrow after reduced intensity conditioning regimens for acute leukemia: a report from the ALWP of the EBMT. Haematologica 101:256–262

Scaradavou A, Brunstein CG, Eapen M et al (2013) Double unit grafts successfully extend the application of umbilical cord blood transplantation in adults with acute leukemia. Blood 121:752–758

Schrezenmeier H, Passweg JR, Marsh JC et al (2007) Worse outcome and more chronic GVHD with peripheral blood progenitor cells than bone marrow in HLA-matched sibling donor transplants for young patients with severe acquired aplastic anemia. Blood 110:1397–1400

Schroeder MA, Rettig MP, Lopez S et al (2017) Mobilization of allogeneic peripheral blood stem cell donors with intravenous plerixafor mobilizes a unique graft. Blood 129:2680–2692

Seita J, Weissman IL (2010) Hematopoietic stem cell: self-renewal versus differentiation. Wiley Interdiscip Rev Syst Biol Med 2:640–653

Spees LP, Martin PL, Kurtzberg J et al (2019) Reduction in mortality after umbilical cord blood transplantation in children over a 20-year period (1995–2014). Biol Blood Marrow Transplant 25:756–763

Sun YQ, Wang Y, Zhang XH et al (2019) Virus reactivation and low dose of CD34+ cell, rather than haploidentical transplantation, were associated with secondary poor graft function within the first 100 days after allogeneic stem cell transplantation. Ann Hematol 98:1877–1883

Switzer GE, Bruce JG, Harrington D et al (2014) Health-related quality of life of bone marrow versus peripheral blood stem cell donors: a prespecified subgroup

analysis from a phase III RCT-BMTCTN protocol 0201. Biol Blood Marrow Transplant 20:118–127
Thomas ED, Lochte HL Jr, Lu WC et al (1957) Intravenous infusion of bone marrow in patients receiving radiation and chemotherapy. N Engl J Med 257:491–496
Thomas ED, Buckner CD, Clift RA et al (1973) Marrow grafting in patients with acute leukemia. Transplant Proc 5:917–922
van Besien K, Childs R (2016) Haploidentical cord transplantation – the best of both worlds. Semin Hematol 53:257–266
Vander Lugt MT, Chen X, Escolar ML et al (2020) Reduced-intensity single-unit unrelated cord blood transplant with optional immune boost for nonmalignant disorders. Blood Adv 4:3041–3052
Wachsmuth LP, Patterson MT, Eckhaus MA et al (2019) Post-transplantation cyclophosphamide prevents graft-versus-host disease by inducing alloreactive T cell dysfunction and suppression. J Clin Invest 129:2357–2373
Wagner JE Jr, Brunstein CG, Boitano AE et al (2016) Phase I/II trial of StemRegenin-1 expanded umbilical cord blood hematopoietic stem cells supports testing as a stand-alone graft. Cell Stem Cell 18:144–155
Wang TF, Wen SH, Chen RL et al (2008) Factors associated with peripheral blood stem cell yield in volunteer donors mobilized with granulocyte colony-stimulating factors: the impact of donor characteristics and procedural settings. Biol Blood Marrow Transplant 14:1305–1311
Wing K, Lindgren S, Kollberg G et al (2003) CD4 T cell activation by myelin oligodendrocyte glycoprotein is suppressed by adult but not cord blood CD25+ T cells. Eur J Immunol 33:579–587
Zhao Y, Gao F, Shi J et al (2019) Incidence, risk factors, and outcomes of primary poor graft function after allogeneic hematopoietic stem cell transplantation. Biol Blood Marrow Transplant 25:1898–1907
Zhuang L, Lauro D, Wang S et al (2022) Addition of plerixafor in poorly mobilized allogeneic stem cell donors. J Clin Apher 37:388–394
Zuckerman T (2017) Allogeneic transplant: does age still matter? Blood 130:1079–1080

Optimal GVHD Prophylaxis

5

Hany Elmariah and Javier Bolaños-Meade

Abstract

Graft-versus-host disease (GVHD) limits the curative potential of allogeneic hematopoietic cell transplant (HCT). This immune-mediated disease can result in multisystem organ morbidity, chronic debility, or mortality. Prevention of GVHD is a primary objective when treating patients with allogeneic HCT. While many factors are considered in transplant planning, the choice of GVHD prophylaxis is a critical and modifiable factor that impacts the risk of GVHD. Each regimen is associated with variable GVHD risk, immune reconstitution, infection risk, graft-versus-tumor effect, and drug-related adverse events. Furthermore, the utility of a specific regimen must be considered in the context of the transplant platform such as conditioning regimen and donor choice. In this chapter, we review the current state of GVHD prophylaxis, highlighting the historical standard approach of calcineurin inhibitor-based regimens, as well as the emerging standard of post-transplant cyclophosphamide.

H. Elmariah
Moffitt Cancer Center, Tampa, FL, USA

J. Bolaños-Meade (✉)
The Sidney Kimmel Comprehensive Cancer Center at Johns Hopkins University, Baltimore, MD, USA
e-mail: fbolano2@jhmi.edu

Keywords

Post-transplant cyclophosphamide · Calcineurin inhibitor · Methotrexate · Antithymocyte globulin · Abatacept · GVHD · GVHD prophylaxis · Matched donor · Mismatched unrelated · Haploidentical

5.1 Mechanisms of GVHD

Graft-versus-host disease (GVHD) is a unique complication of allogeneic hematopoietic cell transplant (HCT) leading to morbidity and transplant-related mortality (TRM). GVHD is subdivided into acute GVHD, characterized by rapid onset, and chronic GVHD, with a more indolent course and generally developing later. GVHD overlap syndrome, with a mix of acute and chronic features, may also occur (Przepiorka et al. 1995; Carpenter et al. 2015). Common to all subgroups of GVHD, the donor immune system recognizes recipient tissue as "non-self" thus initiating an antirecipient immune response. This recognition is highly dependent on the matching of the major histocompatibility complex (MHC), called human leukocyte antigens (HLA) in humans. While treatments are available, prevention is key to optimal transplant outcomes.

© The Author(s), under exclusive license to Springer Nature Switzerland AG 2025

K. Ballen (ed.), *Clinical Management of Challenges in Allogeneic Stem Cell Transplant*, Advances in Experimental Medicine and Biology 1475, https://doi.org/10.1007/978-3-031-84988-6_5

5.2 Pathophysiology

Acute GVHD As indicated by its name, acute GVHD is characterized by acute onset and a more aggressive course. Clinically, acute GVHD presents with a combination of symptoms including maculopapular rashes, nausea, vomiting, diarrhea, and hepatotoxicity (Przepiorka et al. 1995). Although traditionally considered to occur in the first 100 days after transplant, acute GVHD may occur late after transplant in isolation as "late acute GVHD" or in combination with chronic GVHD features as "GVHD overlap syndrome." (Holtan et al. 2016; Pidala et al. 2012a; Lee 2017).

The pathophysiology of acute GVHD begins with an initiating event leading to the activation of donor alloreactive T cells (Malard et al. 2023; Zeiser and Blazar 2017a). In the case of human leukocyte antigen (HLA) mismatching, the donor CD4+ and CD8+ T cells recognize the major HLA mismatch on antigen-presenting cells (APCs) leading to T-cell activation (Neefjes et al. 2011). Independent of HLA matching, minor histocompatibility antigens may also lead to acute GVHD (Goulmy et al. 1996). This begins with injury to host tissue that leads to a cascade of immune activation. The initial injury is most commonly due to the conditioning regimen or infection, leading to the release of cytokines such as tumor necrosis factor (TNF), interleukin (IL)-1, and IL-6, and the release of tissue antigens (Blazar et al. 2012). Furthermore, this tissue injury leads to instability of epithelial surfaces, such as the skin and intestines, leading to translocation of microbial pathogens and release of pathogen-related antigens and an inflammatory milieu including APCs such as macrophages and dendritic cells (Blazar et al. 2012; Jenq et al. 2012; Legoff et al. 2017; Zeiser et al. 2011). The APCs are then activated and present host antigens to donor T cells (Zeiser et al. 2011; Matte et al. 2004). Whether activated by major or minor histocompatibility antigens, the activated alloreactive CD4+ and CD8+ donor T cells release additional cytokines and cause apoptosis of host tissues (Zeiser and Blazar 2017a). Notably, regulatory cells such as T-regulatory (Treg) cells, mesenchymal stem cells, and myeloid-derived suppressor cells function to suppress this inflammatory response and improve immune tolerance (Zeiser and Blazar 2017a; Edinger et al. 2003). However, these cells may be depleted by conditioning regimens or inhibited by immune suppression, thus further promoting acute GVHD (Zeiser and Blazar 2017a).

Chronic GVHD In contrast to the acute inflammatory mechanism of acute GVHD, chronic GVHD has a more indolent progression with a fibrotic phenotype. Manifestations most often include skin, ocular, oral, pulmonary, gastrointestinal, hepatic, and musculoskeletal, though any organ system may be involved (Carpenter et al. 2015).

The pathophysiology of chronic GVHD initiates with tissue injury, which may include delayed effects of conditioning but also may be incited by other factors such as sun exposure or infections (Cooke et al. 2017). This leads to early inflammation and endothelial injury resulting in T cells, B cells, and APCs entering end organs (Cooke et al. 2017; Zeiser and Blazar 2017b). Release of cytokines such as IL-1 and IL-6 supports T-cell differentiation into Th1, Th2, and Th17 cells that can cause tissue injury (Chen et al. 2007). In the absence of adequate immune regulation by Tregs, this acute inflammation may persist and cause chronic inflammation (Zeiser and Blazar 2017b). Chronic inflammation is supported by transforming growth factor (TGF), IL1, and TNFa, all of which mediate fibrosis and scarring that distinguish chronic GVHD from acute GVHD (Cooke et al. 2017; Wynn and Ramalingam 2012).

5.3 General Goals of GVHD Prophylaxis

Given the potential for morbidity and mortality related to GVHD, prevention of both acute and chronic GVHD is an overarching objective following allogeneic HCT. Clinically, donor selection is an initial consideration in transplant planning, which may be influenced by HLA matching, donor age, blood groups (ABO) compatibility, and cytomegalovirus serology (Bray et al. 2008; Dehn et al. 2019). Historically, the use of an HLA-matched sibling offers the lowest risk of GVHD, followed by an HLA-matched unrelated donor, and then alternative donor sources such as HLA haploidentical relative or HLA mismatched unrelated donors (Fuchs et al. 2021; O'Donnell et al. 2022). However, the risk of GVHD varies across transplant platforms and may be influenced by other factors. Graft source is also a significant modifiable factor as bone marrow grafts carry a lower risk of GVHD than peripheral blood stem cell grafts (Anasetti et al. 2012; Bashey et al. 2017). The choice of pharmacologic or ex vivo cell manipulation for GVHD prophylaxis may also significantly alter the risk of GVHD, and more recent advances have successfully mitigated the increased risk of GVHD associated with HLA mismatched donors (Gooptu et al. 2021; Shaffer et al. 2024).

In clinical practice, approaches to prevent GVHD must be balanced against reciprocal risks. Most notably, GVHD occurs in parallel to the beneficial graft-versus-tumor (GVT) effect of an allogeneic transplant that is necessary to provide optimal disease control (Edinger et al. 2003; Horowitz et al. 1990). While prevention of GVHD is achievable with maximal T-cell depletion, loss of immune function prevents GVT and, thus, may precipitate disease relapse. Indeed, prior studies have demonstrated a correlation between GVHD and a lower risk of relapse, highlighting the close relationship between antihost and antitumor immunity (Weiden et al. 1979, 1981). As the risk of GVHD is especially high in the first 3–6 months after transplant, intensive immune suppression is generally continued for up to 6 months after transplant. Beyond this, in the absence of GVHD requiring treatment, the goal is to discontinue immune suppression to mitigate the long-term risk of relapse (Pidala et al. 2020; DeZern et al. 2020). Separating the two immune responses, thus eliminating GVHD while maximizing GVT, remains the "holy grail" of GVHD research.

In addition to preventing GVT, intensive immune suppression also limits immune reconstitution against infectious pathogens (Taplitz et al. 2018). While all HCT recipients are immune compromised, GVHD prophylactic regimens that are used in patients at particularly high risk of GVHD, such as in vivo or ex vivo T-cell depletion, also increase the risk of opportunistic infections such as pneumocystis (PCP) pneumonia, cytomegalovirus (CMV) infections, and invasive fungal infections (Bolaños-Meade et al. 2023; Roex et al. 2020; Wagner et al. 2005).

Overall, in choosing GVHD prophylaxis, the provider must consider the risks of GVHD, relapse, and infection inherent to the specific patient, donor, and disease being treated. Additional considerations may include regimen-specific toxicities and tolerability based on the patient's level of fitness and preexisting comorbidities. Ultimately, the regimen of choice is that which maximizes the long-term potential for disease-free survival (DFS), overall survival (OS), and GVHD-free relapse-free survival (GRFS) (Holtan et al. 2015).

5.4 Pharmacologic Agents

Calcineurin Inhibitors The calcineurin inhibitors (CNI) cyclosporine and tacrolimus are incorporated into most GVHD prevention regimens used in the modern era. Mechanistically, T-cell receptor (TCR) signaling requires the binding of calcium to calmodulin, which then binds to calcineurin, resulting in dephosphorylation and activation of nuclear factor of activated T cells (NF-AT) leading to cytokine release and T-cell activation (Wiederrecht et al. 1993). Cyclosporine and tacrolimus bind to cyclophilins, thus inhibiting calcineurin and preventing the activation of NF-AT and the downstream cascade that leads to GVHD (Zeiser and Blazar 2017a). Notably, the CNIs also suppress Tregs, which are a population of immune cells important for immune tolerance and prevention of GVHD (Coenen et al. 2006).

Structurally, cyclosporine and tacrolimus differ as the former is a cyclic endecapeptide while the latter is a macrocyclic lactone (Barbarino et al. 2013). Tacrolimus is more potent than cyclosporine, inhibiting calcineurin at much lower concentrations (Barbarino et al. 2013). Both drugs are metabolized by cytochrome p450 CYP3A enzymes in the liver (Chen et al. 2017). Thus, metabolism is significantly impacted by drugs that inhibit or induce CYP3A, including azoles commonly used for antifungal prophylaxis, as well as by food interactions such as furanocoumarins (e.g., grapefruit). Because of this, systemic exposure to CNIs is highly variable, and therapeutic drug monitoring from whole blood samples is necessary to ensure appropriate dosing (Kidney Disease: Improving Global Outcomes Transplant Work G 2009). To improve stability in the early posttransplant period, many centers prefer IV administration and later transition to oral formulations as the risk of acute GVHD decreases over time (Ho et al. 2024).

Both CNIs have similar toxicity profiles. Common side effects include hypomagnesemia, hyperkalemia, tremors, hypertension, and nephrotoxicity (Barbarino et al. 2013). In some cases, nephrotoxicity may be associated with transplant-associated thrombotic microangiopathy (TA-TMA), which is characterized by endothelial injury, microangiopathic hemolytic anemia, platelet over activation with microthrombi, acute kidney injury, and proteinuria (Jodele et al. 2014). In rare cases, CNIs may cause posterior reversible encephalopathy syndrome (PRES) (Wong et al. 2003).

Methotrexate Methotrexate (MTX) is an antimetabolite typically combined with a CNI for GVHD prophylaxis (Storb et al. 1986a). MTX functions by inhibiting dihydrofolate reductase (DHFR) (Storb et al. 1970). DHFR is responsible for the metabolism of folic acid to tetrahydrofolates, a step necessary for purine synthesis. By inhibiting DHFR, MTX prevents the proliferation of the activated T cells that contribute to GVHD (Storb et al. 1970, 1986a). In the context of HCT, MTX is administered intravenously (Storb et al. 1986a). The drug is primarily renally cleared. In contrast to CNIs, MTX has few drug interactions. Therapeutic drug monitoring is not necessary.

MTX is associated with a wide range of toxicities and most patients experience at least one such adverse event when receiving MTX for GVHD prophylaxis. Common side effects include nausea/diarrhea, mucositis, liver injury, dermatitis, and nephrotoxicity (Perkins et al. 2010; Cutler et al. 2005). Less commonly, MTX may cause lung injury that may present as hypersensitivity pneumonitis, organizing pneumonia, or pulmonary fibrosis (Rosenow 3rd et al. 1992; Kremer et al. 1997). Due to a high risk of toxicity, administration of MTX is typically combined with leucovorin, a folate analog that can prevent MTX-related toxicity (Ortiz et al. 1998). Otherwise, management is supportive but may include dose adjustment or omission of MTX doses (Nakamura et al. 2023). Notably, MTX is contraindicated in patients with pleural effusions or ascites as MTX may pool in these extravascular fluid collections and result in increased toxicity (Evans and Pratt 1978).

Mycophenolate Mofetil and Mycophenolic Acid Mycophenolate mofetil (MMF) or its analog mycophenolic acid (MPA) are antimetabolites that are included in various regimens in lieu of MTX (Ruutu et al. 2014; Allison and Eugui 2005). Like MTX, MMF inhibits purine synthesis needed for T-cell activation (Allison and Eugui 2005). However, the mechanism is through inhibition of inosine monophosphate dehydrogenase (Allison and Eugui 2005). Notably, this mechanism spares neutrophils, and thus, MMF is less immune suppressive and allows for more rapid neutrophil recovery (Vogelsang and Arai 2001).

MMF is a prodrug of the active ingredient mycophenolic acid (MPA), providing better bioavailability than MPA (Lamba et al. 2014). While available both orally and intravenously, MMF has an oral bioavailability of ~90% (Staatz and Tett 2007). MMF is metabolized primarily in the liver (Lamba et al. 2014). Drug interactions are not significant, and therapeutic drug monitoring is not typically performed.

MMF is generally well tolerated, but is often associated with gastrointestinal adverse events and may cause pathologic changes such as crypt apoptosis that histologically resemble intestinal GVHD (Liapis et al. 2013). Cytopenias are also commonly observed. Rare cases of progressive multifocal leukoencephalopathy have been reported (Pavlovic et al. 2012).

Sirolimus Sirolimus has been adopted by some centers in combination with tacrolimus or as an alternative to tacrolimus in select regimens (Kelly et al. 1997). Sirolimus is a macrolide that inhibits the mammalian target of rapamycin (mTOR). This interferes with cell proliferation by preventing progression from the G1 to S phase of mitosis. Furthermore, sirolimus disrupts the CD28 costimulatory signal required for T-cell activation (Nourse et al. 1994). In contrast to CNIs, sirolimus does not inhibit T-regulatory cell proliferation which adds theoretical benefit for GVHD prevention (Coenen et al. 2006; Singh et al. 2014; Peccatori et al. 2015).

In contrast to CNIs, sirolimus is only available through the oral route. Metabolism is primarily by the liver via the cytochrome p450 system (Kahan and Napoli 1998). As such, like CNIs, drug interactions with inducers and inhibitors of cytochrome p450 significantly impact drug metabolism (Kahan and Napoli 1998). Thus, therapeutic drug monitoring is warranted by whole blood trough levels (Kahan and Napoli 1998).

Adverse events associated with sirolimus differ from those seen with other GVHD agents, and thus, consideration may be given to which drug is better tolerated based on the patient's existing comorbidities (Elmariah et al. 2024a). Commonly, sirolimus may cause cytopenias, dyslipidemia, blood pressure changes, and hepatic toxicity (Kelly et al. 1997). More rarely, sirolimus may cause oral mucositis and pericardial/pleural effusions (Grenier et al. 2021). When combined with myeloablative conditioning, sirolimus may increase the risk of vaso-occlusive disease of the liver (Cutler et al. 2008). When combined with tacrolimus, sirolimus may increase the risk of renal injury and TMA (Jodele et al. 2014; Mirza et al. 2022).

Cyclophosphamide Cyclophosphamide is an alkylating agent that emerged as a preferred option for GVHD prophylaxis in the setting of HLA mismatched donors and has more recently been applied in the HLA-matched setting. The mechanism of action was originally attributed to the selective elimination of rapidly proliferating alloreactive T cells when cyclophosphamide is given on days +3 and + 4 after allogeneic transplant (Eto et al. 1991). However, more recent studies suggest that it is functional impairment, rather than selective depletion, of alloreactive T cells that help decrease this risk of GVHD (Fletcher et al. 2023; Wachsmuth et al. 2019). Furthermore, Treg cells are resistant to cyclophosphamide toxicity due to the expression of aldehyde dehydrogenase, thus promoting long-term immune tolerance and limiting the risk of chronic GVHD (Kanakry et al. 2013a).

When used for GVHD prophylaxis, cyclophosphamide is administered as an IV bolus, generally on days +3 and + 4 after transplant. The drug is metabolized by aldehyde dehydrogenase and the cytochrome P450 into acrolein and the active metabolite, phosphoramide (Hall and Tilby 1992). Because acrolein causes hemorrhagic cystitis, cyclophosphamide is administered with MESNA and aggressive intravenous hydration for bladder protection (Andriole et al. 1987). Drugs that impact cytochrome P450 may impact cyclophosphamide metabolism, though clinical impacts are not well established and drug level monitoring is not standard.

Like other alkylating agents used in conditioning regimens, cyclophosphamide is associated with cytopenias, and this may lead to slower engraftment when used as GVHD prophylaxis (Hall and Tilby 1992). Immune reconstitution may also be slower, leading to a higher risk of opportunistic infections (Goldsmith et al. 2021; Khimani et al. 2021). Otherwise, gastrointestinal toxicity including nausea, vomiting, and diarrhea is common. Hemorrhagic cystitis is also common, both due to direct cyclophosphamide toxicity and indirectly due to BK and adenovirus, though usually self-limited (Andriole et al. 1987; Rice et al. 1985). Cardiotoxicity including arrhythmias, decreased ejection fraction, or pericardial effusions are also significant (Dulery et al. 2021). Pneumonitis is also reported though less common. While there are theoretical risks for secondary donor-derived malignancies due to the administration of an alkylating agent in the posttransplant setting, clinical outcomes have not corroborated this concern (Kanakry et al. 2013b; Majzner et al. 2017).

Antithymocyte Globulin Antithymocyte globulin (ATG) is an animal-derived antibody product used both to reduce GVHD and graft rejection. ATG is produced by immunizing rabbits or horses with human thymocytes. The resultant antithymocyte antibodies are then collected for therapeutic purposes. The final ATG product is, thus, primed to target antigens expressed on human T cells, such as CD3 and CD152 (Raefsky et al. 1986). For allogeneic HCT, rabbit ATG is favored both for efficacy and because it offers more consistent potency than equine ATG. When administered, ATG effectively depletes a broad population of T cells resulting in significant immune suppression (Mohty 2007).

ATG is administered intravenously. The half-life is 30 days, and ATG persists in human circulation beyond 5 weeks after transplant. However, it has been shown that ATG remains in circulation even after it is no longer biologically active (Waller et al. 2003). When administered for GVHD prophylaxis, ATG has a well-established dose-dependent response whereby suppression of T-cell function and the likelihood of GVHD are directly associated with higher exposure to ATG (Waller et al. 2003). ATG does not require dose adjustments for organ function nor drug interactions and similarly does not require therapeutic drug monitoring.

Given the effective T-cell depletion associated with ATG, risks of infections, including viral reactivation such as Epstein–Barr and cytomegalovirus, are high (Figgins et al. 2019; Dominietto et al. 2012). Furthermore, as a xenogenic product, ATG confers a high risk of immune reactions including fevers, cytokine release, hypoxia, myalgias, anaphylaxis, and serum sickness (Pihusch et al. 2002; Nishihori et al. 2016). Thus, premedications with corticosteroids, antihistamines, and acetaminophen are typically administered. ATG frequently causes liver injury with transaminitis, though this is typically mild and self-limited. ATG has also been associated with a higher risk of posttransplant lymphoproliferative disorders (Kalra et al. 2018).

Abatacept Abatacept is the first and only drug approved by the US Food and Drug Administration (FDA) for acute GVHD prophylaxis. Abatacept is a soluble fusion protein with a cytotoxic T-lymphocyte-associated antigen 4 (CTLA-4) domain bound to human immunoglobulin (Moreland et al. 2006). Abatacept binds to the CD80 and CD86 costimulatory domains on APCs, which prevents these ligands from binding to the CD28 costimulatory domains on alloreactive T cells (Picchianti Diamanti et al. 2014). Without this costimulatory signal, T-cell activation is decreased, thus preventing acute GVHD (Picchianti Diamanti et al. 2014).

For GVHD prevention, abatacept is administered intravenously and has a long half-life of 13 days (Lutt 2009). Abatacept is not metabolized by the kidneys or the liver, so dose modifications are not required. Similarly, there are no known significant drug interactions. Therapeutic drug monitoring is not warranted for abatacept.

The most significant toxicity of abatacept is risk of infection. This may include a risk of mycobacterium infections, so tuberculosis testing is indicated prior to treatment (Weinblatt et al. 2013). Headaches, hypertension, dizziness, gastrointestinal toxicity, and rash are also common (Dubois and Cohen 2009). As an antibody fusion protein, abatacept is also associated with risks of anaphylaxis and the development of anti-abatacept antibodies (Dubois and Cohen 2009). There is also concern that abatacept may increase the risk of malignancy, though this has not been demonstrated specifically in the setting of abatacept used for GVHD prophylaxis (Weinblatt et al. 2006).

5.5 Regimens

In 1957, E. Donnall Thomas and colleagues published their groundbreaking research infusing allogeneic bone marrow grafts into five patients with hematologic malignancies (Thomas et al.

1957). Though these patients did not survive past day 100, this experience laid the foundations for allogeneic HCT as a therapeutic approach for hematologic diseases. The feasibility of allogeneic HCT was realized by the discovery that HLA compatibility could reduce the risk of graft rejection and GVHD (Storb et al. 1968; Good et al. 1964). Even still, the risk of GVHD was significant. Recognition that suppressing alloreactive T cells either by pharmacologic methods, such as cyclosporine and methotrexate, or ex vivo T-cell depletion (TCD) could reduce GVHD was a turning point in the success of allogeneic HCT (Thomas et al. 1975; Lochte Jr. et al. 1962; Reisner et al. 1983). Over the last 50 years, optimizing the approaches to GVHD prophylaxis has remained an important sector of research for allogeneic HCT. Commonly used regimens are shown in Table 5.1.

CNI/MTX In the mid-1980s, Storb et al. published results of a prospective trial of patients receiving HLA-matched sibling donor (MSD) bone marrow transplant (BMT) randomized to cyclosporine (n = 50) versus cyclosporine/MTX (n = 43) for GVHD prophylaxis (Storb et al. 1986a). In that study, the combination regimen resulted in a cumulative incidence of acute GVHD of 33% compared to 54% with single-agent cyclosporine (p = 0.014). Grade 4 acute GVHD was only observed in the single-agent cyclosporine group. OS at 15 months was also superior in the combination group (80% vs. 55%, p = 0.042). The same group similarly compared cyclosporine/MTX (n = 22) to single-agent MTX (n = 24) (Storb et al. 1986b). Again, all transplants were from MSDs. Graft rejection occurred in one patient in each cohort. The cumulative incidence of grade II–IV acute GVHD was 18% in the cyclosporine/MTX group, compared to 53% with MTX (p = 0.012), and grade IV acute GVHD was only observed in the MTX group. The 2-year OS was 82% in the combination group compared to 60% in the single-agent group (p = 0.062). These studies established CNI/MTX as a standard GVHD prevention platform, and this regimen remains in use for HLA-matched donor transplants.

The establishment of CNI/MTX as the standard GVHD prevention strategy led next to the question of whether tacrolimus may yield better outcomes than cyclosporine. A large, phase III study compared tacrolimus/MTX (n = 90) to cyclosporine/MTX (n = 90) in patients receiving HLA-matched unrelated donor (MUD) transplants (Nash et al. 2000). The risk of grade II–IV acute GVHD was much lower in the tacrolimus group: 56–74%, p = 0.0002. Survival and toxicity outcomes were similar between both groups. A subsequent phase III trial with MSD donors similarly showed lower rates of grade II–IV acute GVHD with tacrolimus/MTX (32% vs. 44%, p = 0.01) (Ratanatharathorn et al. 1998). However, limiting to only grade III–IV acute GVHD, outcomes were similar. Furthermore, OS was superior in the cyclosporine/MTX arm (50.4% vs. 40.5%, p = 0.01). This difference was attributed to imbalanced patient populations with a higher proportion of patients in the tacrolimus group having more aggressive malignancies. Yet another phase III trial examined this same question and found similar rates of acute GVHD in both groups, with 33.3% in the tacrolimus/MTX group and 39.6% in the cyclosporine group (p = 0.41) (Kanda et al. 2016). Other outcomes including OS, DFS, relapse, and nonrelapse mortality (NRM) were also similar in both groups. Retrospective data have also supported these results (Inamoto et al. 2011). Taken together, tacrolimus/MTX likely offers improved rates of acute GVHD but the absence of a clear and consistent advantage has led to some centers continuing to use cyclosporine/MTX.

The standard regimen currently in use combines MTX 15 mg/m^2 on day 1 and 10 mg/m^2 on days 3, 6, and 11 with a CNI initiated on day −2 and continued through approximately 3–6 months posttransplant. While effective for GVHD prophylaxis, CNI/MTX does have significant adverse effects. MTX doses are often held due to toxicities such as mucositis and hepatotoxicity, thus decreasing the GVHD prophylactic benefits (Hamilton et al. 2015a). To combat these risks, a regimen of tacrolimus with "mini-dose" MTX 5 mg/m^2 IV on days 1, 3, 6, and 11 has been

Table 5.1 Summary of current standard graft-versus-host disease prophylactic regimens

Regimen		Dosing	Indication	Notes
Calcineurin inhibitor (CNI) with antimetabolite	Cyclosporine	Cyclosporine day −1, target 200–300 ng/mL, taper to discontinue at 3–6 months	Matched donor	
	OR			
	Tacrolimus	Or tacrolimus day −1 goal trough 5–10 ng/mL, taper to discontinue at 3–6 months	Matched donor	
	PLUS			
	Methotrexate (MTX)	MTX 15 mg/m^2 IV on day +1, 10 mg/m^2 on day +3, +6, +11	Matched donor	High risk of mucositis, MTX dosing available to reduce toxicity, MTX is contraindicated with effusions
	OR			
	Mycophenolate mofetil (MMF)	MMF 15 mg/kg TID up to 3 g/day from day 0 to 1 month in matched sibling or 3 months in matched unrelated donor	Matched donor, umbilical cord blood	MMF has less GVHD control but less mucositis than MTX
Other CNI-Based Regimens	Tacrolimus/ Sirolimus	Tacrolimus day −3 to trough 5–10 ng/ mL, Sirolimus day −3 to trough 3–12 ng/mL, taper to discontinue both drugs by 1 year	Matched donor	Sirolimus has less mucositis but more renal and hepatic toxicity than MTX
	CNI/Antimetabolite/ rabbit antithymocyte globulin (ATG)	Add ATG to CNI regimen, variable dosing 10–30 mg/ kg/day with conditioning	Add to CNI-based regimen for myeloablative matched unrelated or mismatched donor	ATG increases the risk of viral infections
	CNI/MTX/ Abatacept	Add Abatacept to CNI regimen, dosing 10 mg/kg IV on days −1, +5, +14, +28	Add to CNI-based regimen for matched unrelated or mismatched unrelated donor	
Post-transplant cyclophosphamide (PTCy)-based	PTCy/tacrolimus/ MMF	PTCy 50 mg/kg IV on days +3 and +4	Matched donor	PTCy has a high risk of hemorrhagic cystitis, administer with MESNA
		Tacrolimus day +5 to trough 5–15 ng/ mL, discontinue by 6 months	Haploidentical donor	Shorter duration of tacrolimus may be considered in low GVHD risk patients
		Mycophenolate 15 mg/kg TID up to 3 g/day from day +5 to +35	Mismatched unrelated donor	Sirolimus may be substituted for tacrolimus if contraindication

explored (Przepiorka et al. 1996). In a study of 30 adult patients receiving MUD donor HCT, grade II–IV acute GVHD was observed in 34%, grade III–IV in 17%, and OS at 1 year was 47%. However, risks of infection, kidney injury, and liver injury remained high. This study was duplicated in a cohort of 40 patients receiving a one-antigen (7/8) mismatched unrelated donor (MMUD) HCT with a cumulative incidence of grade II–IV acute GVHD of 59%, grade III–IV acute GVHD of 17%, and 1-year OS of 29% (Przepiorka et al. 1999).

The CNI/MTX combination has remained a standard GVHD prophylactic regimen for matched donor transplants for ~40 years and remains in use at the time of this publication. However, this approach is reliant on a suitable well-matched donor with poor outcomes in the setting of HLA mismatched or haploidentical donors.

Other CNI-Based Regimens A number of CNI-based regimens were developed to try to improve upon the experience with CNI/MTX. While none demonstrated clear superiority, they have demonstrated other benefits leading to preferential use at some centers, or for use when there is a contraindication to MTX.

1. *Tacrolimus/Sirolimus*: The mTOR inhibitor sirolimus garnered interest as a GVHD preventing agent to complement tacrolimus in the early 2000s (Cutler and Antin 2004). In contrast to CNIs, sirolimus spares donor Tregs, thus promoting antihost immune tolerance (Kopf et al. 2007). Additionally, sirolimus was assumed to be more easily tolerated than MTX. Indeed, the tissue injury caused by MTX had created concern for possibly inciting GVHD through the release of tissue antigens and proinflammatory cytokines (Antin and Ferrara 1992). Tacrolimus/sirolimus also offers the subjective benefit of an oral regimen that can be continued in the outpatient setting or modulated in real-time in response to toxicity, infection, or onset of GVHD. In this regimen, tacrolimus is started on day −3 to a goal trough of 5–10 ng/mL and sirolimus on day −3 to a goal trough of 3–12 ng/mL and then tapered off after day +100.

A number of single-center studies suggested promising results with the tacrolimus/sirolimus regimen (Cutler et al. 2004; Rodriguez et al. 2010; Pidala et al. 2012b). Most encouraging, a randomized phase II trial stratified trial with MSD or MUD donors compared tacrolimus/sirolimus to tacrolimus/MTX (Pidala et al. 2012b). The tacrolimus/sirolimus group had significantly lower rates of grade II–IV acute GVHD (43% vs. 89%, $p < 0.001$) and moderate/severe chronic GVHD (24% vs. 64%, $p = 0.008$), as well as similar survival outcomes. However, these results could not be replicated in a subsequent multicenter phase III trial where the probability of grade II–IV acute GVHD was 26% for tacrolimus/sirolimus and 34% for tacrolimus/MTX ($p = 0.48$), and the probability of chronic GVHD was 53% for tacrolimus/sirolimus and 45% for tacrolimus/MTX ($p = 0.06$) (Cutler et al. 2014). Nonetheless, the tacrolimus/sirolimus group did experience more rapid engraftment and lower risks of oral mucositis, though a trend toward higher risks of veno-occlusive disease of the liver or sinosoidal obstruction sindrome (VOD) and TMA. This study concluded that tacrolimus/MTX remained the standard of care for matched donor HCT, but tacrolimus/sirolimus was an acceptable alternative.

In the real-world experience, there are unique practical considerations for using the combination of tacrolimus/sirolimus. First, while this oral regimen allows for dose adjustments, drug–drug interactions impact both drugs and, thus, both medications must be monitored. Further, sirolimus is only available in oral form, so this therapy may require interruption in patients who cannot take oral medications. Overlapping toxicities, particularly increased risk of TMA or (VOD) veno-occlusive disease of the liver or sinosoidal obstruction sindrome when including sirolimus, also can result in interruptions to the regimen (Jodele et al. 2014; Rodriguez et al. 2010). Of significance, patients who are forced to discontinue the combination tacrolimus/sirolimus prior to day 100 have a high risk of GVHD and NRM (Mirza et al. 2022).

2. *CNI/MMF*: MMF offers a direct alternative to MTX as both are antimetabolites that can be combined with CNI, but with differing toxicities. Early animal models of MMF for GVHD prophylaxis suggested synergy in preventing GVHD when combined with cyclosporine (Yu et al. 1998). Furthermore, MMF was able to promote more rapid and sustained engraftment (Storb et al. 1997). These data, combined with a favorable safety profile, led to the investigation of CNI/MMF in humans.

Initial trials with CNI/MMF occurred in the late 1990s, at a time before reduced intensity conditioning (RIC) and tacrolimus for GVHD prophylaxis had become standard. Thus, MMF was first combined with cyclosporine for GVHD prophylaxis following myeloablative conditioning (MAC) HCT (Bornhäuser et al. 1999; Kiehl et al. 2002). A single-center randomized trial compared cyclosporine/MTX to cyclosporine with MMF 2 g/day and cyclosporine with MMF 3 g/day in 45 patients receiving MAC marrow graft HCT. Though no statistically significant differences were observed in this small study, rates of grade II–IV acute GVHD appeared to be higher in the cyclosporine/MMF groups, rates of mucositis were higher in the cyclosporine/MTX group, and survival outcomes were similar (Kiehl et al. 2002). A subsequent randomized trial compared cyclosporine/MMF to cyclosporine/MTX in 40 patients receiving HCT with MAC and MSDs (Bolwell et al. 2004). The risk of severe mucositis was significantly lower in the cyclosporine/MMF group (21% vs. 65%, $p = 0.008$), and time to neutrophil engraftment was faster in the cyclosporine/MMF group (11 days vs. 18 days, $p < 0.001$). GVHD rates were similar between the two groups. Notably, the trial was stopped early due to high rates of toxicity in the cyclosporine/MTX group. In 2014, a larger randomized trial limited also to MAC and MSDs compared cyclosporine/MMF ($n = 174$) to cyclosporine/MTX ($n = 67$). This larger sample size was able to confirm the lower risk of severe mucositis (19% vs. 53%) and faster engraftment (11 days vs. 19 days) in the cyclosporine/MMF group. However, this trial also confirmed a significantly higher risk of grade III–IV acute GVHD with cyclosporine/MMF by multivariate analysis (Hamilton et al. 2015b).

As more centers were choosing tacrolimus/MTX over cyclosporine/MTX, another randomized phase II trial sought to compare tacrolimus/MMF versus tacrolimus/MTX in a cohort of 89 patients receiving MAC and matched donors (Perkins et al. 2010). MMF was started on day 0 and continued for 1 year. Again, mucositis was significantly less in the MMF group. However, no difference was seen in time to neutrophil engraftment. Like the cyclosporine trials, tacrolimus/MMF resulted in higher risks of grade III–IV acute GVHD compared to the tacrolimus/MTX arm (19% vs. 4%, $p = 0.03$). Other outcomes were similar between the groups. In combination, the existing data suggests that CNI/MMF results in more severe acute GVHD than CNI/MTX in the context of MAC regimens (Perkins et al. 2010; Hamilton et al. 2015b).

CNI/MMF has also been studied in the RIC setting with more promising results. A single-center study compared cyclosporine/MMF with cyclosporine/MTX in patients receiving RIC MSD transplants and found the cyclosporine/MMF group had a lower risk of grade 2–4 mucositis (23% vs. 57%, $p = 0.001$), with no differences observed in the risks of acute GVHD, chronic GVHD, NRM, relapse, or OS (Pinana et al. 2010). A registry study from the Center for International Blood and Marrow Transplant Research (CIBMTR) compared tacrolimus/MTX to cyclosporine/MTX to cyclosporine/MMF to tacrolimus/MMF in 1564 patients who received RIC-matched donor HCT (Chhabra et al. 2019). While this study corroborated the findings that CNI/MMF provides comparable GVHD and survival outcomes to CNI/MTX with MSD donors, both cyclosporine/MMF and tacrolimus/MMF led to higher risks of grade II–IV acute GVHD and grade III–IV acute GVHD compared to tacrolimus/MTX for patients receiving MUD transplants. Tacrolimus/MMF was also associated with higher NRM when used for MUD transplants.

3. *CNI/Antimetabolite with Antithymocyte Globulin*: As alternate CNI-based doublets could not universally supplant CNI/MTX, the community also looked to CNI-based triplet regimens to improve GVHD outcomes. Building on the importance of T-lymphocytes in the pathophysiology of GVHD, the addition of rabbit ATG to CNI/MTX was a promising approach to reduce GVHD risk. A multicenter, prospective phase III randomized trial compared cyclosporine/MTX plus ATG to cyclosporine/MTX in 161 patients receiving MAC followed by MSD peripheral blood stem cell grafts (Kroger et al. 2016). The addition of ATG resulted in a significant reduction in the risk of chronic GVHD (32.2% vs. 68.7%, $p < 0.001$). Infection, Acute GVHD, relapse, and survival were all similar between the groups. Another phase III multicenter randomized trial compared cyclosporine/MTX with ATG to cyclosporine/MTX in 201 patients receiving a MAC MUD transplant (Finke et al. 2009; Finke et al. 2017). The addition of ATG led to a reduced risk of grade II–IV acute GVHD (33.0% vs. 51.0%, $p = 0.011$) and chronic GVHD (12.2% vs. 42.6%, $p < 0.0001$) and a higher likelihood of long-term immunosuppression-free survival (47% vs. 11%). The risk of viral infections, including cytomegalovirus (CMV) and Epstein–Barr Virus (EBV), was higher in the ATG group, though the incidence of fatal infections was comparable with both regimens.

Because of the profound T-cell-depleting impact of ATG, there exists concern that ATG could blunt the GVT effect and increase the risk of post-HCT relapse, especially in patients who receive RIC conditioning where relapse rates are higher than MAC regimens. Data supporting the use of ATG for matched donor transplants with RIC regimens is limited. One prospective phase III randomized trial of matched donor transplants included patients receiving both MAC and RIC regimens (Walker et al. 2016). Though only 33% of the patients received RIC regimens, subgroup analysis did demonstrate a higher rate of freedom from immune suppression with ATG, with no difference in relapse. Two large registry studies have offered conflicting findings. The first, from the CIBMTR, included RIC transplants from matched siblings and matched unrelated donors (Soiffer et al. 2011). While ATG was associated with lower chronic GVHD rates (40% vs. 52%), the risk of relapse (51% vs. 38%) and OS (38% vs. 46%) was worse with ATG. In contrast, a registry study of RIC MSD transplants from the European Group for Blood and Marrow Transplantation (EBMT) confirmed lower risks of chronic GVHD (39% vs. 57%) and extensive chronic GVHD (19% vs. 38%) with ATG compared to no ATG, while other outcomes including relapse were similar between groups (Baron et al. 2014).

ATG is now commonly added to CNI-based GVHD regimens when combined with regimens with matched or mismatched donors. This remains a preferred regimen particularly in Europe based on expert opinion and current consensus guidelines (Penack et al. 2024a; Piekarska et al. 2023). ATG has also been explored in the haploidentical donor setting with acceptable grade II–IV acute GVHD rates of 35%, but high rates of chronic GVHD, NRM, and relapse resulting in only 25% OS (Peccatori et al. 2015). Of note, ATG dosing in transplant regimens is variable across transplant platforms.

4. *CNI/MTX/Abatacept*: Another successful addition to the CNI-doublet is the biologic abatacept. This drug was studied in a randomized prospective phase II trial comparing CNI/MTX/abatacept to CNI/MTX in 142 patients receiving MUD HCT (Watkins et al. 2021). The addition of abatacept resulted in a lower risk of grade II–IV acute GVHD (44.8% vs. 63.7%, $p = 0.006$), though chronic GVHD, NRM, relapse, relapse-free survival (RFS), and OS were similar between the two groups. This trial also included a single-arm cohort of 38 patients who received HLA 7/8 MMUD grafts, which were then compared to a historical cohort from the CIBMTR. In this HLA 7/8 MMUD cohort, the addition of abatacept was associated with significantly lower risks of

grade II–IV acute GVHD (39.5% vs. 57.4%, $p = -0.03$), grade III–IV acute GVHD (0% vs. 32.1%, $p < 0.001$), NRM (16.3% vs. 40.3%, $p = 0.010$), and OS (76.9% vs. 45.3%, $p = -0.002$). In response to this study, abatacept became the first drug approved by the US Food and Drug Administration (FDA) specifically for acute GVHD prevention. Multicenter trials are ongoing to evaluate optimal dosing schedules of abatacept, as well as effectiveness in preventing chronic GVHD.

PTCy/Tacrolimus/MMF While CNI-based GVHD prophylaxis led to acceptable outcomes, the need for an HLA-matched donor limited access as only ~30% of patients have an MSD and the likelihood of finding an MUD varies significantly by race and ethnicity (Gragert et al. 2014). Indeed, requiring a full HLA match was a major source of disparities for underrepresented minority patients with hematologic malignancies (Kirtane and Lee 2017). A potential approach to solve this problem was to optimize the safety of HLA haploidentical-related donor HCT. Because of the Mendelian inheritance pattern of HLA haplotypes, parents and offspring are uniformly haploidentical and siblings have a 50% likelihood of being haploidentical. For elderly patients whose siblings and children may be old enough that donating risks worse transplant outcomes, extended relatives of younger age may even be considered (Elmariah et al. 2018; Mehta et al. 2023). Thus, the incorporation of haploidentical-related donors greatly expands access to transplants as >90% of patients have a suitable haploidentical-related donor, regardless of racial/ethnic background (Bolaños-Meade et al. 2012).

Citing evidence that the alkylating agent cyclophosphamide could prevent skin graft rejection in MHC mismatched mice, investigators at the Johns Hopkins University sought to develop a regimen using cyclophosphamide for GVHD prophylaxis with HLA haploidentical donor HCT (Mayumi and Good 1989; Luznik et al. 2001). In murine models, Luznik et al. found that MHC-incompatible marrow transplant followed by posttransplant cyclophosphamide (PTCy) on day 3 led to sustained engraftment and attenuated GVHD (Luznik et al. 2001). Based on these promising results, the regimen was implemented in 68 patients receiving HLA haploidentical-related donor HCT at two centers (Luznik et al. 2008). In this trial, PTCy 50 mg/kg IV was given on days +3 in 28 patients or days +3 and +4 in 40 patients. The regimen was combined with tacrolimus given from day +5 to +180 and MMF from day +5 to +35. The incidence of grade II–IV acute GVHD was 34% and of grade III–IV acute GVHD was 6%. The risk of extensive chronic GVHD was improved in the cohort of patients receiving two doses of PTCy (5% vs. 25%). NRM was only 15% at 1 year, and OS was 46% due to a high 51% risk of relapse. The results of this study were significant in demonstrating the feasibility of haploidentical donor transplant, with rates of GVHD that appeared similar to matched donor transplant with CNI-based regimens. Furthermore, while most prior studies focused on lowering the risk of acute GVHD, this study suggested that PTCy also lowers the risk of chronic GVHD. These results were replicated in subsequent studies demonstrating the feasibility of haploidentical HCT with PTCy using varied conditioning regimens, graft sources, and extended relatives (Bashey et al. 2017; Elmariah et al. 2018; Ciurea et al. 2015; Gaballa et al. 2016; DiStasi et al. 2014). This led to a boom in haploidentical donor HCT, with steadily increasing volumes across the world after 2014 (D'Souza et al. 2020).

The success of PTCy in the haploidentical setting led to the question of whether this regimen could also improve outcomes in combination with HLA-matched donors (Gooptu et al. 2021; Kanakry et al. 2014). In the setting of MAC and marrow grafts, the randomized multicenter phase III BMT Clinical Trials Network (CTN) 1301 trial compared tacrolimus/MTX to single-agent PTCy and to CD34 selected grafts (Luznik et al. 2022). Outcomes were similar between tacrolimus/MTX and PTCy, with no significant differences in GVHD or survival outcomes. The negative trial results were attributed possibly to the omission of tacrolimus/MMF from the PTCy regimen. Separately, the BMT CTN 1203 study evaluated three novel prophylactic regimens for

RIC and matched donor peripheral blood stem cell (PBSC) grafts: PTCy/tacrolimus/MMF versus tacrolimus/MTX/bortezomib versus tacrolimus/MTX/maraviroc (Bolaños-Meade et al. 2019a). The PTCy/tacrolimus/MMF arm led to the most promising outcomes and thus was chosen as the investigational arm for comparison against tacrolimus/MTX in the subsequent multicenter, randomized phase III trial BMT CTN 1703 (Bolaños-Meade et al. 2023). This trial met the primary endpoint with PTCy/tacrolimus/MMF yielding superior GVHD-free, relapse-free survival (GRFS) at 1-year post-transplant: 52.7% versus 34.9%. This benefit was attributable to lower incidences of grade III–IV acute GVHD (6.3% vs. 14.7%) and chronic GVHD (21.9% vs. 35.1%) with similar OS, DFS, relapse, and NRM (Fig. 5.1). Furthermore, the likelihood of immune-suppression-free survival at 1 year was higher with PTCy/tacrolimus/MMF (50% vs. 39.7%). Notably, the PTCy arm was associated with a higher risk of infections, though this was attributed specifically to a higher incidence of grade 2 infections, with no significant increase in grade 3–4 infections. The results of BMT CTN 1703 were practice-changing, heralding widespread adoption of PTCy/tacrolimus/MMF as a standard GVHD prophylactic option for matched donor transplants.

Beyond haploidentical and matched donor transplants, PTCy is also being studied in the setting of MMUD transplants. Based on research from Johns Hopkins that showed feasibility, a national multicenter phase II trial by the (NMDP) National Marrow Donor Program evaluated PTCy/tacrolimus/MMF in 80 patients receiving MMUD bone marrow HCT, of whom 48% were ethnic minorities (Shaw et al. 2021; Kasamon et al. 2017). In this study, rates of grade II–IV acute GVHD, severe chronic GVHD, and OS at 1 year were 47.5%, 7.6%, and 72.3%, respectively. A multicenter prospective trial studying PTCy with MMUD PBSC grafts is ongoing. The PTCy experience has reframed the conventional thinking regarding the significance of HLA matching in unrelated donor HCT. Whereas donor search algorithms with CNI-based GVHD prophylaxis heavily weigh HLA matching, emerging data suggests that PTCy mitigates the importance of an 8/8 HLA-matched donor (Shaffer et al. 2024). Instead, other factors such as donor age, time to transplant, and blood groups (ABO) compatibility may warrant more emphasis in the PTCy setting.

As the original PTCy regimen with tacrolimus/MMF gains wide acceptance, ongoing studies seek to elucidate optimal immune suppression and modifications with this regimen. In the setting of nonmyeloablative (NMA) conditioning, continuing tacrolimus for only 2 months may be as effective as 3–6 months (DeZern et al. 2020; Kasamon et al. 2018). While MMF was initially included to promote engraftment in combination with NMA haploidentical bone marrow HCT, more recent studies suggest that lower doses or shortened duration of MMF may result in improved immune reconstitution and more favorable outcomes (Mehta et al. 2022; Elmariah et al. 2024b). Even further, some studies seek to completely omit tacrolimus or MMF and replace them with novel agents with synergistic mechanisms or preferable safety profiles (Elmariah et al. 2024a; Al-Homsi et al. 2023; Morozova et al. 2021). The addition of ATG to PTCy regimens has proven effective, especially in patients with nonmalignant diseases (Bolaos-Meade et al. 2019b; DeZern et al. 2023). Ongoing trials are also studying lower doses of PTCy with the goal of reducing toxicity without increasing GVHD.

Ex vivo T-Cell Depletion To avoid the adverse events associated with pharmacologic prophylaxis, some centers have instead attempted ex vivo T-cell depletion (TCD) to mitigate the risks of GVHD.

A multicenter phase III trial randomized patients receiving MUD or MMUD transplants to either TCD with cyclosporine ($n = 201$) or cyclosporine/MTX ($n = 204$). TCD was performed by either centrifugal elutriation or antibody targeting to achieve a prescribed cell dose of 5×10^5 CD3 cells/kg. TCD successfully lowered the risks of grade III–IV acute GVHD (18% vs. 37%, $p < 0.0001$), though not chronic GVHD (Wagner et al. 2005). However, TCD was also associated with significantly higher risks of relapse (20% vs.

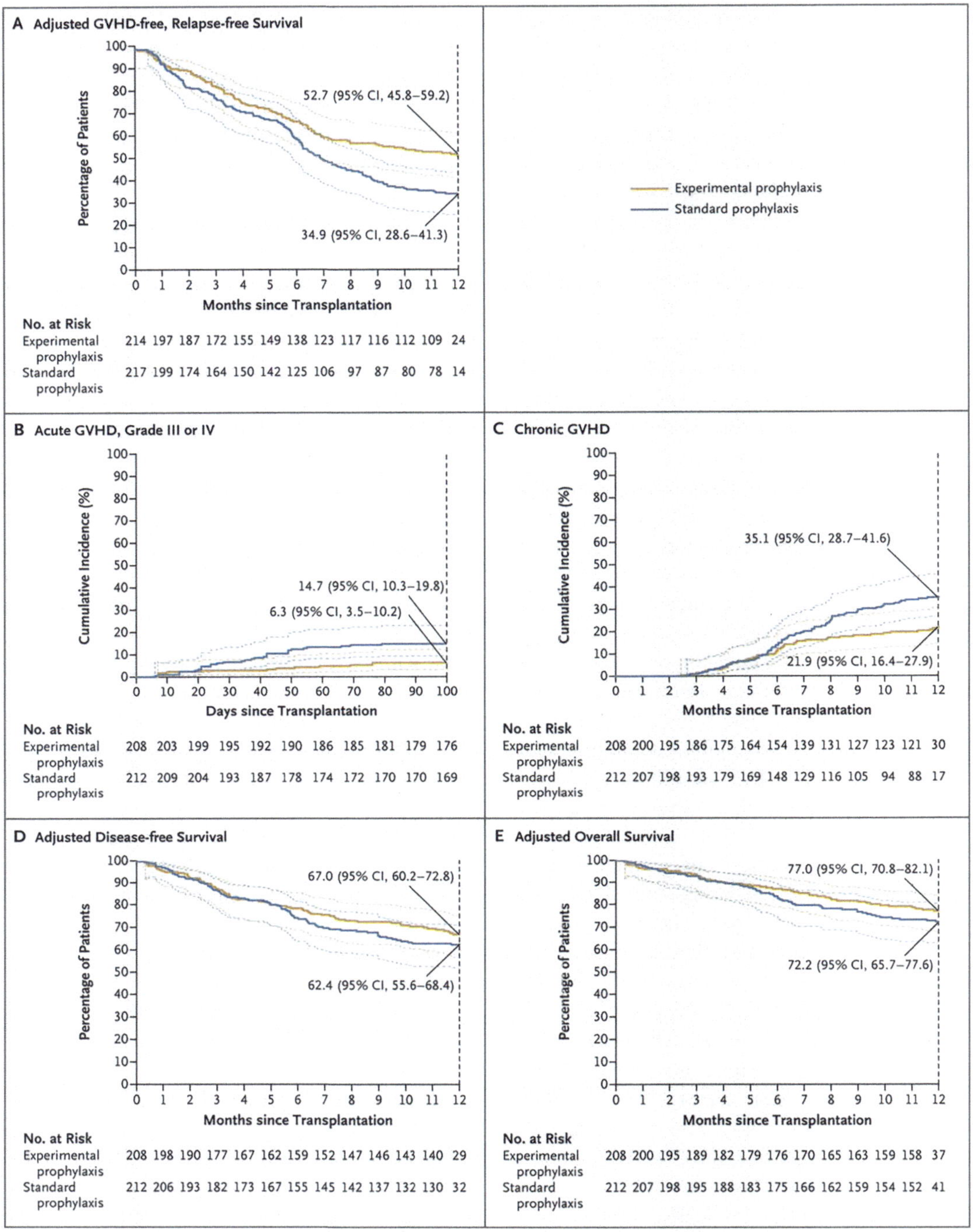

Fig. 5.1 Results of BMTCTN 1703, establishing post-transplant cyclophosphamide as a new standard of care for GVHD prevention for matched donor transplants.(Source: Courtesy of the New England Journal of Medicine)

7%, $p = 0.009$) and CMV infection (28% vs. 17%, $p = 0.023$).

The subsequent BMT Clinical Trials Network (CTN) 0303 multicenter trial limiting to MSD transplants compared TCD without pharmacologic prophylaxis ($n = 44$) to CNI-based GVHD prophylaxis ($n = 84$) (Pasquini et al. 2012). In this trial, TCD was achieved by CD34+ stem cell selection and capping the T-cell (CD3+) cell dose to <1 × 10^5 cells/kg. Incidence of grade II–IV acute GVHD was 23% in the TCD group and 39% in the CNI group, though this did not reach statistical significance ($p = 0.07$). The incidence of chronic GVHD was significantly lower in the TCD group (19% vs. 50%, $p < 0.001$), resulting in improved GVHD-free survival at 2 years post-HCT (41% vs. 19%, $p = 0.006$). Notably, graft rejection, relapse, NRM, DFS, and OS were similar between the two groups.

The results of BMTCTN 0303 created optimism that GVHD prevention without a pharmacologic agent was feasible, leading to the BMTCTN 1301 trial (Luznik et al. 2022). This phase III trial included three arms: CD34+ selected grafts, single-agent PTCy, or tacrolimus/MTX in patients receiving MAC-matched donor bone marrow grafts. Compared to the tacrolimus/MTX control arm, CD34 selection resulted in lower grade II–IV acute GVHD (16.3% vs. 29.8%, $p = 0.036$) and moderate to severe chronic GVHD (8.9% vs. 33.7%, $p < 0.001$). However, CD34 selection also led to a 2-year NRM of 21.5%, compared to 7.9% in the tacrolimus/MTX arm ($p = 0.008$), attributable to high rates of infection. Thus, CD34 selection was associated with inferior OS (60.1% vs. 76.1%, $p = 0.02$).

Ex vivo TCD has garnered interest as a strategy for GVHD prevention in the haploidentical donor setting. A European trial of 101 adult patients with acute leukemia receiving TCD haploidentical peripheral blood stem cell grafts demonstrated rates of grade II–IV acute GVHD < 10% and chronic GVHD < 20% (Aversa et al. 2005). However, this benefit was offset by high rates of NRM, due mostly to infection.

An alternative approach to TCD in haploidentical transplants is the depletion specifically of αβ T cells (Schumm et al. 2013). The rationale of this approach is based on evidence that αβ T cells contribute to GVHD, while other cell populations including γδ T cells and natural killer cells contribute to immune reconstitution and GVT without causing GVHD (Airoldi et al. 2015; Perko et al. 2015). Thus, depletion of αβ T cells may reduce GVHD without high risks of infection, relapse, or graft rejection seen with other TCD platforms. This approach was evaluated in a phase II trial of 80 pediatric patients with acute leukemia receiving haploidentical transplants (Locatelli et al. 2017). No patients developed severe acute GVHD or extensive chronic GVHD. The incidence of NRM was 5% and of relapse was 24%, resulting in 5-year leukemia-free survival (LFS) and OS of 71% and 72%, respectively.

5.6 Donor-Specific Recommendations and Guidelines for Use

The optimal approach to GVHD prevention for a specific patient depends upon a range of factors including conditioning regimens and donor sources. Interpretation of existing literature should carefully consider the inclusion criteria of practice-changing trials, with attention to the population for whom a specific regimen was validated, and whether the results may be generalized to other populations.

Matched Siblings Historically, MSD transplants have the lowest risk of GVHD and the need for aggressive immune suppression is minimized. A CNI-based doublet has remained a standard GVHD prevention regimen for ~40 years. The phase III BMT CTN 1703 trial included patients receiving RIC MSD or MUD peripheral blood grafts (Bolaos-Meade et al. 2023). The combination of PTCy/tacrolimus/MMF proved superior to tacrolimus/MTX with an improvement in GVHD-free, relapse-free survival. Thus, in the setting of RIC-matched donor PBSC grafts, PTCy/tacrolimus/MMF is considered the current standard of care.

For patients receiving a MAC regimen or bone marrow graft with MSD, the optimal GVHD prophylaxis is less clear. In prospective randomized trials, no regimen has proven superior to CNI-based doublet GVHD prophylaxis. Thus, CNI-based regimens remain an acceptable standard of care. Nonetheless, PTCy-based regimens have now been adopted by many centers as the preferred regimen for all MSD transplants. This is largely based on the extrapolation of BMT CTN 1703 data. The equivalent outcomes of CNI-based regimens to single-agent PTCy in the setting of MAC bone marrow grafts with matched donors have also been cited by many as a justification to use PTCy in this setting, though the trial did not meet the primary endpoint (Luznik et al. 2022).

MUD The decision to use CNI-based versus PTCy-based GVHD prophylaxis with MUD transplants is supported by the same data and recommendations as MSD. In summary, PTCy-based regimens are widely preferred for RIC MUD peripheral blood stem cell transplants. However, as BMTCTN 1703 did not include ATG for MUD transplants, some centers continue to prefer CNI-based regimens with ATG. Thus, CNI-based or PTCy-based regimens may be considered (Penack et al. 2024a).

For those patients receiving a PTCy-based regimen, a combination of PTCy with tac and MMF is preferred over single-agent PTCy given a lack of data to support the latter regimen. For CNI-based regimens for MUD transplants with MAC regimens, a CNI-based triplet should be considered. Both ATG and abatacept have demonstrated efficacy for lowering risks of GVHD when added to a CNI-based doublet without significant negative implications (Walker et al. 2016; Baron et al. 2014; Watkins et al. 2021; Theurich et al. 2012).

Haploidentical Relative There are currently no large prospective trials comparing GVHD prophylactic regimens for haploidentical-related donor transplants. Due to a lack of prospective data, PTCy-based regimens, CNI/MTX with ATG, and ex vivo TCD are all considered suitable options for haploidentical transplants. However, a registry trial from the EBMT compared PTCy ($n = 2999$) versus ATG ($n = 358$) in adult patients who received haploidentical donor HCT for acute myeloid leukemia and showed the ATG group had a higher risk of NRM, worse DFS, and worse OS (Bazarbachi et al. 2024). PTCy has emerged as the most widely accepted regimen for haploidentical donor transplants, particularly in adult patients.

Mismatched Unrelated Donors Currently accepted regimens for MMUD HCT include PTCy-based regimens, CNI/MTX with ATG, and tacrolimus/MTX/abatacept. Prospective phase III data demonstrated the superiority of PTCy/tacrolimus/MMF over tacrolimus/MTX in a cohort that included HLA 7/8 MMUD transplants. A registry study from the EBMT compared PTCy ($n = 583$) to ATG (1540) in patients who received HLA 9/10 MMUD transplants and showed that ATG was associated with worse NRM (18% vs. 24.9%, $p = 0.028$) and worse OS (65.7% vs. 55.7%, $p < 0.001$) (Penack et al. 2024b). Another study from the CIBMTR showed that HLA 7/8 MMUD transplants with PTCy resulted in similar outcomes to HLA 8/8 MUD HCT, suggesting data with this regimen in MUDs can be extrapolated to HLA 7/8 mismatched transplants (Shaffer et al. 2024). However, with no prospective comparisons between PTCy and regimens containing ATG or abatacept, more data is needed to deem one regimen as preferable over the others. Furthermore, more studies are warranted to determine the optimal GVHD prophylactic regimen for patients receiving HLA < 7/8 matched donor HCT.

Umbilical Cord Blood Data supporting GVHD prophylactic strategies in related and unrelated donor HCT cannot be applied to umbilical cord blood transplants (UCBT) as this platform has unique immune reconstitution and an inherently low risk of GVHD (Fuchs et al. 2021; Cunha et al. 2014; Hiwarkar et al. 2017). While CNI-based regimens are standard for GVHD prophylaxis with UCBTs, the incorporation of MTX has been shown to decrease the likelihood of neutrophil engraftment (Herr et al. 2010; Yoshida et al. 2019). Instead, CNI/MMF is more commonly used with UCBT (Bejanyan et al. 2018).

Other Considerations Published trials are often studying outcomes under ideal situations with eligibility restrictions based on comorbidities and organ function. In practice, it is important to consider patient-related factors and comorbidities that may interfere with favorable outcomes. When feasible, substitutions may be considered. For patients receiving CNI-based regimens, factors such as older age, baseline liver or kidney disease, or presence of extravascular fluid collections (e.g., pleural effusions) would warrant consideration for dose-reduced MTX or substitution with MMF (Hamilton et al. 2015b). As PTCy is associated with cardiac toxicities, some providers avoid PTCy in patients with baseline cardiac comorbidities (Dulery et al. 2021). Similarly, in the presence of a relative contraindication to tacrolimus, such as chronic kidney disease, PTCy/sirolimus/MMF may be considered (Elmariah et al. 2024a; Bejanyan et al. 2021). For patients who received a T-cell-depleting regimen prior to transplant, such as alemtuzumab, the addition of ATG to the GVHD prophylactic regimen should be used cautiously since both agents have a high risk of opportunistic infections (Ueda Oshima et al. 2023).

5.7 Future Directions

Janus Kinase Inhibitors The Janus kinase signal transducer and activator of transcription proteins (JAK-STAT) pathway is now recognized as a critically important target in the pathophysiology of both acute and chronic GVHD. JAK2/STAT3 activity is required for the interleukin signaling that facilitates T-cell activation needed for GVHD (Betts et al. 2015; Betts et al. 2018). Clinically, the JAK1/JAK2 inhibitor ruxolitinib has demonstrated effectiveness for the treatment of steroid-refractory acute and chronic GVHD, with response rates of ~60% and ~50%, respectively (Zeiser et al. 2020; Zeiser et al. 2021). For GVHD prevention, a pilot study of 20 patients receiving matched donor transplants for myelofibrosis combined PTCy with ruxolitinib continued through day +100 (Morozova et al. 2021). The incidence of grade II–IV acute GVHD was 25%, of grade III–IV acute GVHD was 15%, and of severe chronic GVHD was 0%. A more recent multicenter combined phase I/II trial studied the novel JAK2-specific inhibitor pacritinib in combination with tacrolimus/sirolimus for patients receiving matched donor transplants. Although correlative studies confirmed suppression of the STAT3 pathway and subsequent inhibition of Th1/Th17 lymphocytes, risks of GVHD were not decreased compared to historic controls (Pidala et al. 2024). Studies evaluating JAK2 inhibition for GVHD prevention are ongoing.

Ustekinumab Ustekinumab is a monoclonal antibody that targets IL-12 and IL-23, currently in use to treat Crohn's disease and psoriasis. As IL-12 contributes to Th1 differentiation and IL-23 contributes to Th17 differentiation, suppression of these cytokines by ustekinumab offers potential for GVHD prevention (Pidala et al. 2018). A single-center, randomized, blinded placebo-controlled trial compared CNI-based GVHD prophylaxis with or without ustekinumab and found ustekinumab significantly suppressed IL-12, IL-23, and led to improved overall survival (85.7% vs. 46.7%, $p = 0.027$). Ustekinumab in combination with CNI-based GVHD prophylaxis is now being studied in a randomized, multicenter phase II trial (Pidala et al. 2018).

ORCA-T The use of ex vivo T-cell depletion has been limited by complications such as relapse and infection that are secondary to the removal of essential T-cell populations. The Orca-T platform exploits immunomagnetic selection and high-speed cell sorting technology to create a graft with a prespecified dose of stem cells progenitor cells to promote engraftment, Tregs to prevent GVHD, and conventional T cells to mitigate infection and relapse (Meyer et al. 2019). In a trial of patients receiving MUD ORCA-T grafts with MAC and single-agent tacrolimus for GVHD prevention, rates of NRM, DFS, and OS were 3%, 77%, and 94%, respectively (Gomez-

Arteaga et al. 2024). These results were superior to a historical cohort of patients receiving PTCy-based conditioning. However, there are currently no prospective comparative trials with ORCA, and data with RIC regimens, haploidentical donors, and MMUDs are also needed.

Vedolizumab Vedolizumab is a monoclonal antibody that targets gut-selective $\alpha_4\beta_7$ integrin (Wyant et al. 2016). This drug can block T-lymphocyte migration to intestinal tissue and, thus, prevent gastrointestinal (GI) GVHD (Soler et al. 2009; Chen et al. 2009). A randomized, controlled phase III trial compared CNI-based GVHD prophylaxis with vedolizumab versus placebo in 343 patients receiving unrelated donor HCT (Chen et al. 2024). The trial met the primary endpoint of improving GI acute GVHD-free survival at 6 months after transplant (85.5% vs. 70.9%, $p < 0.001$), and the drug was well tolerated with no differences in NRM, relapse, or OS.

Sitagliptin Commonly used for the treatment of type 2 diabetes mellitus, sitagliptin inhibits dipeptidyl peptidase 4 (DPP-4), which contributes to the costimulation of T cells by upregulating CD86 on APCs (Hatano et al. 2013). A phase II trial evaluated sitagliptin from day −1 to +14 combined with tacrolimus/sirolimus in 36 patients receiving MAC-matched donor transplants. The incidence of grade II acute GVHD was only 5%, much lower than expected with tacrolimus/sirolimus alone and 1-year OS was 94%. However, sitagliptin did not add any perceived benefit for the prevention of chronic GVHD.

5.8 Conclusions

GVHD is a major driver of morbidity and mortality after allogeneic HCT. From the 1980s, CNI/MTX remained unbeaten for 40 years as the standard of care for GVHD prophylaxis in most settings. Modern approaches have finally offered significantly lower risks of GVHD, and PTCy particularly has emerged as a new standard for matched donors, haploidentical-related donors, and MMUDs. Still, over half of patients develop GVHD during their posttransplant course. Future studies should continue to focus on reducing this challenging complication, while also separating and augmenting the critical GVT effect to reduce relapse.

References

Airoldi I, Bertaina A, Prigione I, Zorzoli A, Pagliara D, Cocco C et al (2015) Gammadelta T-cell reconstitution after HLA-haploidentical hematopoietic transplantation depleted of TCR-alphabeta+/CD19+ lymphocytes. Blood 125(15):2349–2358

Al-Homsi AS, Cirrone F, Wo S, Cole K, Suarez-Londono JA, Gardner SL et al (2023) PTCy, abatacept, and a short course of tacrolimus for GVHD prevention after haploidentical transplantation. Blood Adv 7(14):3604–3611

Allison AC, Eugui EM (2005) Mechanisms of action of mycophenolate mofetil in preventing acute and chronic allograft rejection. Transplantation 80(2 Suppl):S181–S190

Anasetti C, Logan B, Lee S, Waller E, Weisdorf D, Wingard J et al (2012) Peripheral-blood stem cells versus bone marrow from unrelated donors. NEJM 367:1487–1496

Andriole G, Sandlund J, Miser J, Arasi V, Linehan M, Magrath I (1987) The efficacy of mesna (2-mercaptoethane sodium sulfonate) as a uroprotectant in patients with hemorrhagic cystitis receiving further oxazaphosphorine chemotherapy. JCO 5(5):799

Antin JH, Ferrara JL (1992) Cytokine dysregulation and acute graft-versus-host disease. Blood 80(12):2964–2968

Aversa F, Terenzi A, Tabilio A, Falzetti F, Carotti A, Ballanti S et al (2005) Full haplotype-mismatched hematopoietic stem-cell transplantation: a phase II study in patients with acute leukemia at high risk of relapse. J Clin Oncol 23(15):3447–3454

Barbarino JM, Staatz CE, Venkataramanan R, Klein TE, Altman RB (2013) PharmGKB summary: cyclosporine and tacrolimus pathways. Pharmacogenet Genomics 23(10):563–585

Baron F, Labopin M, Blaise D, Lopez-Corral L, Vigouroux S, Craddock C et al (2014) Impact of in vivo T-cell depletion on outcome of AML patients in first CR given peripheral blood stem cells and reduced-intensity conditioning Allo-SCT from a HLA-identical sibling donor: a report from the Acute Leukemia Working Party of the European Group for Blood and Marrow Transplantation. Bone Marrow Transplant 49(3):389–396

Bashey A, Zhang M, McCurdy S, StMartin A, Argall T, Anasetti C et al (2017) Mobilized peripheral blood stem cells versus unstimulated bone marrow as a

graft source for T-cell-replete Haploidentical donor Transplantation using post-Transplant cyclophosphamide. JCO. 35(26):3002–3009
Bazarbachi AH, Labopin M, Raiola AM, Blaise D, Arcese W, Santarone S et al (2024) Posttransplant cyclophosphamide versus anti-thymocyte globulin versus combination for graft-versus-host disease prevention in haploidentical transplantation for adult acute myeloid leukemia: a report from the European Society for Blood and Marrow Transplantation Acute Leukemia Working Party. Cancer 130:3123
Bejanyan N, Rogosheske J, Cao Q, Lazaryan A, Holtan S, Ustun C et al (2018) Weight-based mycophenolate mofetil dosing predicts acute Gvhd and relapse after allogeneic hematopoietic cell transplantation. BBMT. 24(3):s291
Bejanyan N, Pidala JA, Wang X, Thapa R, Nishihori T, Elmariah H et al (2021) A phase 2 trial of GVHD prophylaxis with PTCy, sirolimus, and MMF after peripheral blood haploidentical transplantation. Blood Adv 5(5):1154–1163
Betts B, Sagatys E, Veerapathran A, Lloyd M, Beato F, Lawrence H et al (2015) CD4+ T cell STAT3 phosphorylation precedes acute GVHD, and subsequent Th17 tissue invasion correlates with GVHD severity and therapeutic response. J Leukoc Biol 97(4):807–819
Betts B, Bastian D, Iamsawat S, Nguyen H, Heinrichs J, Wu Y et al (2018) Targeting JAK2 reduces GVHD and xenograft rejection through regulation of T cell differentiation. PNAS 115(7):1582–1587
Blazar BR, Murphy WJ, Abedi M (2012) Advances in graft-versus-host disease biology and therapy. Nat Rev Immunol 12(6):443–458
Bolaños-Meade J, Fuchs E, Luznik L, Lanzkron S, Gamper C, Jones R et al (2012) HLA-haploidentical bone marrow transplantation with posttransplant cyclophosphamide expands the donor pool for patients with sickle cell disease. Blood 120:4285–4291
Bolaños-Meade J, Reshef R, Fraser R, Fei M, Abhyankar S, Al-Kadhimi Z et al (2019a) Three prophylaxis regimens (tacrolimus, mycophenolate mofetil, and cyclophosphamide; tacrolimus, methotrexate, and bortezomib; or tacrolimus, methotrexate, and maraviroc) versus tacrolimus and methotrexate for prevention of graft-versus-host disease with haemopoietic cell transplantation with reduced-intensity conditioning: a randomised phase 2 trial with a non-randomised contemporaneous control group (BMT CTN 1203). Lancet Haematol 6(3):e132–ee43
Bolaños-Meade J, Cooke KR, Gamper CJ, Ali SA, Ambinder RF, Borrello IM et al (2019b) Effect of increased dose of total body irradiation on graft failure associated with HLA-haploidentical transplantation in patients with severe haemoglobinopathies: a prospective clinical trial. Lancet Haematol. 6(4):e183–ee93
Bolaños-Meade J, Hamadani M, Wu J, Al Malki MM, Martens MJ, Runaas L et al (2023) Post-Transplantation cyclophosphamide-based graft-versus-host disease prophylaxis. N Engl J Med 388(25):2338–2348
Bolwell B, Sobecks R, Pohlman B, Andresen S, Rybicki L, Kuczkowski E et al (2004) A prospective randomized trial comparing cyclosporine and short course methotrexate with cyclosporine and mycophenolate mofetil for GVHD prophylaxis in myeloablative allogeneic bone marrow transplantation. Bone Marrow Transplant 34(7):621–625
Bornhäuser M, Schuler U, Pörksen G, Naumann R, Geissler G, Thiede C et al (1999) Mycophenolate mofetil and cyclosporine as graft-versus-host disease prophylaxis after allogeneic blood stem cell transplantation. Transplantation 67(4):499–504
Bray R, Hurley C, Kamani N, Woolfrey A, Müller C, Spellman S et al (2008) National Marrow Donor Program HLA matching guidelines for unrelated adult donor hematopoietic cell transplants. BBMT 14(9):45–53
Carpenter PA, Kitko CL, Elad S, Flowers ME, Gea-Banacloche JC, Halter JP et al (2015) National Institutes of Health consensus development project on criteria for clinical trials in chronic graft-versus-host Disease: V. The 2014 ancillary therapy and supportive care working group report. Biol Blood Marrow Transplant 21(7):1167–1187
Chen X, Vodanovic-Jankovic S, Johnson B, Keller M, Komorowski R, Drobyski WR (2007) Absence of regulatory T-cell control of TH1 and TH17 cells is responsible for the autoimmune-mediated pathology in chronic graft-versus-host disease. Blood 110(10):3804–3813
Chen YB, Kim HT, McDonough S, Odze RD, Yao X, Lazo-Kallanian S et al (2009) Up-regulation of alpha4beta7 integrin on peripheral T cell subsets correlates with the development of acute intestinal graft-versus-host disease following allogeneic stem cell transplantation. Biol Blood Marrow Transplant 15(9):1066–1076
Chen P, Li J, Li J, Deng R, Fu Q, Chen J et al (2017) Dynamic effects of CYP3A5 polymorphism on dose requirement and trough concentration of tacrolimus in renal transplant recipients. J Clin Pharm Ther 42(1):93–97
Chen YB, Mohty M, Zeiser R, Teshima T, Jamy O, Maertens J et al (2024) Vedolizumab for the prevention of intestinal acute GVHD after allogeneic hematopoietic stem cell transplantation: a randomized phase 3 trial. Nat Med 30:2277
Chhabra S, Liu Y, Hemmer MT, Costa L, Pidala JA, Couriel DR et al (2019) Comparative analysis of calcineurin inhibitor-based methotrexate and mycophenolate mofetil-containing regimens for prevention of graft-versus-host disease after reduced-intensity conditioning allogeneic transplantation. Biol Blood Marrow Transplant 25(1):73–85
Ciurea S, Zhang M, Bacigalupo A, Bashey A, Appelbaum F, Aljitawi O et al (2015) Haploidentical transplant with post-transplant cyclophosphamide versus matched unrelated donor transplant for acute myeloid leukemia. Blood 126(8):1033–1040

Coenen JJ, Koenen HJ, van Rijssen E, Hilbrands LB, Joosten I (2006) Rapamycin, and not cyclosporin A, preserves the highly suppressive CD27+ subset of human CD4+CD25+ regulatory T cells. Blood 107(3):1018–1023

Cooke KR, Luznik L, Sarantopoulos S, Hakim FT, Jagasia M, Fowler DH et al (2017) The biology of chronic graft-versus-host disease: a task force report from the National Institutes of Health consensus development project on criteria for clinical trials in chronic graft-versus-host disease. Biol Blood Marrow Transplant 23(2):211–234

Cunha R, Loiseau P, Ruggeri A, Sanz G, Michel G, PaolaIori A et al (2014) Impact of HLA mismatch direction on outcomes after umbilical cord blood transplantation for hematological malignant disorders: a retrospective Eurocord-EBMT analysis. Bone Marrow Transplant 49:24–29

Cutler C, Antin JH (2004) Sirolimus for GVHD prophylaxis in allogeneic stem cell transplantation. Bone Marrow Transplant 34(6):471–476

Cutler C, Kim HT, Hochberg E, Ho V, Alyea E, Lee SJ et al (2004) Sirolimus and tacrolimus without methotrexate as graft-versus-host disease prophylaxis after matched related donor peripheral blood stem cell transplantation. Biol Blood Marrow Transplant 10(5):328–336

Cutler C, Li S, Kim HT, Laglenne P, Szeto KC, Hoffmeister L et al (2005) Mucositis after allogeneic hematopoietic stem cell transplantation: a cohort study of methotrexate- and non-methotrexate-containing graft-versus-host disease prophylaxis regimens. Biol Blood Marrow Transplant 11(5):383–388

Cutler C, Stevenson K, Kim HT, Richardson P, Ho VT, Linden E et al (2008) Sirolimus is associated with veno-occlusive disease of the liver after myeloablative allogeneic stem cell transplantation. Blood 112(12):4425–4431

Cutler C, Logan B, Nakamura R, Johnston L, Choi S, Porter D et al (2014) Tacrolimus/sirolimus vs tacrolimus/methotrexate as GVHD prophylaxis after matched, related donor allogeneic HCT. Blood 124(8):1372–1377

Dehn J, Spellman S, Hurley C, Shaw B, Barker J, Burns L et al (2019) Selection of unrelated donors and cord blood units for hematopoietic cell transplantation: guidelines from the NMDP/CIBMTR. Blood 134(12):924–934

DeZern AE, Elmariah H, Zahurak M, Rosner GL, Gladstone DE, Ali SA et al (2020) Shortened-duration immunosuppressive therapy after Nonmyeloablative, related HLA-Haploidentical or unrelated peripheral blood grafts and post-Transplantation cyclophosphamide. Biol Blood Marrow Transplant 26(11):2075–2081

DeZern AE, Zahurak M, Symons HJ, Cooke KR, Huff CA, Jain T et al (2023) Alternative donor BMT with posttransplant cyclophosphamide as initial therapy for acquired severe aplastic anemia. Blood 141(25):3031–3038

DiStasi A, Milton D, Poon L, Hamdi A, Rondon G, Chen J et al (2014) Similar transplantation outcomes for acute myeloid leukemia and myelodysplastic syndrome patients with haploidentical versus 10/10 human leukocyte antigen–matched unrelated and related donors. BBMT 20(12):1975–1981

Dominietto A, Tedone E, Soracco M, Bruno B, Raiola AM, Van Lint MT et al (2012) In vivo B-cell depletion with rituximab for alternative donor hemopoietic SCT. Bone Marrow Transplant 47(1):101–106

D'Souza A, Fretham C, Lee SJ, Arora M, Brunner J, Chhabra S et al (2020) Current use of and trends in hematopoietic cell transplantation in the United States. Biol Blood Marrow Transplant 26:e177

Dubois EA, Cohen AF (2009) Abatacept. Br J Clin Pharmacol 68(4):480–481

Dulery R, Mohty R, Labopin M, Sestili S, Malard F, Brissot E et al (2021) Early cardiac toxicity associated with post-transplant cyclophosphamide in allogeneic stem cell transplantation. JACC Cardio Oncol 3(2):250–259

Edinger M, Hoffmann P, Ermann J, Drago K, Fathman CG, Strober S et al (2003) CD4+CD25+ regulatory T cells preserve graft-versus-tumor activity while inhibiting graft-versus-host disease after bone marrow transplantation. Nat Med 9(9):1144–1150

Elmariah H, Kasamon YL, Zahurak M, Macfarlane KW, Tucker N, Rosner GL et al (2018) Haploidentical bone marrow transplantation with post-transplant cyclophosphamide using non-first-degree related donors. Biol Blood Marrow Transplant 24(5):1099–1102

Elmariah H, Otoukesh S, Kumar A, Ali H, Arslan S, Shouse G et al (2024a) Sirolimus is an acceptable alternative to tacrolimus for graft-versus-host disease prophylaxis after Haploidentical peripheral blood stem cell transplantation with post-transplantation cyclophosphamide. Transplant Cell Ther. 30(2):229.e1–229e11

Elmariah H, Otoukesh S, Kumar A, Haris A, Shukaib A, Dimaggio E et al (2024b) Lower weight-based mycophenolate mofetil dosing is associated with superior outcomes after haploidentical hematopoietic cell transplant with post-transplant cyclophosphamide. Transplant Cell Ther. 30:1019.e1

Eto M, Mayumi H, Tomita Y, Yoshikai Y, Nishimura Y, Maeda T et al (1991) Specific destruction of host-reactive mature T cells of donor origin prevents graft-versus-host disease in cyclophosphamide-induced tolerant mice. J Immunol 146(5):1402–1409

Evans WE, Pratt CB (1978) Effect of pleural effusion on high-dose methotrexate kinetics. Clin Pharmacol Ther 23(1):68–72

Figgins B, Hammerstrom A, Ariza-Heredia E, Oran B, Milton DR, Yeh J (2019) Characterization of viral infections after antithymocyte globulin-based conditioning in adults undergoing allogeneic hematopoietic stem cell transplantation. Biol Blood Marrow Transplant 25(9):1837–1843

Finke J, Bethge WA, Schmoor C, Ottinger HD, Stelljes M, Zander AR et al (2009) Standard graft-versus-host disease prophylaxis with or without anti-T-cell globulin in haematopoietic cell transplantation from matched unrelated donors: a randomised, open-label, multicentre phase 3 trial. Lancet Oncol 10(9):855–864

Finke J, Schmoor C, Bethge WA, Ottinger H, Stelljes M, Volin L et al (2017) Long-term outcomes after standard graft-versus-host disease prophylaxis with or without anti-human-T-lymphocyte immunoglobulin in haemopoietic cell transplantation from matched unrelated donors: final results of a randomised controlled trial. Lancet Haematol. 4(6):e293–e301

Fletcher RE, Nunes NS, Patterson MT, Vinod N, Khan SM, Mendu SK et al (2023) Posttransplantation cyclophosphamide expands functional myeloid-derived suppressor cells and indirectly influences Tregs. Blood Adv 7(7):1117–1129

Fuchs EJ, O'Donnell PV, Eapen M, Logan B, Antin JH, Dawson P et al (2021) Double unrelated umbilical cord blood vs HLA-haploidentical bone marrow transplantation: the BMT CTN 1101 trial. Blood 137(3):420–428

Gaballa S, Ge I, El Fakih R, Brammer JE, Kongtim P, Tomuleasa C et al (2016) Results of a 2-arm, phase 2 clinical trial using post-transplantation cyclophosphamide for the prevention of graft-versus-host disease in haploidentical donor and mismatched unrelated donor hematopoietic stem cell transplantation. Cancer 122(21):3316–3326

Goldsmith SR, Abid MB, Auletta JJ, Bashey A, Beitinjaneh A, Castillo P et al (2021) Posttransplant cyclophosphamide is associated with increased cytomegalovirus infection: a CIBMTR analysis. Blood 137(23):3291–3305

Gomez-Arteaga A, Oliai C, Patel SS, Pantin J, Gandhi AP, Salhotra A et al (2024) A retrospective analysis comparing Orca-T to post-Transplant cyclophosphamide based allogeneic hematopoietic cell Transplant in patients with matched unrelated donors receiving Myeloablative conditioning. Transplant Cell Ther 30(2):S51–S52

Good RA, Martinez C, Gabrielsen AE (1964) Progress toward transplantation of tissues in man. Adv Pediatr Infect Dis 13:93–127

Gooptu M, Romee R, St Martin A, Arora M, Al Malki MM, Antin JH et al (2021) HLA Haploidentical versus matched unrelated donor transplants with post-Transplant cyclophosphamide based prophylaxis. Blood 138:273

Goulmy E, Schipper R, Pool J, Blokland E, Falkenburg JH, Vossen J et al (1996) Mismatches of minor histocompatibility antigens between HLA-identical donors and recipients and the development of graft-versus-host disease after bone marrow transplantation. N Engl J Med 334(5):281–285

Gragert L, Eapen M, Williams E, Freeman J, Spellman S, Baitty R et al (2014) HLA match likelihoods for hematopoietic stem-cell grafts in the U.S. Registry. NEJM 371:339–348

Grenier PO, McCormack L, Alshamekh SA, Adams DM, Trenor CC 3rd, O'Hare MJ et al (2021) Mucocutaneous adverse events associated with oral sirolimus for the treatment of vascular anomalies. JAMA Dermatol 157(2):233–235

Hall AG, Tilby MJ (1992) Mechanisms of action of, and modes of resistance to, alkylating agents used in the treatment of haematological malignancies. Blood Rev 6(3):163–173

Hamilton BK, Rybicki L, Haddad H, Abounader D, Yurch M, Majhail NS et al (2015a) Does day 11 omission of methotrexate due to toxicity influence the outcome in myeloablative hematopoietic cell transplant? Results from a single-center retrospective cohort study. Blood Cancer J 5(8):e344

Hamilton BK, Rybicki L, Dean R, Majhail NS, Haddad H, Abounader D et al (2015b) Cyclosporine in combination with mycophenolate mofetil versus methotrexate for graft versus host disease prevention in myeloablative HLA-identical sibling donor allogeneic hematopoietic cell transplantation. Am J Hematol 90(2):144–148

Hatano R, Ohnuma K, Yamamoto J, Dang NH, Morimoto C (2013) CD26-mediated co-stimulation in human CD8(+) T cells provokes effector function via pro-inflammatory cytokine production. Immunology 138(2):165–172

Herr AL, Kabbara N, Bonfim CM, Teira P, Locatelli F, Tiedemann K et al (2010) Long-term follow-up and factors influencing outcomes after related HLA-identical cord blood transplantation for patients with malignancies: an analysis on behalf of Eurocord-EBMT. Blood 116(11):1849–1856

Hiwarkar P, Hubank M, Qasim W, Chiesa R, Gilmour KC, Saudemont A et al (2017) Cord blood transplantation recapitulates fetal ontogeny with a distinct molecular signature that supports CD4(+) T-cell reconstitution. Blood Adv 1(24):2206–2216

Ho TT, Perkins JB, Gonzalez R, Hicks JK, Martinez RA, Duranceau K et al (2024) Association between CYP3A4, CYP3A5 and ABCB1 genotype and tacrolimus treatment outcomes among allogeneic HSCT patients. Pharmacogenomics 25(1):29–40

Holtan S, DeFor T, Lazaryan A, Bejanyan N, Arora M, Brunstein C et al (2015) Composite end point of graft-versus-host disease-free, relapse-free survival after allogeneic hematopoietic cell transplantation. Blood 125:1333–1338

Holtan SG, Khera N, Levine JE, Chai X, Storer B, Liu HD et al (2016) Late acute graft-versus-host disease: a prospective analysis of clinical outcomes and circulating angiogenic factors. Blood 128(19):2350–2358

Horowitz M, Gale R, Sondel P, Goldman J, Kersey J, Kolb H et al (1990) Graft-versus-leukemia reactions after bone marrow transplantation. Blood 75(3):555–562

Inamoto Y, Flowers ME, Appelbaum FR, Carpenter PA, Deeg HJ, Furlong T et al (2011) A retrospective comparison of tacrolimus versus cyclosporine with methotrexate for immunosuppression after allogeneic hematopoietic cell transplantation with mobilized blood cells. Biol Blood Marrow Transplant 17(7):1088–1092

Jenq RR, Ubeda C, Taur Y, Menezes CC, Khanin R, Dudakov JA et al (2012) Regulation of intestinal inflammation by microbiota following allogeneic bone marrow transplantation. J Exp Med 209(5):903–911

Jodele S, Davies SM, Lane A, Khoury J, Dandoy C, Goebel J et al (2014) Diagnostic and risk criteria for HSCT-associated thrombotic microangiopathy: a study in children and young adults. Blood 124(4):645–653

Kahan BD, Napoli KL (1998) Role of therapeutic drug monitoring of rapamycin. Transplant Proc 30(5):2189–2191

Kalra A, Roessner C, Jupp J, Williamson T, Tellier R, Chaudhry A et al (2018) Risk factors for post-transplant lymphoproliferative disorder after thymoglobulin-conditioned hematopoietic cell transplantation. Clin Transpl 32(1)

Kanakry C, Ganguly S, Zahurak M, Bolaños-Meade J, Thoburn C, Perkins B et al (2013a) Aldehyde dehydrogenase expression drives human regulatory T cell resistance to Posttransplantation cyclophosphamide. Sci Transl Med 5(211):211ra157

Kanakry J, Kasamon Y, Bolaños-Meade J, Borrello I, Brodsky R, Fuchs E et al (2013b) Absence of post-transplantation lymphoproliferative disorder after allogeneic blood or marrow transplantation using post-transplantation cyclophosphamide as graft-versus-host disease prophylaxis. BBMT. 19(10):1514–1517

Kanakry C, Tsai H, Bolaños-Meade J, Smith B, Gojo I, Kanakry J et al (2014) Single-agent GVHD prophylaxis with posttransplantation cyclophosphamide after myeloablative, HLA-matched BMT for AML, ALL, and MDS. Blood 124:3817–3827

Kanda Y, Kobayashi T, Mori T, Tanaka M, Nakaseko C, Yokota A et al (2016) A randomized controlled trial of cyclosporine and tacrolimus with strict control of blood concentrations after unrelated bone marrow transplantation. Bone Marrow Transplant 51(1):103–109

Kasamon Y, Ambinder R, Fuchs E, Zahurak M, Rosner G, Bolaños-Meade J et al (2017) Prospective study of nonmyeloablative, HLA-mismatched unrelated BMT with high-dose posttransplantation cyclophosphamide. Blood Adv 1:288–292

Kasamon Y, Fuchs E, Zahurak M, Rosner G, Symons H, Gladstone D et al (2018) Shortened-duration tacrolimus after nonmyeloablative, HLA-haploidentical bone marrow transplantation. BBMT 24(5):1022–1028

Kelly PA, Gruber SA, Behbod F, Kahan BD (1997) Sirolimus, a new, potent immunosuppressive agent. Pharmacotherapy 17(6):1148–1156

Khimani F, Ranspach P, Elmariah H, Kim J, Whiting J, Nishihori T et al (2021) Increased infections and delayed CD4+ T-cell but faster B-cell immune reconstitution after Posttransplant cyclophosphamide compared to conventional GVHD prophylaxis in allogeneic transplantation. Transplant Cell Ther. 27:940

Kidney Disease: Improving Global Outcomes Transplant Work G (2009) KDIGO clinical practice guideline for the care of kidney transplant recipients. Am J Transplant 9(Suppl 3):S1–S155

Kiehl MG, Schafer-Eckart K, Kroger M, Bornhauser M, Basara N, Blau IW et al (2002) Mycophenolate mofetil for the prophylaxis of acute graft-versus-host disease in stem cell transplant recipients. Transplant Proc 34(7):2922–2924

Kirtane K, Lee S (2017) Racial and ethnic disparities in hematologic malignancies. Blood 130(15):1699–1705

Kopf H, de la Rosa GM, Howard OM, Chen X (2007) Rapamycin inhibits differentiation of Th17 cells and promotes generation of FoxP3+ T regulatory cells. Int Immunopharmacol 7(13):1819–1824

Kremer JM, Alarcon GS, Weinblatt ME, Kaymakcian MV, Macaluso M, Cannon GW et al (1997) Clinical, laboratory, radiographic, and histopathologic features of methotrexate-associated lung injury in patients with rheumatoid arthritis: a multicenter study with literature review. Arthritis Rheum 40(10):1829–1837

Kroger N, Solano C, Wolschke C, Bandini G, Patriarca F, Pini M et al (2016) Antilymphocyte globulin for prevention of chronic graft-versus-host disease. N Engl J Med 374(1):43–53

Lamba V, Sangkuhl K, Sanghavi K, Fish A, Altman RB, Klein TE (2014) PharmGKB summary: mycophenolic acid pathway. Pharmacogenet Genomics 24(1):73–79

Lee SJ (2017) Classification systems for chronic graft-versus-host disease. Blood 129(1):30–37

Legoff J, Resche-Rigon M, Bouquet J, Robin M, Naccache SN, Mercier-Delarue S et al (2017) The eukaryotic gut virome in hematopoietic stem cell transplantation: new clues in enteric graft-versus-host disease. Nat Med 23(9):1080–1085

Liapis G, Boletis J, Skalioti C, Bamias G, Tsimaratou K, Patsouris E et al (2013) Histological spectrum of mycophenolate mofetil-related colitis: association with apoptosis. Histopathology 63(5):649–658

Locatelli F, Merli P, Pagliara D, Li Pira G, Falco M, Pende D et al (2017) Outcome of children with acute leukemia given HLA-haploidentical HSCT after alphabeta T-cell and B-cell depletion. Blood 130(5):677–685

Lochte HL Jr, Levy AS, Guenther DM, Thomas ED, Ferrebee JW (1962) Prevention of delayed foreign marrow reaction in lethally irradiated mice by early administration of methotrexate. Nature 196:1110–1111

Lutt JR (2009) Efficacy, safety, and tolerability of abatacept in the management of rheumatoid arthritis. Open Access Rheumatol 1:17–35

Luznik L, Jalla S, Engstrom L, Iannone R, Fuchs E (2001) Durable engraftment of major histocompatibility complex–incompatible cells after nonmyeloablative conditioning with fludarabine, low-dose total body irradiation, and posttransplantation cyclophosphamide. Blood 98:3456–3464

Luznik L, O'Donnell P, Symons H, Chen A, Leffell M, Zahurak M et al (2008) HLA-haploidentical bone marrow transplantation for hematologic malignancies using nonmyeloablative conditioning and high-dose, posttransplantation cyclophosphamide. BBMT. 14(6):641–650

Luznik L, Pasquini MC, Logan B, Soiffer RJ, Wu J, Devine SM et al (2022) Randomized phase III BMT CTN trial of calcineurin inhibitor-free chronic graft-versus-host disease interventions in myeloablative hematopoietic cell transplantation for hematologic malignancies. J Clin Oncol 40(4):356–368

Majzner RG, Mogri H, Varadhan R, Brown P, Cooke KR, Bolaños-Meade J et al (2017) Post-transplantation cyclophosphamide after bone marrow transplantation is not associated with an increased risk of donor-derived malignancy. Biol Blood Marrow Transplant 23(4):612–617

Malard F, Holler E, Sandmaier BM, Huang H, Mohty M (2023) Acute graft-versus-host disease. Nat Rev Dis Primers 9(1):27

Matte CC, Liu J, Cormier J, Anderson BE, Athanasiadis I, Jain D et al (2004) Donor APCs are required for maximal GVHD but not for GVL. Nat Med 10(9):987–992

Mayumi H, Good R (1989) Long-lasting skin allograft tolerance in adult mice induced across fully allogeneic (multimajor H-2 plus multiminor histocompatibility) antigen barriers by a tolerance-inducing method using cyclophosphamide. J Exp Med 169(1):213–238

Mehta RS, Saliba RM, Hayase E, Jenq RR, Abraham S, Rashid A et al (2022) Mycophenolate mofetil: a friend or a foe with post-transplantation cyclophosphamide and tacrolimus prophylaxis in HLA-matched donors? Transplant Cell Ther. 28(8):500.e1–500e10

Mehta RS, Ramdial J, Marin D, Alousi A, Kanakry CG, Champlin RE et al (2023) Impact of donor age in Haploidentical-post-transplantation cyclophosphamide versus matched unrelated donor post-transplantation cyclophosphamide hematopoietic stem cell transplantation in patients with acute myeloid leukemia. Transplant Cell Ther. 29(6):377.e1–377.e7

Meyer EH, Laport G, Xie BJ, MacDonald K, Heydari K, Sahaf B et al (2019) Transplantation of donor grafts with defined ratio of conventional and regulatory T cells in HLA-matched recipients. JCI Insight 4(10):e127244

Mirza AS, Tandon A, Jenneman D, Cao S, Brimer T, Kumar A et al (2022) Outcomes following intolerance to tacrolimus/sirolimus graft-versus-host disease prophylaxis for allogeneic hematopoietic cell transplantation. Transplant Cell Ther. 28(4):185.e1–185.e7

Mohty M (2007) Mechanisms of action of antithymocyte globulin: T-cell depletion and beyond. Leukemia 21(7):1387–1394

Moreland L, Bate G, Kirkpatrick P (2006) Abatacept. Nat Rev Drug Discov 5(3):185–186

Morozova EV, Barabanshikova MV, Moiseev IS, Shakirova AI, Barhatov IM, Ushal IE et al (2021) A prospective pilot study of graft-versus-host disease prophylaxis with post-transplantation cyclophosphamide and ruxolitinib in patients with myelofibrosis. Acta Haematol 144(2):158–165

Nakamura N, Wada F, Kondo T, Aoki K, Arai Y, Mizumoto C et al (2023) Significance of omitting day 11 mini-dose methotrexate for GVHD prophylaxis after unrelated bone marrow transplantation. Transplant Cell Ther 29(2):119.e1–119.e7

Nash RA, Antin JH, Karanes C, Fay JW, Avalos BR, Yeager AM et al (2000) Phase 3 study comparing methotrexate and tacrolimus with methotrexate and cyclosporine for prophylaxis of acute graft-versus-host disease after marrow transplantation from unrelated donors. Blood 96(6):2062–2068

Neefjes J, Jongsma ML, Paul P, Bakke O (2011) Towards a systems understanding of MHC class I and MHC class II antigen presentation. Nat Rev Immunol 11(12):823–836

Nishihori T, Al-Kadhimi Z, Hamadani M, Kharfan-Dabaja MA (2016) Antithymocyte globulin in allogeneic hematopoietic cell transplantation: benefits and limitations. Immunotherapy 8(4):435–447

Nourse J, Firpo E, Flanagan WM, Coats S, Polyak K, Lee MH et al (1994) Interleukin-2-mediated elimination of the p27Kip1 cyclin-dependent kinase inhibitor prevented by rapamycin. Nature 372(6506):570–573

O'Donnell PV, Brunstein CG, Fuchs EJ, Zhang MJ, Allbee-Johnson M, Antin JH et al (2022) Umbilical cord blood or HLA-Haploidentical Transplantation: real-world outcomes versus randomized trial outcomes. Transplant Cell Ther. 28(2):109.e1–109.e8

Ortiz Z, Shea B, Suarez-Almazor ME, Moher D, Wells GA, Tugwell P (1998) The efficacy of folic acid and folinic acid in reducing methotrexate gastrointestinal toxicity in rheumatoid arthritis. A metaanalysis of randomized controlled trials. J Rheumatol 25(1):36–43

Pasquini MC, Devine S, Mendizabal A, Baden LR, Wingard JR, Lazarus HM et al (2012) Comparative outcomes of donor graft CD34+ selection and immune suppressive therapy as graft-versus-host disease prophylaxis for patients with acute myeloid leukemia in complete remission undergoing HLA-matched sibling allogeneic hematopoietic cell transplantation. J Clin Oncol 30(26):3194–3201

Pavlovic AM, Bonaci-Nikolic B, Kozic D, Ostojic J, Abinun M, Svabic-Medjedovic T et al (2012) Progressive multifocal leukoencephalopathy associated with mycophenolate mofetil treatment in a woman with lupus and CD4+ T-lymphocyte deficiency. Lupus 21(1):100–102

Peccatori J, Forcina A, Clerici D, Crocchiolo R, Vago L, Stanghellini MT et al (2015) Sirolimus-based graft-versus-host disease prophylaxis promotes the in vivo expansion of regulatory T cells and permits peripheral blood stem cell transplantation from haploidentical donors. Leukemia 29(2):396–405

Penack O, Marchetti M, Aljurf M, Arat M, Bonifazi F, Duarte RF et al (2024a) Prophylaxis and management of graft-versus-host disease after stem-cell transplantation for haematological malignancies: updated consensus recommendations of the European Society for Blood and Marrow Transplantation. Lancet Haematol. 11(2):e147–ee59

Penack O, Abouqateb M, Peczynski C, Boreland W, Gulbas Z, Gedde-Dahl T et al (2024b) PTCy versus ATG as graft-versus-host disease prophylaxis in mismatched unrelated stem cell transplantation. Blood Cancer J 14(1):45

Perkins J, Field T, Kim J, Kharfan-Dabaja MA, Fernandez H, Ayala E et al (2010) A randomized phase II trial comparing tacrolimus and mycophenolate mofetil to tacrolimus and methotrexate for acute graft-versus-host disease prophylaxis. Biol Blood Marrow Transplant 16(7):937–947

Perko R, Kang G, Sunkara A, Leung W, Thomas PG, Dallas MH (2015) Gamma delta T cell reconstitution is associated with fewer infections and improved event-free survival after hematopoietic stem cell transplantation for pediatric leukemia. Biol Blood Marrow Transplant 21(1):130–136

Picchianti Diamanti A, Rosado MM, Scarsella M, Germano V, Giorda E, Cascioli S et al (2014) Abatacept (cytotoxic T lymphocyte antigen 4-immunoglobulin) improves B cell function and regulatory T cell inhibitory capacity in rheumatoid arthritis patients non-responding to anti-tumour necrosis factor-alpha agents. Clin Exp Immunol 177(3):630–640

Pidala J, Vogelsang G, Martin P, Chai X, Storer B, Pavletic S et al (2012a) Overlap subtype of chronic graft-versus-host disease is associated with an adverse prognosis, functional impairment, and inferior patient-reported outcomes: a chronic graft-versus-host disease consortium study. Haematologica 97(3):451–458

Pidala J, Kim J, Jim H, Kharfan-Dabaja MA, Nishihori T, Fernandez HF et al (2012b) A randomized phase II study to evaluate tacrolimus in combination with sirolimus or methotrexate after allogeneic hematopoietic cell transplantation. Haematologica 97(12):1882–1889

Pidala J, Beato F, Kim J, Betts B, Jim H, Sagatys E et al (2018) In vivo IL-12/IL-23p40 neutralization blocks Th1/Th17 response after allogeneic hematopoietic cell transplantation. Haematologica 103(3):531–539

Pidala J, Martens M, Anasetti C, Carreras J, Horowitz M, Lee SJ et al (2020) Factors associated with successful discontinuation of immune suppression after allogeneic hematopoietic cell Transplantation. JAMA Oncol 6(1):e192974

Pidala JA, Holtan SG, Walton K, Kim J, Cao B, Elmariah H et al (2024) JAK2/mTOR inhibition fails to prevent acute GVHD despite reduced Th1/Th17 cells: final phase II trial results. Blood 144:2295

Piekarska A, Czyz A, Peczynski C, Ambron P, Polge E, Moiseev I et al (2023) ATG or no ATG? – survey of clinical practice in EBMT centers on behalf of the Transplant Complications Working Party of EBMT. Bone Marrow Transplant 58(3):337–339

Pihusch R, Holler E, Muhlbayer D, Gohring P, Stotzer O, Pihusch M et al (2002) The impact of antithymocyte globulin on short-term toxicity after allogeneic stem cell transplantation. Bone Marrow Transplant 30(6):347–354

Pinana JL, Valcarcel D, Fernandez-Aviles F, Martino R, Rovira M, Barba P et al (2010) MTX or mycophenolate mofetil with CsA as GVHD prophylaxis after reduced-intensity conditioning PBSCT from HLA-identical siblings. Bone Marrow Transplant 45(9):1449–1456

Przepiorka D, Weisdorf D, Martin P, Klingemann H, Beatty P, Hows J et al (1995) 1994 consensus conference on acute GVHD grading. Bone Marrow Transplant 15(6):825

Przepiorka D, Ippoliti C, Khouri I, Woo M, Mehra R, Le Bherz D et al (1996) Tacrolimus and minidose methotrexate for prevention of acute graft-versus-host disease after matched unrelated donor marrow transplantation. Blood 88(11):4383–4389

Przepiorka D, Khouri I, Ippoliti C, Ueno NT, Mehra R, Korbling M et al (1999) Tacrolimus and minidose methotrexate for prevention of acute graft-versus-host disease after HLA-mismatched marrow or blood stem cell transplantation. Bone Marrow Transplant 24(7):763–768

Raefsky EL, Gascon P, Gratwohl A, Speck B, Young NS (1986) Biological and immunological characterization of ATG and ALG. Blood 68(3):712–719

Ratanatharathorn V, Nash RA, Przepiorka D, Devine SM, Klein JL, Weisdorf D et al (1998) Phase III study comparing methotrexate and tacrolimus (prograf, FK506) with methotrexate and cyclosporine for graft-versus-host disease prophylaxis after HLA-identical sibling bone marrow transplantation. Blood 92(7):2303–2314

Reisner Y, Kapoor N, Kirkpatrick D, Pollack MS, Cunningham-Rundles S, Dupont B et al (1983) Transplantation for severe combined immunodeficiency with HLA-A,B,D,DR incompatible parental marrow cells fractionated by soybean agglutinin and sheep red blood cells. Blood 61(2):341–348

Rice S, Bishop J, Apperley J, Gardner S (1985) BK virus as cause of haemorrhagic cystitis after bone marrow transplantation. Lancet 2(8459):844

Rodriguez R, Nakamura R, Palmer JM, Parker P, Shayani S, Nademanee A et al (2010) A phase II pilot study of tacrolimus/sirolimus GVHD prophylaxis for sibling donor hematopoietic stem cell transplantation using 3 conditioning regimens. Blood 115(5):1098–1105

Roex MCJ, Wijnands C, Veld SAJ, van Egmond E, Bogers L, Zwaginga JJ et al (2020) Effect of alemtuzumab-based T-cell depletion on graft compositional change in vitro and immune reconstitution early after allogeneic stem cell transplantation. Cytotherapy 23:46

Rosenow EC 3rd, Myers JL, Swensen SJ, Pisani RJ (1992) Drug-induced pulmonary disease. An update. Chest 102(1):239–250

Ruutu T, Gratwohl A, de Witte T, Afanasyev B, Apperley J, Bacigalupo A et al (2014) Prophylaxis and treatment of GVHD: EBMT-ELN working group recommendations for a standardized practice. Bone Marrow Transplant 49(2):168–173

Schumm M, Lang P, Bethge W, Faul C, Feuchtinger T, Pfeiffer M et al (2013) Depletion of T-cell receptor alpha/beta and CD19 positive cells from apheresis products with the CliniMACS device. Cytotherapy 15(10):1253–1258

Shaffer BC, Gooptu M, DeFor TE, Maiers M, Bolaños-Meade J, Abboud R et al (2024) Post-transplant

cyclophosphamide-based graft-versus-host disease prophylaxis attenuates disparity in outcomes between use of matched or mismatched unrelated donors. J Clin Oncol 42:JCO2400184
Shaw BE, Jimenez-Jimenez AM, Burns LJ, Logan BR, Khimani F, Shaffer BC et al (2021) National Marrow Donor Program-sponsored multicenter, phase II trial of HLA-mismatched unrelated donor bone marrow transplantation using post-transplant cyclophosphamide. J Clin Oncol 39(18):1971–1982
Singh K, Stempora L, Harvey RD, Kirk AD, Larsen CP, Blazar BR et al (2014) Superiority of rapamycin over tacrolimus in preserving nonhuman primate Treg half-life and phenotype after adoptive transfer. Am J Transplant 14(12):2691–2703
Soiffer RJ, Lerademacher J, Ho V, Kan F, Artz A, Champlin RE et al (2011) Impact of immune modulation with anti-T-cell antibodies on the outcome of reduced-intensity allogeneic hematopoietic stem cell transplantation for hematologic malignancies. Blood 117(25):6963–6970
Soler D, Chapman T, Yang LL, Wyant T, Egan R, Fedyk ER (2009) The binding specificity and selective antagonism of vedolizumab, an anti-alpha4beta7 integrin therapeutic antibody in development for inflammatory bowel diseases. J Pharmacol Exp Ther 330(3):864–875
Staatz CE, Tett SE (2007) Clinical pharmacokinetics and pharmacodynamics of mycophenolate in solid organ transplant recipients. Clin Pharmacokinet 46(1):13–58
Storb R, Epstein RB, Bryant J, Ragde H, Thomas ED (1968) Marrow grafts by combined marrow and leukocyte infusions in unrelated dogs selected by histocompatibility typing. Transplantation 6(4):587–593
Storb R, Epstein RB, Graham TC, Thomas ED (1970) Methotrexate regimens for control of graft-versus-host disease in dogs with allogeneic marrow grafts. Transplantation 9(3):240–246
Storb R, Deeg HJ, Whitehead J, Appelbaum F, Beatty P, Bensinger W et al (1986a) Methotrexate and cyclosporine compared with cyclosporine alone for prophylaxis of acute graft versus host disease after marrow transplantation for leukemia. N Engl J Med 314(12):729–735
Storb R, Deeg HJ, Farewell V, Doney K, Appelbaum F, Beatty P et al (1986b) Marrow transplantation for severe aplastic anemia: methotrexate alone compared with a combination of methotrexate and cyclosporine for prevention of acute graft-versus-host disease. Blood 68(1):119–125
Storb R, Yu C, Wagner JL, Deeg HJ, Nash RA, Kiem HP et al (1997) Stable mixed hematopoietic chimerism in DLA-identical littermate dogs given sublethal total body irradiation before and pharmacological immunosuppression after marrow transplantation. Blood 89(8):3048–3054
Taplitz RA, Kennedy EB, Bow EJ, Crews J, Gleason C, Hawley DK et al (2018) Antimicrobial prophylaxis for adult patients with cancer-related immunosuppression: ASCO and IDSA clinical practice guideline update. J Clin Oncol 36(30):3043–3054
Theurich S, Fischmann H, Shimabukuro-Vornhagen A, Chemnitz JM, Holtick U, Scheid C et al (2012) Polyclonal anti-thymocyte globulins for the prophylaxis of graft-versus-host disease after allogeneic stem cell or bone marrow transplantation in adults. Cochrane Database Syst Rev 9:CD009159
Thomas ED, Lochte HL Jr, Lu WC, Ferrebee JW (1957) Intravenous infusion of bone marrow in patients receiving radiation and chemotherapy. N Engl J Med 257(11):491–496
Thomas E, Storb R, Clift RA, Fefer A, Johnson FL, Neiman PE et al (1975) Bone-marrow transplantation (first of two parts). N Engl J Med 292(16):832–843
Ueda Oshima M, Xie H, Zamora D, Flowers ME, Hill GR, Mielcarek MB et al (2023) Impact of GVHD prophylaxis on CMV reactivation and disease after HLA-matched peripheral blood stem cell transplantation. Blood Adv 7(8):1394–1403
Vogelsang GB, Arai S (2001) Mycophenolate mofetil for the prevention and treatment of graft-versus-host disease following stem cell transplantation: preliminary findings. Bone Marrow Transplant 27(12):1255–1262
Wachsmuth LP, Patterson MT, Eckhaus MA, Venzon DJ, Gress RE, Kanakry CG (2019) Post-transplantation cyclophosphamide prevents graft-versus-host disease by inducing alloreactive T cell dysfunction and suppression. J Clin Invest 129(6):2357–2373
Wagner JE, Thompson JS, Carter SL, Kernan NA, Unrelated Donor Marrow Transplantation T (2005) Effect of graft-versus-host disease prophylaxis on 3-year disease-free survival in recipients of unrelated donor bone marrow (T-cell depletion trial): a multi-centre, randomised phase II-III trial. Lancet 366(9487):733–741
Walker I, Panzarella T, Couban S, Couture F, Devins G, Elemary M et al (2016) Pretreatment with anti-thymocyte globulin versus no anti-thymocyte globulin in patients with haematological malignancies undergoing haemopoietic cell transplantation from unrelated donors: a randomised, controlled, open-label, phase 3, multicentre trial. Lancet Oncol 17(2):164–173
Waller EK, Langston AA, Lonial S, Cherry J, Somani J, Allen AJ et al (2003) Pharmacokinetics and pharmacodynamics of anti-thymocyte globulin in recipients of partially HLA-matched blood hematopoietic progenitor cell transplantation. Biol Blood Marrow Transplant 9(7):460–471
Watkins B, Qayed M, McCracken C, Bratrude B, Betz K, Suessmuth Y et al (2021) Phase II trial of costimulation blockade with abatacept for prevention of acute GVHD. J Clin Oncol 39(17):1865–1877
Weiden PL, Flournoy N, Thomas ED, Prentice R, Fefer A, Buckner CD et al (1979) Antileukemic effect of graft-versus-host disease in human recipients of allogeneic-marrow grafts. N Engl J Med 300(19):1068–1073
Weiden PL, Sullivan KM, Flournoy N, Storb R, Thomas ED, Seattle Marrow Transplant T (1981) Antileukemic effect of chronic graft-versus-host disease: contribution to improved survival after allogeneic marrow transplantation. N Engl J Med 304(25):1529–1533

Weinblatt M, Combe B, Covucci A, Aranda R, Becker JC, Keystone E (2006) Safety of the selective costimulation modulator abatacept in rheumatoid arthritis patients receiving background biologic and nonbiologic disease-modifying antirheumatic drugs: a one-year randomized, placebo-controlled study. Arthritis Rheum 54(9):2807–2816

Weinblatt ME, Moreland LW, Westhovens R, Cohen RB, Kelly SM, Khan N et al (2013) Safety of abatacept administered intravenously in treatment of rheumatoid arthritis: integrated analyses of up to 8 years of treatment from the abatacept clinical trial program. J Rheumatol 40(6):787–797

Wiederrecht G, Lam E, Hung S, Martin M, Sigal N (1993) The mechanism of action of FK-506 and cyclosporin A. Ann N Y Acad Sci 696:9–19

Wong R, Beguelin GZ, de Lima M, Giralt SA, Hosing C, Ippoliti C et al (2003) Tacrolimus-associated posterior reversible encephalopathy syndrome after allogeneic haematopoietic stem cell transplantation. Br J Haematol 122(1):128–134

Wyant T, Fedyk E, Abhyankar B (2016) An overview of the mechanism of action of the monoclonal antibody Vedolizumab. J Crohns Colitis 10(12):1437–1444

Wynn TA, Ramalingam TR (2012) Mechanisms of fibrosis: therapeutic translation for fibrotic disease. Nat Med 18(7):1028–1040

Yoshida S, Ohno Y, Nagafuji K, Yoshimoto G, Sugio T, Kamimura T et al (2019) Comparison of calcineurin inhibitors in combination with conventional methotrexate, reduced methotrexate, or mycophenolate mofetil for prophylaxis of graft-versus-host disease after umbilical cord blood transplantation. Ann Hematol 98(11):2579–2591

Yu C, Seidel K, Nash RA, Deeg HJ, Sandmaier BM, Barsoukov A et al (1998) Synergism between mycophenolate mofetil and cyclosporine in preventing graft-versus-host disease among lethally irradiated dogs given DLA-nonidentical unrelated marrow grafts. Blood 91(7):2581–2587

Zeiser R, Blazar BR (2017a) Acute graft-versus-host disease – biologic process, prevention, and therapy. N Engl J Med 377(22):2167–2179

Zeiser R, Blazar BR (2017b) Pathophysiology of chronic graft-versus-host disease and therapeutic targets. N Engl J Med 377(26):2565–2579

Zeiser R, Penack O, Holler E, Idzko M (2011) Danger signals activating innate immunity in graft-versus-host disease. J Mol Med (Berl) 89(9):833–845

Zeiser R, vonBubnoff N, Butler J, Mohty M, Niederwieser D, Or R et al (2020) Ruxolitinib for glucocorticoid-refractory acute graft-versus-host disease. NEJM 382:1800–1810

Zeiser R, Polverelli N, Ram R, Hashmi SK, Chakraverty R, Middeke JM et al (2021) Ruxolitinib for glucocorticoid-refractory chronic graft-versus-host disease. N Engl J Med 385(3):228–238

Modern-Era Challenges in the Clinical Management of Graft-Versus-Host Disease

6

Catherine J. Lee and Paul A. Carpenter

Abstract

For several decades, graft-versus-host disease (GVHD) has been a long-standing barrier to successful allogenic hematopoietic cell transplantation and a significant cause of post-transplant morbidity and mortality. Initially described as secondary disease or wasting syndrome in transplanted mice, the pathobiology of GVHD is increasingly understood as a dynamic interplay between innate and adaptive immunity in response to initial tissue damage, leading to inflammation and end-organ damage. In parallel, more uniform symptom capture, diagnosis, and response criteria have facilitated rigorous clinical trial design and conduct; together, these advancements have facilitated the development of novel GVHD prevention and treatment strategies. While these advancements have improved the GVHD treatment paradigm, new questions arise within this complex patient population. This chapter discusses several of the most pertinent current clinical practice challenges in GVHD, including its earlier diagnosis, risk stratification, initial and more advanced stage management, as well as a renewed focus on supportive care, given our increased understanding of key roles played by the human microbiome.

Keywords

Hematopoietic cell transplantation · Graft versus host disease · Acute GVHD · Chronic GVHD · Steroid-refractory · Corticosteroids · Immunosuppression · Treatment · Gastrointestinal · Fibrotic · Supportive care · Microbiome · Nutrition

6.1 Introduction

Graft-versus-host disease (GVHD) is a significant complication following allogeneic hematopoietic cell transplantation (HCT) that can be broadly characterized by donor immune cells attacking recipient tissues due to histocompatibility differences (Holtan et al. 2014). Its acute and chronic forms have distinct clinical presentations and management strategies. For decades, GVHD has been the main complication of allogenic HCT and the leading cause of post-transplant morbidity and nonrelapse mortality (NRM). Significant progress has been made in understanding its pathophysiology and improving diagnostic criteria and clinical staging. This has led to its improved recognition, more uniform symptom capture, and development of

C. J. Lee (✉) · P. A. Carpenter
Fred Hutchinson Cancer Center, Seattle, WA, USA

University of Washington, Seattle, WA, USA
e-mail: cjlee@fredhutch.org

© The Author(s), under exclusive license to Springer Nature Switzerland AG 2025

K. Ballen (ed.), *Clinical Management of Challenges in Allogeneic Stem Cell Transplant*, Advances in Experimental Medicine and Biology 1475, https://doi.org/10.1007/978-3-031-84988-6_6

novel prevention and treatment strategies. While GVHD rates are declining, the clinical complexity of acute GVHD (aGVHD) and chronic (cGVHD) syndromes remains a challenge.

This chapter will specifically discuss the key contemporary challenges of early GVHD diagnosis, risk stratification, and treatment strategies for new onset and steroid-refractory GVHD. We will also address challenges in supportive care following a diagnosis of GVHD. While recognizing that the best GVHD strategy is prevention, we defer this discussion and refer readers to Chap. 6.

6.2 Acute GVHD

Acute GVHD is a frequent complication of HCT involving the activation of donor T lymphocytes against host tissues. Despite prophylactic immunosuppression (IST), up to 50% of HCT recipients experience aGVHD and ~14% experience severe aGVHD (grade III–IV) (Martin et al. 2012; Zeiser and Blazar 2017a; Malard et al. 2020). The aGVHD syndrome most frequently targets the skin (maculopapular rash), gastrointestinal (GI) tract (nausea, vomiting, and/or diarrhea), or liver (traditionally with cholestatic liver injury ± transaminase elevation), alone or in combination (Glucksberg et al. 1974). *Classic* aGVHD develops before 100 days post-HCT but since 2005, *late* aGVHD has been used to categorize aGVHD that occurs after day 100, more often after nonmyeloablative conditioning (Filipovich et al. 2005). Several commonly used aGVHD grading systems include Glucksberg (Glucksberg et al. 1974), International Bone Marrow Transplant Registry (IBMTR) (Rowlings et al. 1997), Minnesota (MacMillan et al. 2002a, 2015), and Mount Sinai Acute GVHD International Consortium (MAGIC) (Harris et al. 2016). MAGIC grading is the most current and detailed approach that is recommended by the EBMT–NIH–CIBMTR Task Force (Table 6.1) (Schoemans et al. 2018a).

The choice of initial therapy depends on aGVHD severity and initial organ involvement (Malard et al. 2023a). Mild (MAGIC grade 1) aGVHD is most often managed by continuing aGVHD prophylaxis (or resuming or dose-escalating prophylaxis when it was tapering or already discontinued) and adding organ-directed topical corticosteroids (CS). Treatment of grade 2 or higher aGVHD begins with high-dose CS, traditionally at a prednisone-equivalent dose (PED) of 1–2 mg/kg/day (Martin et al. 2012; Penack et al. 2024). The Seattle group has shown that for adults without liver involvement, with <50% body surface area rash and <1 L per day of diarrhea, therapy can safely and effectively begin with a PED of 0.5–1 mg/kg/day, often supported with organ-directed topical CS (Mielcarek et al. 2009, 2015).

While approximately 60% of grade II aGVHD responds to CS, 30–40% of grade III/IV aGVHD is CS-nonresponsive, leading to shortened overall survival (OS) (Deeg 2007; Westin et al. 2011; MacMillan et al. 2010; Yu et al. 2020; Holtan et al. 2022). The incomplete efficacy of prolonged CS therapy is compounded by significant associated opportunistic infections, metabolic and mood disturbances, bone loss, and osteonecrosis.

6.3 Steroid-Refractory Acute GVHD

Steroid-refractory aGVHD (SR-aGVHD) is most often defined as any organ that is progressing 3–5 days after starting 2 mg/kg/day of PED or no response after 7 days (Pavletic and Fowler 2012; Schoemans et al. 2018b). Other poor prognosis aGVHD scenarios include an initial response followed by recurrence, as well as persistent or worsening visceral aGVHD despite skin improvement (MacMillan et al. 2015; Axt et al. 2019). Patients may also experience steroid-dependent aGVHD (SD-aGVHD), defined as the inability to taper PED to <2 mg/kg/day after at least 7 days of successful initial treatment, or recurrent aGVHD in the same or new organs during CS tapering (Schoemans et al. 2018b). Prognosis in SD-aGVHD has been less studied but a retrospective study suggests a similar 2-year OS and NRM compared to steroid-sensitive aGVHD (El Jurdi et al. 2021). However, the prolonged IST

Table 6.1 MAGIC aGVHD target organ staging and overall severity grading

Stage	Skin (active erythema only)	Liver (total bilirubin)	Upper GI	Lower GI (stool output/day)
0	No rash	<2 mg/dL	No or intermittent[a] anorexia or nausea or vomiting	Adult: diarrhea[b] <500 mL/day or <3 episodes/day Child: diarrhea[b] <10 mL/kg/day or <4 episodes/day
1	Rash <25% of BSA	2–3 mg/dL	Persistent[a] anorexia or nausea or vomiting	Adult: diarrhea[b] 500–999 mL/day or 3–4 episodes/day Child: diarrhea[b] 10–19.9 mL/kg/day or 4–6 episodes/day
2	Rash 25–50% of BSA	3.1–6 mg/dL		Adult: diarrhea[b] 1000–1500 mL/day or 5–7 episodes/day Child: diarrhea[b] 20–30 mL/kg/day or 7–10 episodes/day
3	Rash >50% of BSA	6.1–15 mg/dL		Adult: diarrhea[b] >1500 mL/day or >7 episodes/day Child: >diarrhea[b] 30 mL/kg/day or >10 episodes/day
4	Generalized erythroderma (>50% BSA) plus bullous formation and desquamation >5% of BSA	>15 mg/dL		Severe abdominal pain with or without ileus or grossly bloody stool (regardless of stool volume)
Overall clinical severity grade (based on most severe target organ involvement)				
0 No organ involvement (skin = 0; and liver = 0; and GI=0) corresponds to the absence of aGVHD				
I Skin = 1 or 2, without liver, upper or lower GI involvement				
II Skin = 3; and/or liver = 1; and/or upper GI=1; and/or lower GI=1				
III Liver = 2 or 3; and/or lower GI=2 or 3				
IV Skin = 4; and/or liver = 4; and/or lower GI=4				

Adapted from Schoemans et al. (2018b)

[a]To be suggestive of GVHD, anorexia should be accompanied by weight loss, nausea should last at least 3 days, or be accompanied by at least 2 vomiting episodes per day for at least 2 days

[b] One episode of diarrhea is considered to be about 200 mL for an adult and 3 mL/kg for a child (<50 kg)

commonly used in SD-aGVHD is associated with more frequent serious infections (El Jurdi et al. 2021). Last, the inability to tolerate acute high-dose CS toxicities (e.g., uncontrolled infections, hyperglycemia, anxiety) is another indication for second-line therapy (Schoemans et al. 2018b).

Until the US Food and Drug Administration (FDA) approval of the dual JAK1/2 inhibitor, ruxolitinib in May 2019, there was no accepted standard-of-care treatment for SR-aGVHD (Zeiser and Blazar 2017a). The approval for ruxolitinib rested upon sequential phase 2 (REACH1) (Jagasia et al. 2020) and phase 3 (REACH2) studies (Zeiser et al. 2020) with REACH2 showing superior ~2.6-fold higher overall response rate (ORR) at day 28 for ruxolitinib versus 39% for the "investigator's choice" control arm of nine commonly available options (ORR 62% vs. 39%); at day 28, the complete response (CR) rate was

34% for ruxolitinib versus 19% for controls. The estimated cumulative 6-month incidence of loss of response was 10% (95% CI, 4–17) for ruxolitinib versus 39% (95% CI, 26–52) for controls. Recipients of ruxolitinib had a significantly higher incidence of thrombocytopenia and a modestly higher incidence of anemia, infection, and cytomegalovirus infection/reactivation compared to recipients of investigator's choice therapy.

Multiple other agents for second-line therapy have been investigated without an identified effective therapy (Martin et al. 2012; Martin 2020). In REACH2, the most common second-line approach was extracorporeal photopheresis (ECP), which accounted for 27% of the investigator's choice arm. Response endpoints for ECP have varied from day 28 to the best overall response, which includes beyond day 28, ranging from 25% to 100% but usually most effective in the skin compared to viscera (Kitko et al. 2022; Drexler et al. 2020). ECP may be preferred over ruxolitinib as second-line therapy for patients with severe infection (Martin 2020).

Despite the landmark FDA approval of ruxolitinib, responses are mostly incomplete and sometimes transient. SR-aGVHD has been associated with high GVHD-associated mortality (Martin et al. 2012; Westin et al. 2011; Abedin et al. 2019; Levine et al. 2010) resulting from cumulative damage to target tissues which can outpace augmented systemic IST attempts, together with a higher incidence of bloodstream infections and other infection-related mortality (Garcia-Cadenas et al. 2017).

Beyond ruxolitinib, there is no established next-line treatment for patients who are intolerant or refractory to ruxolitinib. Enrollment in a clinical trial is recommended but if unavailable, third-line treatment is selected based on center experience with available therapy options and relevance of particular medication safety profiles to individual patients. Treatment options include emerging pharmacologic therapies such as mesenchymal stromal cells (Kebriaei et al. 2020) and fecal microbiota transplantation (Malard et al. 2023b) (Table 6.2), but optimizing treatment for severe SR-aGVHD that has failed second-line therapy remains an unmet medical need.

6.4 Identifying Patients at High Risk of Steroid-Refractory Acute GVHD

Two of the challenges in the clinical management of aGVHD are as follows: (A) identifying who is at high risk (HR) for developing SR-aGVHD and (B) whether more aggressive first-line therapy might mitigate the inferior outcomes observed for HR patients. Conversely, who is more likely to have mild aGVHD, and can this group potentially be spared prolonged high-dose CS? Both the Glucksberg (Glucksberg et al. 1974) and IBMTR (Przepiorka et al. 1995) aGVHD grading approaches fall well short of this challenge. The Minnesota group subsequently developed a refined, clinical-based risk score model, applied at aGVHD onset, that dichotomizes adults (MacMillan et al. 2015) and children (MacMillan et al. 2020) into HR or standard-risk (SR) aGVHD groups who have, respectively, higher and lower risks for NRM. Building on initial observations from 1990 that single-organ involvement, as well as organ severity, were more associated with CR than multiorgan involvement (Weisdorf et al. 1990), the refined Minnesota grading defines SR aGVHD as single-organ involvement (stage 1–3 skin or stage 1–2 GI) or two-organ involvement (stage 1–3 skin plus stage 1 GI; or stage 1–3 skin plus stage 1–4 liver) and is associated with an ORR rate of 69% at day 28 compared to 43% for the HR group, which includes all other patients (MacMillan et al. 2015).

Most recently, the MAGIC algorithm used serum biomarker concentrations of suppressor of tumorigenesis 2 (ST2) and regenerating islet-derived 3α (REG3α) to categorize patients into three distinct 6-month NRM risk groups: *low* (Ann Arbor 1), *intermediate* (Ann Arbor 2), and *high* (Ann Arbor 3) (Hartwell et al. 2017; Levine et al. 2015). ST2 and REG3α both signify gastrointestinal crypt damage (Zhao et al. 2018; Reichenbach et al. 2015) and may identify patients already with, or about to develop, high-risk, lower GI GVHD. In fact, patients without GI symptoms yet deemed high-risk based on Ann Arbor 3 biomarker panel results

Table 6.2 Treatment options for SR-acute GVHD after ruxolitinib

Agent	Function
Alemtuzumab (Gomez-Almaguer et al. 2008; Schnitzler et al. 2009)	Anti-CD52 monoclonal antibody that depletes T and B cells
Alpha-1 antitrypsin (Magenau et al. 2018)	Circulating protease inhibitor with immunomodulatory activity
ATG (MacMillan et al. 2002b)	T-cell depleting antibody
Basiliximab (Schmidt-Hieber et al. 2005)	IL-2 receptor antibody that inhibits T-cell proliferation and activation
CNIs (tacrolimus, cyclosporine)	Inhibits T-cell activation
Etanercept (Busca et al. 2007)	Fusion protein that inactivates TNF-α
Extracorporeal photopheresis (Abu-Dalle et al. 2014)	Involves exposing leukocytes to a photosensitizing drug and ultraviolet A light that leads to immunomodulatory activity
Fecal microbiota transplantation	Re-establishes gut microbiota (NCT04711967, NCT04280471, NCT03819803, NCT05094765, NCT03812705 and NCT04769895)
Infliximab (Couriel et al. 2004)	Monoclonal antibody that blocks TNF-α activity
Mesenchymal stromal cells (Kurtzberg et al. 2020)	Immunomodulatory activity (NCT05333029, NCT04744116 and NCT04629833)
Mycophenolate mofetil (Pidala et al. 2010; Furlong et al. 2009)	Inhibits IMPDH proliferation of T and B cells
Pentostatin (Ragon et al. 2018; Bolanos-Meade et al. 2005)	Inhibits adenosine deaminase and induces lymphocyte apoptosis
Sirolimus (Hoda et al. 2010; Pidala et al. 2009)	Inhibitor of mTOR and inhibits T-cell proliferation and differentiation
Tocilizumab (Roddy et al. 2016; Drobyski et al. 2011; Ganetsky et al. 2019)	Anti-IL6 receptor antibody and blocks IL-6, a proinflammatory cytokine
Vedolizumab (Li et al. 2022) 22	Inhibits activation of α4β7 integrin and trafficking of T cells to the GI mucosa

Abbreviations: *ATG* antithymocyte globulin, *CNI* calcineurin inhibitor, *IMPDH* inosine-5′-monophosphate dehydrogenase

had high NRM rates due to subsequently developing lower GI SR-aGVHD (Hartwell et al. 2017; Levine et al. 2015; Major-Monfried et al. 2018). Consequently, the MAGIC aGVHD algorithm has been developed into a commercially available test for predicting severe aGVHD, treatment resistance, and NRM (Eurofins-Viracor n.d.).

While Minnesota (clinical) and MAGIC (biomarker) risk categorizations both identify patient groups more or less likely to respond to CS, and with higher or lower NRM rates, outside of clinical trials, these approaches have not been established as standard of care. Perhaps the biggest concern relates to inevitably exposing a proportion of patients who are deemed high-risk to intensified IST who did not stand to benefit due to imperfect performance characteristics of these risk categories.

6.5 Gastrointestinal (GI) GVHD

Approximately 50% of patients with aGVHD have GI tract involvement making it the second most affected organ behind skin (MacMillan et al. 2015); upper (UGI) and mid-to-lower GI tract (LGI) may be involved separately or together. Approximately two-thirds of the GI involvement (or 30% overall) is UGI, typically manifesting as persistent nausea, vomiting, and anorexia beyond the time when conditioning toxicities should have resolved (Nikiforow et al. 2018). UGI aGVHD without LGI involvement typically responds well to systemic CS ± oral topical CS, such as beclomethasone dipropionate and budesonide, and has a lower risk for NRM (Nikiforow et al. 2018; Hockenbery et al. 2007). LGI aGVHD is characterized by watery, secretory, protein-rich diarrhea >500 mL per day. With advanced aGVHD, the entire GI tract becomes inflamed and edematous, leading to voluminous diarrhea, malabsorption, mucosal ulceration, sloughing, bleeding, and severe abdominal pain (Naymagon et al. 2017). NRM for Stage 3–4 LGI aGVHD is higher than for stage 1–2 (MacMillan et al. 2015). While <10% present with de novo stage 3–4 LGI aGVHD, the vast majority who develop severe disease do so while already on first-line therapy. Although rates of severe GI GVHD have declined over the past decades, severe GI GVHD remains a major cause of post-HCT mortality (Harris et al. 2016; Aasebo et al. 2022). Standard first-line therapy remains a PED of 2 mg/kg/day, with second-line IST only being added when manifestations are deemed SR-aGVHD. Outcomes are poor for these patients who do not completely respond to high-dose CS; accordingly, reducing aGVHD mortality is contingent upon improving responses in patients with lower GI GVHD.

Vedolizumab recently emerged as a potential preventative option for intestinal GVHD. It is a humanized monoclonal antibody that inhibits α4β7 interaction on T cells with their ligand MAdCAM-1 on endothelium within the GI tract and associated lymphoid tissues thereby inhibiting trafficking of donor T lymphocytes to these target tissues (Floisand et al. 2019). This mechanism of action supported the evaluation of vedolizumab added to standard GVHD prophylaxis after unrelated HCT in an international double-blind, randomized, placebo-controlled phase 3 trial (Chen et al. 2024). The primary endpoint of lower-GI aGVHD-free survival by day+180 after HCT was met: 85.5% (79.2–90.1) with vedolizumab versus 70.9% (63.2–77.2) with placebo (hazard ratio, 0.45; 95% confidence interval, 0.27–0.73; $P < 0.001$) (Chen et al. 2024).

The role of vedolizumab for GI aGVHD treatment is yet to be defined. A meta-analysis that included seven studies showed vedolizumab to yield a pooled ORR of 61% at 14 days after its initiation (pooled risk ratio, RR = 14.14, 95% CI: 2.95–67.71), 50% at 28 days (pooled RR = 7.36, 95% CI: 2.14–25.37), and 77% at 12 months (pooled RR = 13.66, 95% CI: 3.50–53.35) (Li et al. 2022), yet OS remained low with pooled OS of 34% at 6 months and 29% at 12 months. Significantly increased risks such as infections, impaired metabolism, and nutritional complications occurred in the treatment but not prophylactic setting. Further research is required to delineate its role in GI aGVHD treatment.

6.6 Future Directions in Gastrointestinal (GI) GVHD

Strategies beyond pharmacological immunosuppressants are under investigation that recognize the relative importance of the intestinal microbiome and mucosal barrier in immune-mediated inflammatory diseases like GI GVHD. We learned from murine models that butyrate, a short chain fatty acid, was decreased in intestinal epithelial cells (IECs) of mice experiencing GVHD (Goldberg et al. 2003). Restoring butyrate levels by increasing intestinal butyrate-producing bacteria reduced experimental aGVHD severity and mortality (Goldberg et al. 2003). Administration of resistant potato starch (RPS) to normal healthy human volunteers promotes increased butyrogenic bacteria and intestinal butyrate levels (Couriel et al. 2005). A subsequent feasibility study where patients (age ≥10 years) were given RPS from D−7 to D+100 after HCT showed feasibility of RPS administration and RPS significantly increasing salutary intestinal butyrate levels (Gomez-Almaguer et al. 2008). A phase 2 placebo-controlled study is underway to see if aGVHD rates can be reduced (NCT02763033). Interestingly, this approach might not be wise if treating non-intact GI mucosa after GI aGVHD onset, based on other observations where increased butyrogenic bacteria were associated with subsequent SR-GVHD or cGVHD (Hamilton et al. 2024), reminding us that context can be important.

Another prophylactic approach involves manipulation of the gut microbiome by capsular administration of a commercially produced, uniform bacterial consortia investigational product (IP), before and after HCT (Couriel et al. 2004). The IP was designed to decolonize GI pathogens, improve epithelial barrier integrity, and induce immune tolerance to prevent blood stream infections, antimicrobial resistance, and other pathogen-associated negative clinical outcomes. Data from a small, yet randomized Phase Ib study of adults demonstrated engraftment of bacterial strains in the IP; blood stream infection rates were 10% for the IP vs 43% with placebo (OR: 0.15; 95% CI: 0.01, 1.13, $p = 0.04$), representing a 77% relative risk reduction; systemic antibiotic exposure, and incidence of fever and neutropenia were also lower with the IP (NCT04995653) (Harper and McMillen 2021).

FMT for treatment and prophylaxis of aGVHD has also been evaluated in patients but the best source of FMT product remains unclear (Harris et al. 2016). Screened third party FMT is commonly used and can be delivered as oral capsules. In one placebo-controlled phase II study of HCT recipients or AML patients undergoing induction chemotherapy, FMT timed with neutrophil recovery was safe and ameliorated intestinal dysbiosis but did not meet its primary end-point of reducing all-cause infection rate (Hartwell et al. 2017). However, a *post-hoc* analysis of the FMT treated group showed a lower rate of grade II-IV aGVHD in patients who had greater donor microbiota engraftment, suggesting that FMT may protect against aGVHD (Hashmi et al. 2015). Subsequently, a placebo-controlled phase 2 double blind trial of oral encapsulated third-party FMT to prevent severe aGVHD is ongoing at Fred Hutchinson Cancer Center (FHCC) (NCT06026371) (Hashmi et al. 2018). Lingering concerns with third-party FMT are the transfer of drug-resistant microorganisms that can be life-threatening (Hills et al. 2019). Large, randomized studies of FMT or FMT-like (i.e., bacterial consortia) will be required before these therapies can be offered routinely.

When intestinal crypt loss is present, sometimes with a denuded mucosal barrier, GI recovery can be exceedingly slow despite potentially effective novel IST. Lithium carbonate can attenuate GVHD via effects on the intestinal stem cell niche (Cutler et al. 2021). It is thought to do this by inhibiting the normal glycogen synthase-3 (GSK-3) antagonism of WNT/b-catenin signaling which then drives cell proliferation and epithelial regeneration; fortunately, long-term lithium use is not associated with increased risk of colorectal cancer (Hockenbery et al. 2007). In a pilot study at FHCC, adults with GI GVHD and mucosal denudation at endoscopy received oral lithium plus second-line IST; 50% had CR by intention to treat, but 8 of 12 who started lithium (with therapeutic drug monitoring) ≤3 days from baseline endoscopy showing denuded mucosa had a CR and survived >1 year. Compared to historical controls, 2-year survival was 35% vs 10% (Hoda et al. 2010). While fatigue, dulled sensorium, and flattened affect occurred commonly this was difficult to assess in a nonrandomized critically ill cohort.

Lastly, there has been recent interest in teduglutide and apraglutide, which are both GLP-2 analogs. GLP-2 helps maintain and protect GI epithelial cells, including enterocytes, intestinal stem cells, and Paneth Cells, which are direct targets of aGVHD (Deeg 2007; Holler et al. 2014). Teduglutide is approved for once daily SQ injection in patients age ≥1 year with short bowel syndrome (SBS). A pediatric case series with pre- and post-treatment correlative immunohistochemistry data suggested efficacy in 3 children with refractory GI aGVHD (Deeg et al. 1995). Apraglutide is a once weekly SQ injectable that has undergone successful phase 3 testing in adults with SBS with US regulatory submission expected in 2025. Results of a proof-of-concept phase 2 trial (NCT05415410) where apraglutide was added to ruxolitinib as second-line treatment for SR-lower-GI aGVHD in patients age ≥12 years showed apraglutide was well-tolerated with encouraging efficacy in a severely ill population (Cuvelier et al. 2022).

6.7 Nutritional Support in Acute GI GVHD

In addition to IST, aGVHD of the GI tract management includes medical nutrition therapy and supportive care for nausea, early satiety, vomiting, dysphagia, anorexia, diarrhea, abdominal pain, and bleeding to effectively prevent and/or manage malnutrition (Table 6.3). Malnutrition results from inadequate intake compounded by

Table 6.3 Medical nutrition therapies and supportive care in GI GVHD (Harper and McMillen 2021; Akbulut and Yesildemir 2021)

Nutrition impact symptom	Medical nutrition therapy
Xerostomia	Diet plan that includes highly moist foods Normal saline rinses Saliva stimulants (e.g., citric acid drinks, sugar-free candies, mints, gum, cevimeline, or pilocarpine) Dry mouth products (e.g., Biotene (rinse/spray/gel, Xerostom (rinse/spray/gel), mouth Kote (rinse), adhesive tablets (Zylimelts), or adhesive strips (Xerostrips) Gastronomy tube for EN if symptoms preclude adequate calorie-protein intake for weight maintenance and support
Dysgeusia	Diet plan that includes a variety of spicy or flavorful foods (as tolerated) Promotion of good oral hygiene
Nausea/vomiting	Diet plan that includes salty or sour foods, cold and clear liquids, small frequent meal portions, low-fat and low-fiber foods Antiemetics Starting PN or postpyloric EN if symptoms are severe
Anorexia	Consume smaller food portions at more frequent intervals Consumption of high-calorie snacks
Dysphagia	Referral to a speech pathologist for swallowing evaluation Esophageal dilatation for severe symptoms Gastronomy tube for EN if symptoms preclude adequate calorie-protein intake for weight maintenance and support
Diarrhea	Antidiarrheal medications Low lactose diet Avoid GI irritants/simulants and irritating disaccharides Diet plan that includes low-fat, low-insoluble fiber Discontinuing oral intake for large volume, watery diarrhea >500 mL/day
Pancreatic enzyme insufficiency or atrophy	Screening tests (fecal elastase or serum trypsinogen) to confirm fat malabsorption Oral pancreatic enzyme replacement therapy Dietary fat restriction Supplementation with products containing medium-chain triglycerides

Abbreviations: *EN* enteral nutrition, *GI* gastrointestinal, *PN* parenteral nutrition

increased metabolic demands, alterations in protein-calorie intake, nutrient absorption, and utilization leading to macronutrient (e.g., protein catabolism or protein-losing enteropathy) and micronutrient (e.g., electrolytes, vitamin D, B12, zinc) deficiencies, all which can be exacerbated by systemic inflammation and medication toxicities (van der Meij et al. 2013; McMillen et al. 2021; Harper and McMillen 2021). Because malnutrition is a well-established factor associated with poor survival in HCT (Deeg et al. 1995; Beckerson et al. 2019), registered dietitians with HCT expertise are critical to any aGVHD management team. Formal initial and longitudinal nutrition assessments are necessary to determine patients' needs for diet modifications and/or nutrition support (enteral (EN) and/or parenteral nutrition (PN). Although patients with intractable nausea, severe abdominal pain, diarrhea, and/or overt GI bleeding may be placed on bowel rest when they meet the criteria for having a nonfunctional GI tract, oral intake (or EN) should be resumed as soon as possible to prevent gut atrophy (August et al. 2009). Adequate EN during the early transplantation course has been associated with reduced early NRM, improved survival, and GVHD-free/relapse-free survival at 5 years compared to PN and inadequate nutrition (Beckerson et al. 2019). EN protects against overall and gut aGVHD (Zama et al. 2021) perhaps because of its ability to maintain mucosal integrity, modulate the immune response to the effects of intensive chemo/radiotherapy, and support intestinal microbiota homeostasis after HCT (Beckerson et al. 2019; Arai et al. 2015; Seguy et al. 2012).

A healthy (diverse) gut microbiota plays a key role in the immune system, with links to both reduced aGVHD and GVHD-related mortality

(Burgos da Silva et al. 2022; Jenq et al. 2015; Holler et al. 2014). This is not surprising because GVHD-associated disruption of GI mucosal integrity decreases nutrient assimilation and promotes systemic infection through bacterial translocation and enteropathy (Seguy et al. 2012). As maintaining a variety of enteral nutrients is essential to the gut microbiota diversity (Hills Jr. et al. 2019), having a variety of food choices plus good safety practices is more important than a "neutropenic" or "low microbial" diet, which is not supported by evidence in this patient population (Arends et al. 2017). It follows that while PN might be the only option for patients with the most severe manifestations of GI aGVHD or inability to maintain placement of an enteral feeding tube (Arends et al. 2017), if PN is initiated, early reintroduction of oral intake should iteratively be assessed because it leads to improved nutritional outcomes (Imataki et al. 2006).

Whereas the importance of oral intake (or EN) in HCT is well established, the role of micronutrients, their requirements in HCT, and their association with HCT outcomes is less clear (Segon et al. 2022). Vitamins A and D both appear to have immunomodulatory roles, including suppression of host antigen-presenting cells and donor T-cell activation, inhibition of proinflammatory cytokines, and increases in regulatory T cells (Chen and Mayne 2018). These preclinical findings are supported by emerging clinical data—children with lower vitamin A levels had more frequent and more severe intestinal aGVHD (Lounder et al. 2017) which led to a randomized, double-blinded, placebo-controlled phase II trial that compared 4000 IU/kg vitamin A versus placebo given within one week before conditioning (Khandelwal et al. 2024). The cumulative incidence of GI aGHVD and cGVHD was lower in the vitamin A-treated group in the "as-treated" analysis.

The role of vitamin D has been more variable with studies either showing a positive or no association of vitamin D deficiency with GVHD incidence and/or severity [reviewed in Rodriguez-Gil et al. (2022)]. Until these results are validated in well-designed randomized studies, clinicians should use clinical judgment to provide supplementation where appropriate. If micronutrient deficiencies are found or suspected before or after transplantation, a multivitamin (without iron) supplement may be considered. Additionally, EN and PN formulations should be selected carefully to ensure appropriate vitamin and mineral provision.

6.8 Chronic GVHD

Chronic GVHD remains the leading cause of late post-HCT morbidity and mortality (Lee et al. 2003; Pidala et al. 2011). It is a polymorphic, multiorgan syndrome with clinical manifestations having inflammatory and/or fibrotic characteristics but often characterized by the latter as the syndrome gains chronicity. Experimental models suggest that the syndrome arises after early onset inflammation and host tissue damage from pretransplant conditioning and stimulation of alloreactive donor T cells) (Zeiser and Blazar 2017b). Subsequently, thymic injury, loss of peripheral tolerance, emergence of alloreactive B and T (B/T) cells, and loss of regulatory B/T cells together lead to chronic inflammation. The final common pathways appear to involve activated IL-17 and colony-stimulating factor-1 (CSF-1)-dependent, donor-derived macrophages that produce profibrotic cytokines, ultimately leading to the deposition of extracellular collagen by activated fibroblasts (Alexander et al. 2014; MacDonald et al. 2017a, b). The incidence of cGVHD ranges from 30% to 70% (Arai et al. 2015; Arora et al. 2016; Bachier et al. 2021), with increasing rates observed in the early 2000s, attributed to increasing recipient age and the use of unrelated donor and peripheral blood stem cell grafts (Arai et al. 2015). However, recent data suggests that cGVHD is decreasing among adults, children, and patients who underwent HCT for either malignant or nonmalignant disease (Carpenter et al. 2024).

Before the initial National Institutes of Health (NIH) consensus criteria of 2005 (Filipovich et al. 2005), a cGVHD diagnosis was challenging because of its clinical heterogeneity, the coexistence of aGVHD manifestations, and ambiguities in traditional classification approaches (Lee et al. 2003; Shulman et al. 1980). Assignment of cGVHD severity was similarly challenging because of the lack of validated tools to measure cGVHD in individual organs. The 2005 NIH cGVHD consensus effort harmonized methods and definitions for diagnosis, organ scoring, and global severity assignment, as well as treatment response criteria for clinical trials. Importantly, late aGVHD (after day 100) could occur with classic cGVHD manifestations, in which case it was termed overlap subtype, or could occur alone, in which case it has been associated with worse overall and failure-free survival (Holtan et al. 2016). Furthermore, *mild*, *moderate*, and *severe* global severity scores were associated with progressively worse OS and increased NRM (Arai et al. 2011). Iterative refinements in the 2014 NIH criteria (Jagasia et al. 2015) provided clarifications and more reliable tools that were used in a clinical trial that led to the first FDA approval of a cGVHD therapy (Miklos et al. 2017; Jaglowski and Blazar 2018). The third NIH consensus (Pavletic et al. 2021) reported cGVHD basic research and clinical practice gaps relevant to early detection and had the goal to avoid highly morbid cGVHD forms by detecting them (hopefully) before advanced fibrosis causes irreversible morbidity. While these NIH efforts have been undoubtedly pivotal, together with progress in understanding cGVHD biology, several key challenges remain in modern cGVHD management.

6.9 Treatment Challenges in Chronic GVHD

Aside from prevention, cGVHD therapy might ideally rebalance a dysregulated immune system and reverse and prevent fibrosis, leading to the complete resolution of all cGVHD manifestations. This would also be accomplished without significant patient toxicities and would ultimately enable durable and complete withdrawal of IST. While this may be possible for a minority of patients, treatment success might also be considered if there is a reduction in symptom burden, preserved activities of daily living, and improved quality of life. The ideal agent remains elusive, but the cGVHD treatment paradigm has clearly shifted from baseline long-term broadly active CS-based IST toward agents that target specific pathways involved in cGVHD pathogenesis.

6.9.1 First-Line Therapy

Systemic treatment with IST has been used to control moderate-to-severe cGVHD, and so far, multiple phase 3 studies (Sullivan et al. 1988; Koc et al. 2000, 2002; Arora et al. 2001; Martin et al. 2009; Gilman et al. 2012; Carpenter et al. 2018; Miklos et al. 2023) have not identified more efficacious first-line therapy than CS beginning at the PED of 0.5–1 mg/kg/day ± calcineurin inhibitor (CNI) although inclusion of CNI led to less avascular necrosis in a prospective phase 3 study (Koc et al. 2002). After ~14 days, a taper usually can begin and per results of one prospective study, the average reaches 0.2–0.25 mg/kg/day (or 0.4–0.5 mg/kg every other day) within 3 months after starting treatment (Martin et al. 2009). Roughly half of transplant centers prescribe alternate-day prednisone tapers based on old literature, primarily case series in non-HCT populations, supporting fewer side effects and lower total cumulative prednisone exposure. A randomized phase 2 study of daily versus alternate-day prednisone for cGVHD in children was deemed a high-priority trial at the third State of the Science Symposium but did not move forward and so this question remains unanswered for all age groups (Appelbaum et al. 2015). In clinical practice, when cGVHD is not well controlled on an alternating day regimen, some patients can be better controlled simply by switching to an equivalent-dose daily regimen, possibly reflecting the shorter biological half-life of prednisone in some patients. In another scenario, blood sugar levels may swing more erratically in diabetic patients who are on alternate-day versus daily prednisone regimens.

Eventually, after cycles of attempted prednisone tapering and dose reescalation, the dose reaches 2.5 mg/day (5.7 mg/m^2/day of hydrocortisone) in adults and 1.2–2.2 mg/m^2/day (~5–9 mg/m^2/day of hydrocortisone) in children and adolescents, which approximates daily cortisol production (Esteban et al. 1991; Linder et al. 1990). At this point, in non-Cushingoid patients, prednisone is often converted to a split-dose regimen of hydrocortisone (two-thirds taken at ~8 AM and one-third taken at 3–5 PM; no later than 6 PM) to facilitate adrenal recovery during the final taper. Formal testing for secondary adrenal insufficiency is often done when a patient remains Cushingoid on low-dose prednisone and if confirmed, the final taper generally is much slower. Moreover, GVHD activity may recur even when tapering below *physiological* PED, which may also dictate length of the final prednisone (or hydrocortisone) taper.

6.9.2 Prednisone Dependence, Refractoriness, Unresponsiveness, or Intolerance

Phase 3 trials have shown that by 6 months into therapy, only ~50% of patients had a CR or partial response (PR) and just over 25% had a CR. As a result, insufficient response to first-line therapy or recurrent cGVHD after tapering initial therapy means that >50% of patients receive ≥3 lines of therapy and remain on IST for a median of 2–3.5 years (Bachier et al. 2021; Lee et al. 2018); for a minority of patients, time to discontinue all IST may extend beyond 7 years. Indications for second-line therapy include worsening cGVHD manifestations in a previously affected organ, development of signs and symptoms of cGVHD in a previously unaffected organ or absence of improvement after 1 month of standard primary treatment (*refractory*), inability to decrease prednisone below 1 mg/kg per day within 2 months (*dependent*), or significant treatment-related toxicity (*intolerance*). Providers must be aware that some organs are irreversibly damaged (*unresponsive*) like lacrimal or salivary glands or fascia (contractures).

Beyond second-line therapy, additional patience is required before concluding treatment failure. For example, manifestations like sclerosis might be responding even if there is no improvement by 3 months if the prior sclerosis disease trajectory has plateaued, and prednisone can be tapered meaningfully. Well-documented, methodical, and serial skin assessments are necessary to be fully certain of this determination. The recently FDA-approved, more targeted cGVHD therapeutics discussed below had a median time to first response ranging from 4 to 12 weeks, ≥50% ORR at or through 24 weeks, and generally even higher best ORRs. Therefore, in the absence of clear disease progression or intolerance, a reasonable trial is 6 months before declaring efficacy failure.

Finally, it can be helpful for patients to set expectations regarding the duration of cGVHD therapy. Whereas solid organ transplant recipients must take IST for life, a registry study suggested that 32% of HCT recipients *durably* stop IST by 10 years after cGVHD diagnosis, noting that just under one-half were ineligible to achieve this goal due to competing risks of relapse or NRM (Chen et al. 2023). Several factors can shorten the time to durably discontinue IST, with recipients of cord blood, naive T-cell depletion, or post-transplant cyclophosphamide developing cGVHD and/or morbidity less frequently and discontinuing IST much sooner.

6.9.3 Steroid-Refractory cGVHD

Until 2017, there was no standard of care for second-line therapy but commonly used were ECP (Abu-Dalle et al. 2014), low-dose methotrexate (Giaccone et al. 2005; Huang et al. 2005), mycophenolate mofetil (Lopez et al. 2005), rituximab (Kharfan-Dabaja et al. 2009), sirolimus (Couriel et al. 2005; Johnston et al. 2005; Jurado et al. 2007), imatinib (Arai et al. 2016), pentostatin (Goldberg et al. 2003; Jacobsohn et al. 2007), and low-dose interleukin-2 (IL-2) (Koreth et al. 2016) based on mostly heterogeneous, uncontrolled retrospective or prospective study data. The aforementioned NIH consensus effort, which included important clinical trial endpoints, has now facilitated FDA approval of four targeted SR-cGVHD therapeutics (ibrutinib, ruxolitinib, belumosudil, and axatilimab).

Ibrutinib selectively and irreversibly inhibits Bruton's tyrosine kinase (BTK) and interleukin-2-inducible T-cell kinase (ITK), which are important for signaling and activation of B cells and T cells (Miklos et al. 2017). Its 2017 FDA approval was based on the phase Ib/II study results of 42 subjects with SD/SR-cGVHD. Notably, subjects had to have at least an erythematous rash or cGVHD with oral erythema. The best ORR was 67% (21% CR) with a median time to response of 12.4 weeks (range, 4.1 to 42.1 weeks), and ~ 71% of responses were sustained for ≥20 weeks. High response rates were seen in skin (88%), mouth (88%), and GI tract (91%). Eleven of 18 patients with baseline sclerosis showed sclerotic improvement. Patient counseling about side effects should include mention of fatigue, diarrhea, muscle spasms, and bruising. The latter is thought to be due to impaired platelet aggregation and while no major hemorrhages were reported in GVHD trials, the product label advises holding ibrutinib for ±7 days around major surgery and ±3 days around minor surgery. Arrhythmias including atrial fibrillation deserve mention but occurred infrequently in GVHD trials. In 2022, the ibrutinib FDA cGVHD indication was revised to age ≥ 1 year, based on the results of the prospective multicenter iMAGINE study. The study determined pediatric dosing, found no new safety signals, and showed a similar 6-month ORR and a numerically higher best ORR compared to the pivotal adult trial (Carpenter et al. 2022).

Ruxolitinib selectively inhibits Janus-associated kinase (JAK) 1 and 2, thereby inhibiting the JAK-signal transducer and activator of the transcription (JAK/STAT) pathway. Because the activation of intracellular JAK 1/2 leads to transcription of proinflammatory cytokines, mediation of inflammatory neutrophil migration, and upregulation of major histocompatibility class II expression on antigen-presenting cells (Zeiser and Blazar 2017b), JAK/STAT inhibition has opposite effects, as well as diminishing alloreactive T-cell infiltration into target organs mediated by IFN-γ and CXCR3. In 2021, the FDA-approved ruxolitinib based on the results from REACH3, a phase 3, randomized controlled trial (Zeiser et al. 2021), which compared the efficacy and safety of ruxolitinib versus the best available therapy (BAT) in patients with SD or SR-cGVHD. ORR at 24 weeks was 50% for ruxolitinib versus 26% for BAT ($p \leq 0.001$). ORR was 53% for the LGI tract, 50% for the esophagus, 50% for the mouth, 41% for the skin, and 40% for the UGI tract (40%). Responses were ~3-fold higher for ruxolitinib, and partial responses (PR) were more likely in both arms. Patients who crossed over from BAT to ruxolitnib after week 24 subsequently still had a 50% ORR (95% CI: 37.8, 62.2) by 24 weeks, with the best ORR reaching 81.4% (95% CI: 70.3, 89.7), and 74.3% of these being PRs (Zeiser et al. 2023). Ruxolitinib led to longer median failure-free survival than BAT (hazard ratio, 0.37; $P < 0.001$), higher symptom response (odds ratio, 2.62; $P = 0.001$), and spared CS slightly better than BAT. Reversible anemia and thrombocytopenia were more common in the ruxolitinib versus BAT arm; these can be managed with dose reductions and supportive care. Bacterial and viral infections occurred with similar frequency in both arms, but fungal infections were numerically higher in the ruxolitinib arm (11.5% vs. 5.7%). Real-world retrospective studies evaluating ruxolitinib in SR-cGVHD found similar best ORRs and adverse events (AE) (Redondo et al. 2022; Ferreira et al. 2021; Wu et al. 2021; Zeiser et al. 2015), and median times to response were 3 weeks to 2 months (range, 0.5–8 months), which is helpful when counseling patients (Redondo et al. 2022; Zeiser et al. 2015).

Belumosudil (KD025) selectively inhibits rho-associated coiled-coil-containing protein kinase-2 (ROCK2). Because ROCK2 activation results in sequential upregulation of STAT3 phosphorylation and then elaboration of Th17-specific transcription factors, ROCK2 inhibition leads to downregulation of proinflammatory cytokines including IL-17 and IL-21 in T cells (Zanin-Zhorov et al. 2014), with a shift toward more regulatory T cells. ROCK2 activation also results in the polymerization of globular actin to fibrous actin (Riches et al. 2015; Knipe et al. 2015), which in turn leads to further expression of profibrotic genes, promoting myofibroblast and collagen production, in a profibrotic environment. Preclinical cGVHD models showed that inhibition of ROCK2 by belumosudil interferes with this profibrotic process. In 2021, the FDA-approved belumosudil based on the results of ROCKstar, the phase II, open-label, randomized, multicenter study that compared belumosudil 200 mg once daily versus 200 mg twice daily in patients who had received 2–5 previous lines of cGVHD therapy (Cutler et al. 2021). Because the best ORR, at a median follow-up of 14 months, was similar (74% vs. 77%) without differences in individual organ response rates, the recommended dose is 200 mg once daily for age $\geq$ 12 years, unless on a proton pump inhibitor, in which case 200 mg twice daily is recommended. Despite being a more heavily pretreated study population, ORRs were similar if previously treated with ruxolitinib (68%) or ibrutinib (74%); similar to ibrutinib and ruxolitinib, most responses were PR. The median time to response was 5 weeks (range, 4–66 weeks), and the median duration of response was 54 weeks. Belumosudil appeared to provide clinically meaningful benefit in patients with $\geq$4 involved organs, in patients who were refractory to their last line of treatment, and in patients with severe cGVHD. CS dose decreased by a median of 45% from baseline in most patients; 31% discontinued CS therapy, and 22% discontinued CNI therapy over 12 months. Belumosudil was generally well tolerated with a 12% AE-attributable discontinuation rate. The most common AEs of any grade were fatigue, diarrhea, nausea, cough, and upper respiratory tract infection. Reactivation of cytomegalovirus infection was not observed.

Axatilimab inhibits colony-stimulating factor-1 receptor (CSF-1R) on monocytes and macrophages and blocks receptor activation by CSF1 and interleukin-34 (IL-34) (Kitko et al. 2023; Ordentlich et al. 2016), thereby playing key roles in regulating tissue macrophages. CSF1R signaling-dependent monocytes and macrophages are essential mediators of inflammation and fibrosis in cGVHD and a range of autoimmune diseases (Alexander et al. 2014; Jardine et al. 2020). Preclinical studies showed that early post-transplantation antibody blockade of CSF1R could target donor-derived macrophages and ameliorate cGVHD (Alexander et al. 2014). After a successful initial phase I/II clinical SR-cGVHD study, axatilimab showed promising preliminary efficacy and safety results; preferential elimination of CSF1R-dependent nonclassical monocytes and tissue macrophages was demonstrated (Kitko et al. 2023). Efficacy and safety of axatilimab were further established in AGAVE 201, the pivotal phase 2, multinational study that preceded its FDA approval in 2021 (Wolff et al. 2024). A total of 241 patients were randomized (1:1:1) to 0.3-mg, 1-mg, and 3-mg dose groups. While all dose groups met the primary end point, the 6-month ORR was best (74%) for the 0.3-mg group (95% CI: 63–83), 67% (95% CI, 55–77) for the 1-mg group, and 50% (95% CI, 39–61) for the 3-mg group. At least 5-point improvements on the modified Lee Symptom Scale were reported in 60%, 69%, and 41% of the patients in the three dose groups, respectively. In parallel, AE-attributable discontinuations were only 6% for the 0.3-mg group versus 22% for the 1-mg group and 18% for the 3-mg group. Patients should be counseled about dose-dependent transient laboratory abnormalities related to CSF1R blockade (elevated AST, ALT, CPK, LDH, amylase, and lipase) and periorbital edema, albeit this was infrequent at the approved intravenous dose of 0.3 mg every 2 weeks.

6.10 Tailoring Therapy in Chronic GVHD

Although the FDA approvals of ibrutinib, ruxolitinib, belumosudil, and axatilimab reflect an improved understanding of cGVHD biology, harmonized diagnostic, response criteria, and clinical trial endpoints, the existence of these multiple treatment options prompts a requestioning of how best to sequence next-in-line therapy for SR-cGVHD. A corollary question is whether there is a preferred agent for specific organ involvement (e.g., oral vs. skin vs. lung) or clinical phenotype (e.g., inflammatory or fibrotic manifestations). In practice, FDA-labeling indications largely direct the answer to question number one, but the corollary question was never addressed by any of these pivotal clinical trials. While ibrutinib and ruxolitinib are FDA-approved for the failure of ≥1 systemic therapy, the use of belumosudil or axatilimab requires the failure of ≥2 lines of therapy. In Europe, the European Medicines Agency (EMA) has similarly approved ruxolitinib and ibrutinib; however, belumosudil and axatilimab have orphan drug designations. The recognition that ruxolitinib is the only drug approved based on having a comparator treatment arm of the best available therapy may influence some providers to select ruxolitinib before other treatments.

Beyond FDA labeling and pivotal trial study design, selection of next-in-line therapy often considers the agent's mechanism of action, underlying disease indication for HCT (e.g., ruxolitinib with a history of myelofibrosis or ibrutinib with CLL), and relevance of its AE profile to an individual patient's comorbidities inclusive of infections, availability of an oral formulation (e.g., axatilimab being only intravenous), feasibility of patient co-pays, provider experience, and patient preference (Table 6.4). Taking all the above, next-in-line therapy commonly involves shared clinical decision-making.

Establishing patient-tailored approaches to cGVHD treatment would require validated bio-

Table 6.4 Selection of treatment considerations

	Ibrutinib (Imbruciva (ibrutinib) [package insert] 2024)	Ruxolitinib (Jakafi (ruxolitinib) [package insert] 2021)	Belumosudil (Rezurock (belumosudil) [package insert] 2024)	Axatilimab (Niktimvo (axatilimab) [package insert] 2024)
FDA indication	Refractory cGVHD after ≥1 LOT in patients age 1 year and older	Refractory cGVHD after ≥1 LOT in patients age 12 years and older	Refractory cGVHD after ≥2 LOTs in patients age 12 years and older	Refractory cGVHD after ≥2 LOTs in patients with weight ≥ 40 kg
Mode of administration	Oral	Oral	Oral	IV
Most common (≥20%) non-laboratory AEs of all grades	Fatigue, bruising, diarrhea, stomatitis, nausea, muscle spasms, hemorrhage, and pneumonia	Infections	Infections, asthenia, edema, nausea, diarrhea, abdominal pain, dyspnea, cough, hemorrhage, hypertension. MSK pain, headache	Infections, MSK pain, fatigue, nausea, headache
Most common (≥10%) grade ≥ 3 laboratory AEs	Neutropenia	Anemia, neutropenia, thrombocytopenia, hypercholesterolemia, elevated ALT, GGT and lipase	Elevated GGT, lymphocytes decreased	None
Common AEs leading to discontinuation of agent	Fatigue, pneumonia	Pneumonia	Nausea	Unknown
Other considerations	Avoid with other strong CYP3A inhibitors (e.g., CSA)	Avoid with fluconazole doses >200 mg/day	Twice daily dosing requiring when co-administered with PPI (may increase patient OOP costs)	Monitor for infusion-related reactions, including fever, chills, Rash, flushing, dyspnea, and hypertension

Abbreviations: *AEs* adverse events, *CSA* cyclosporin, *GGT* gamma glutamyl transferase, *IV* intravenous, *LOT* line of therapy, *MSK* musculoskeletal, *OOP* out-of-pocket, *PPI* proton pump inhibitor

markers. In the future, predictive biomarkers might assist with the sequencing of CS-sparing therapies, but this information is currently lacking. Although not formally validated, in practice, for some patients, serum creatine kinase (and/or aldolase) can serve as helpful response biomarkers for patients with the atypical cGVHD manifestation of myositis (Couriel et al. 2002; Shahzad et al. 2022). Beyond anecdotal case series, validated biomarkers are not in routine clinical use (Bidgoli et al. 2022; Ji et al. 2023). To address this need, the Chronic GVHD Consortium initiated a prospective, observational cohort study in June 2020, "Predicting the Quality of Response to Specific Treatments" (Hamilton et al. 2024) (NCT04431479), to identify clinical and biological predictors of treatment response among patients receiving first-, second-, or third-line systemic therapy for cGVHD. Detailed clinical data pertaining to cGVHD phenotype and activity and whole blood samples are collected at defined timepoints with the hope of defining clinical and biological subgroups of patients and their response to treatments.

6.11 Should Chronic GVHD Therapies Be Added or Substituted?

The argument for *substituting* rather than simply *adding* next-in-line therapy is that to date, no controlled first-line therapy trials have demonstrated superior efficacy of two-versus-one CS-sparing agent(s) with some trials showing worse outcomes (Sullivan et al. 1988; Koc et al. 2000, 2002; Arora et al. 2001; Martin et al. 2009; Gilman et al. 2012; Carpenter et al. 2018; Miklos et al. 2023). Adding IST is also concerning simply because infection with or without cGVHD is a major cause of NRM, with the hazard for late fatal infections in one large registry study being ~4-fold for adults with cGVHD while still on IST and ~9-fold for children (Norkin et al. 2019; DeFilipp et al. 2021). These data have informed the predominant FHCC practice to substitute next-in-line therapy usually by discontinuing the failed CS-therapeutic within 2 weeks from the substitution.

Whether substituting should still be the preferred approach in the era of recent FDA-approved therapies is arguably best addressed in clinical trials. Combining these newer therapeutics has gained attention because they might have greater CS-sparing potential and be better suited for combination, ideally in first-line therapy with little or no CS. Each of the four approved agents has demonstrated overall responses in SR-cGVHD but most of the responses were partial. Only ibrutinib has been evaluated in first-line therapy, but like the above-mentioned controlled studies, the ORR to ibrutinib plus prednisone was no better than prednisone plus placebo (Miklos et al. 2023). Until new single-agent CS-sparing studies are reported, combining new-era CS-therapeutics should be done sparingly because supporting data is limited to case series mostly in the context of cGVHD that was refractory to multiple therapies (Caputo et al. 2024; Chin et al. 2024; Pusic et al. 2024). In summary, while cGVHD pathophysiology continues to be elucidated, it is logical to study new-era CS-agent combinations that target different biological pathways, with minimal overlapping toxicities, in preclinical models and well-designed prospective trials.

6.12 Early Detection and Preemption

The third NIH cGVHD consensus (Pavletic et al. 2021) highlighted the goal to avoid highly morbid, advanced fibrotic forms of cGVHD through early clinical detection supplemented with relevant validated biomarkers. Feasibility of earlier cGVHD recognition requires multiple stakeholders, including nontransplant providers (e.g., primary care provider, referring hematologist/oncologist), patients, and caregivers. At FHCC, patients and their local health providers have phone or email access to a team of survivorship-dedicated support staff, including long-term follow-up (LTFU) telemedicine registered nurses who answer questions and triage questions needing additional expertise to an LTFU Attending physician (Lee et al. 2021). Coordinating care with the patient's original transplant center is ideal; however, a directory of GVHD specialists

(BMTinfonet.org n.d.) accepting patients regardless of where the patient received their transplant is available.

Encouraging patients and their caregivers to be proactive in cGVHD monitoring might also facilitate early diagnosis and intervention and validated tool development to accomplish this is ongoing. Meanwhile, patient and caregiver education about key reportable signs and symptoms of cGVHD is recommended before discharge from the transplant center back to the referring provider.

Beyond early symptom reporting by patients, clinic visits by a provider with cGVHD expertise should incorporate methodical, iterative (every 1–3 months) organ-specific screenings during the first 1–2 years post-HCT to facilitate early cGVHD diagnosis, perhaps even in the absence of symptoms (Kitko et al. 2021). Pre-HCT baseline examination of the skin, joint range of motion, eyes, oral cavity, lungs, and genitals is important to document baseline tissue abnormalities that may confound subsequent cGVHD interpretations. At these visits, cGVHD symptom screening can be efficiently delegated to a clinical staff member; any positive sign or symptom can be made readily available to the expert provider. Electronic health (e-Health) tools like the "eGVHD app," which has been used worldwide, can facilitate cGVHD assessment and diagnosis because it encourages systematic patient physical evaluations and decreases diagnosis and scoring errors (Schoemans et al. 2016, 2018a, c).

Susceptibility or (subclinical) diagnostic biomarkers might further aid in the early diagnosis of cGVHD but are still in the research phase (Bidgoli et al. 2022; Kitko et al. 2021; Martens et al. 2023). Finally, whether subsequent preemptive therapy for the earliest clinical manifestations of cGVHD can prevent highly morbid forms of cGVHD is now being examined in clinical trials (NCT05996627).

6.13 Highly Morbid Forms and Atypical Manifestations

Highly morbid cGVHD manifests when irreversible fibrosis leads to sequelae that negatively impact physical functioning, quality of life, and survival; particularly common organ syndromes include severe dry eyes, bronchiolitis obliterans, and sclerotic skin and joint/fascia involvement (Wolff et al. 2021). Less commonly, very severe oral and genital manifestations occur, and more rarely, a wasting syndrome not explained by GI causes occurs. Atypical cGVHD manifestations outside the eight commonly affected cGVHD organs (eyes, mouth, skin, liver, lung, GI and GI tracts, joints/fascia) include the hematopoietic system (immune-mediated cytopenias), endothelium, serous membranes, central and peripheral nervous systems, musculoskeletal system, and kidneys (Table 6.5) [reviewed in Cuvelier et al. (2022)].

Prevention and treatment of highly morbid or atypical manifestations remain problematic because of incomplete understanding of their biology, under-recognition, relative infrequency of atypical manifestations, lack of predictive and

Table 6.5 Target organs and manifestations in atypical versus NIH-defined cGVHD

Atypical target organs and manifestations	
Central nervous system	Cognitive deficits, meningoencephalitis, demyelinating diseases, CNS vasculitis
Peripheral nervous system	Neuropathy, myasthenia gravis
Lungs	COP, nonspecific interstitial pneumonia, PPFE
Serositis	Pericardial effusion, pleural effusion, ascites
Renal	Proteinuria, nephrotic syndrome, tubular, glomerular, or interstitial disease, vascular disease
Hematologic	Autoimmune hemolytic anemia, immune thrombocytopenia purpura, autoimmune neutropenia
Muscles, fascia, joints	Edema, muscle cramps, arthralgia, arthritis, myositis

Adapted from Cuvelier et al. (2022)

Abbreviations: *COP* cryptogenic organizing pneumonia, *PPFE* pleuroparenchymal pulmonary fibroelastosis diagnostic biomarkers, and lack of well-established, efficacious organ-specific treatment approaches. Clinical trials focused on organ-specific treatments are additionally challenging to conduct because of the protean nature of cGVHD and the small number of patients with single-organ involvement of interest.

Upon recognition of cGVHD, collaborative involvement of relevant subspecialists can be beneficial. Colleagues in pulmonology, ophthalmology, gynecology, dermatology, nephrology, and rheumatology are often consulted for their subspecialty expertise and may provide additional management insights based on their experience with either GVHD or analogous immune-mediated conditions. Methodical care coordination is required when multiple subspecialists are involved, and it is important that the transplant team remains central to ensure that patients are receiving optimal cGVHD management, including supportive and ancillary care, as well as surveillance for graft function, relapse (if malignant HCT indication), and late effects.

6.14 Ancillary and Supportive Care in Chronic GVHD

The cGVHD burden cannot be overemphasized given that manifestations may evolve and/or recur over several years, yet never in isolation from other HCT late effects. Late effects result from varying degrees of interaction between pre-transplant exposures, post-transplant complications, and positive or negative influences of age, gender, genetics, and lifestyle. Given that clinical immunological tolerance may take years rather than months, toxicities arising from IST and medications used to treat side effects of IST can contribute to chronic health conditions per se, for example, hypertension, dyslipidemia, and chronic renal insufficiency to name a few. It follows that comprehensive LTFU care involves not only monitoring and management of cGVHD, but also monitoring for late effects [reviewed in Carpenter (2023)]. However, a major challenge is that LTFU attenuates over time and compliance with LTFU guidelines is suboptimal by survivors and healthcare providers especially when patients are adolescents or young adults, transplanted for standard-risk malignancy or nonmalignant disease, and perceived to be doing well (Miyamura et al. 2019; Armenian et al. 2012; Khera et al. 2011). An American Society of Transplantation and Cellular Therapy (ASTCT) survey found that respondents agreed on allo-HCT survivors having needs separate from GVHD and that complications can arise during transitions of care from pediatric to adult or from transplant center to primary medical provider. However, only 45% had dedicated LTFU clinics; 84% of individual practitioners preferred to provide their own survivorship care. Other challenges include a lack of subspecialty LTFU expertise, logistics of the LTFU model of care, and financial issues (Hashmi et al. 2015, 2018). More work is needed to optimize LTFU, particularly for young adult survivors who are more prone to becoming lost to follow-up.

Two key components of supportive care for cGVHD include optimizing nonsystemic IST (usually topically directed) for individual organ sites, and broad infection prophylaxis and timely revaccination [reviewed in Carpenter and Englund (2016)] given that a registry study found infection to be a primary or contributing factor for ~30% of late deaths (Norkin et al. 2019) and another large study (Foord et al. 2020) showed vaccine-preventable late infections were >30-fold more frequent than in the general population. Baseline antimicrobial prophylaxis directed against shingles, pneumocystis *jirovecii* pneumonia, encapsulated organisms, and often molds is recommended during cGVHD therapy. Routine post-transplant vaccinations are recommended and detailed elsewhere (Carpenter and Englund 2016; Rubin et al. 2014; Reynolds et al. 2023). Immunoglobulin replacement therapy is prescribed judiciously per ASTCT "Choosing Wisely" unless profoundly low IgG (often with immeasurable IgA) (Bhella et al. 2018).

Additional challenges can be medication burden and monitoring for complications of polypharmacy and noncompliance, plus frequent care coordination around issues of mood disturbance, and ensuring that there is adequate psychosocial support for patients and caregivers; referral to appropriate subspecialists who can assist with symptom control or potentially help with multifactorial problems that often can include contribution(s) from other HCT late effects.

Survivors are at risk for depression, fatigue, pain, social withdrawal, educational problems, and vocational problems including unemployment and dependent living (Armenian et al. 2012). Among allogeneic recipients, cGVHD severity, being female, and having current pain were associated with depression and fatigue (Jim et al. 2016). A registry study showed that return to work by adolescent young adults improved over time but many became or remained unemployed or on medical disability (Bhatt et al. 2021). Return-to-work programs are unavailable at most centers despite 87% endorsing this as a problem (Salit et al. 2020), and ASTCT-endorsed return-to-work guidance for providers of HCT is available in Open Access and can be used as a tool to counsel and support these patients (Salit et al. 2022).

6.15 Future Directions and Conclusions

Remarkable progress has been made in our understanding of GVHD biology, and we are better able to categorize, diagnose and grade GVHD manifestations, and capture cGVHD symptoms. Clinical frameworks now exist to facilitate the design and implementation of clinical trials, as well as increased industry and government collaborations. Novel therapies are more available after failure of initial therapy and additional investigational agents are undergoing evaluation. Challenges still remain, with the priority being the study of first-line agents that hopefully will spare chronic CS exposure. More effective treatments for severe GI aGVHD and highly morbid forms of cGVHD will still be needed when GVHD prevention or early therapy approaches fail. Whether patient-tailored over one-size-fits-all approaches can be achieved remains a stretch goal. Serum biomarkers are more developed for clinical use in aGVHD but still not widely utilized in practice, and cGVHD biomarkers to assign risk, predict, diagnose, and assess treatment response remain in development.

Artificial intelligence-based, machine-learning algorithms may have a role in the management of GVHD, given the preliminary data showing its potential to predict aGVHD and cGVHD (Martens et al. 2023; Chandra et al. 2024; McNeil et al. 2022; Arai et al. 2019), as well as to recognize cutaneous cGVHD (McNeil et al. 2022).

Despite these challenges, the future of GVHD appears bright. Collaborations between transplant clinicians, community providers, and subspecialists are growing stronger to provide optimal GVHD care, as academia, industry, government, and patient advocacy groups also work together toward eradicating this debilitating HCT complication.

Acknowledgments We thank K.K. McMillen MS, RD (Medical Nutrition Therapy Services, Fred Hutchinson Cancer Cancer) for her suggestions and review of Table 6.3.

References

Aasebo AT, Gedde-Dahl T, Reims HM, Baekkevold ES, Jahnsen FL (2022) Calprotectin expressing donor-derived macrophages increase in acute gastrointestinal graft-versus-host disease. Transplant Cell Ther 28(5):248.e1–248.e8. https://doi.org/10.1016/j.jtct.2022.01.028

Abedin S, McKenna E, Chhabra S et al (2019) Efficacy, toxicity, and infectious complications in Ruxolitinib-treated patients with corticosteroid-refractory graft-versus-host disease after hematopoietic cell transplantation. Biol Blood Marrow Transplant 25(8):1689–1694. https://doi.org/10.1016/j.bbmt.2019.04.003

Abu-Dalle I, Reljic T, Nishihori T et al (2014) Extracorporeal photopheresis in steroid-refractory acute or chronic graft-versus-host disease: results of a systematic review of prospective studies. Biol Blood Marrow Transplant 20(11):1677–1686. https://doi.org/10.1016/j.bbmt.2014.05.017

Akbulut G, Yesildemir O (2021) Overview of nutritional approach in hematopoietic stem cell transplantation: COVID-19 update. World J Stem Cells 13(10):1530–1548. https://doi.org/10.4252/wjsc.v13.i10.1530

Alexander KA, Flynn R, Lineburg KE et al (2014) CSF-1-dependant donor-derived macrophages mediate chronic graft-versus-host disease. J Clin Invest 124(10):4266–4280. (In eng). https://doi.org/10.1073/pnas.1414189111

Appelbaum FR, Anasetti C, Antin JH et al (2015) Blood and marrow transplant clinical trials network state of the science symposium 2014. Biol Blood Marrow Transplant 21(2):202–224. https://doi.org/10.1016/j.bbmt.2014.10.003

Arai S, Jagasia M, Storer B et al (2011) Global and organ-specific chronic graft-versus-host disease severity according to the 2005 NIH Consensus Criteria. Blood 118(15):4242–4249. https://doi.org/10.1182/blood-2011-03-344390

Arai S, Arora M, Wang T et al (2015) Increasing incidence of chronic graft-versus-host disease in allogeneic transplantation: a report from the Center for International Blood and Marrow Transplant Research. Biol Blood Marrow Transplant 21(2):266–274. (In eng). https://doi.org/10.1016/j.bbmt.2014.12.001

Arai S, Pidala J, Pusic I et al (2016) A randomized phase II crossover study of Imatinib or rituximab for cutaneous sclerosis after hematopoietic cell transplantation. Clin Cancer Res 22(2):319–327. https://doi.org/10.1158/1078-0432.CCR-15-1443

Arai Y, Kondo T, Fuse K et al (2019) Using a machine learning algorithm to predict acute graft-versus-host disease following allogeneic transplantation. Blood Adv 3(22):3626–3634. (In eng). https://doi.org/10.1182/blood.2019003125

Arends J, Bachmann P, Baracos V et al (2017) ESPEN guidelines on nutrition in cancer patients. Clin Nutr 36(1):11–48. https://doi.org/10.1016/j.clnu.2016.07.015

Armenian SH, Sun CL, Francisco L et al (2012) Health behaviors and cancer screening practices in long-term survivors of hematopoietic cell transplantation (HCT): a report from the BMT Survivor Study. Bone Marrow Transplant 47(2):283–290. https://doi.org/10.1038/bmt.2011.60

Arora M, Wagner JE, Davies SM et al (2001) Randomized clinical trial of thalidomide, cyclosporine, and prednisone versus cyclosporine and prednisone as initial therapy for chronic graft-versus-host disease. Biol Blood Marrow Transplant 7(5):265–273. https://doi.org/10.1053/bbmt.2001.v7.pm11400948

Arora M, Cutler CS, Jagasia MH et al (2016) Late acute and chronic graft-versus-host disease after allogeneic hematopoietic cell transplantation. Biol Blood Marrow Transplant 22(3):449–455. (In eng). https://doi.org/10.1016/j.bbmt.2015.12.020

August DA, Huhmann MB, American Society for P, Enteral Nutrition Board of D. A.S.P.E.N (2009) Clinical guidelines: nutrition support therapy during adult anticancer treatment and in hematopoietic cell transplantation. JPEN J Parenter Enteral Nutr 33(5):472–500. https://doi.org/10.1177/0148607109341804

Axt L, Naumann A, Toennies J et al (2019) Retrospective single center analysis of outcome, risk factors and therapy in steroid refractory graft-versus-host disease after allogeneic hematopoietic cell transplantation. Bone Marrow Transplant 54(11):1805–1814. https://doi.org/10.1038/s41409-019-0544-y

Bachier CR, Aggarwal SK, Hennegan K et al (2021) Epidemiology and treatment of chronic graft-versus-host disease post-allogeneic hematopoietic cell transplantation: a US claims analysis. Transplant Cell Ther 27(6):504.e1–504.e6. https://doi.org/10.1016/j.jtct.2020.12.027

Beckerson J, Szydlo RM, Hickson M et al (2019) Impact of route and adequacy of nutritional intake on outcomes of allogeneic haematopoietic cell transplantation for haematologic malignancies. Clin Nutr 38(2):738–744. https://doi.org/10.1016/j.clnu.2018.03.008

Bhatt NS, Brazauskas R, Salit RB et al (2021) Return to work among Young Adult Survivors of allogeneic hematopoietic cell transplantation in the United States. Transplant Cell Ther 27(8):679.e1–679.e8. https://doi.org/10.1016/j.jtct.2021.04.013

Bhella S, Majhail NS, Betcher J et al (2018) Choosing wisely BMT: American Society for Blood and Marrow Transplantation and Canadian Blood and Marrow Transplant Group's list of 5 tests and treatments to question in blood and marrow transplantation. Biol Blood Marrow Transplant 24(5):909–913. https://doi.org/10.1016/j.bbmt.2018.01.017

Bidgoli A, DePriest BP, Saatloo MV, Jiang H, Fu D, Paczesny S (2022) Current definitions and clinical implications of biomarkers in graft-versus-host disease. Transplant Cell Ther 28(10):657–666. https://doi.org/10.1016/j.jtct.2022.07.008

BMTinfonet.org. Directory of Graft-versus-Host Disease (GVHD) Clinics and Physicians. https://bmtinfonet.org/gvhd-directory

Bolanos-Meade J, Jacobsohn DA, Margolis J et al (2005) Pentostatin in steroid-refractory acute graft-versus-host disease. J Clin Oncol 23(12):2661–2668. https://doi.org/10.1200/JCO.2005.06.130

Burgos da Silva M, Ponce DM, Dai A et al (2022) Preservation of the fecal microbiome is associated with reduced severity of graft-versus-host disease. Blood 140(22):2385–2397. https://doi.org/10.1182/blood.2021015352

Busca A, Locatelli F, Marmont F, Ceretto C, Falda M (2007) Recombinant human soluble tumor necrosis factor receptor fusion protein as treatment for steroid refractory graft-versus-host disease following allogeneic hematopoietic stem cell transplantation. Am J Hematol 82(1):45–52. https://doi.org/10.1002/ajh.20752

Caputo J, Peddireddi A, Bhatta S et al (2024) Combination ruxolitinib and belumosudil is tolerable and induces responses despite treatment failure as monotherapies. Leuk Lymphoma 66:1–8. https://doi.org/10.1080/10428194.2024.2409876

Carpenter PA (2023) Chapter 108: late effects after hematopoietic cell transplantation. In: Means TT, Arber DA, Glader BE et al (eds) Wintrobe's clinical hematology, 15th edn. Wolters Kluwer, pp 2371–2386

Carpenter PA, Englund JA (2016) How I vaccinate blood and marrow transplant recipients. Blood 127(23):2824–2832. https://doi.org/10.1182/blood-2015-12-550475
Carpenter PA, Logan BR, Lee SJ et al (2018) A phase II/III randomized, multicenter trial of prednisone/sirolimus versus prednisone/ sirolimus/calcineurin inhibitor for the treatment of chronic graft-versus-host disease: BMT CTN 0801. Haematologica 103(11):1915–1924. (In eng). https://doi.org/10.1172/jci99261
Carpenter PA, Kang HJ, Yoo KH et al (2022) Ibrutinib treatment of pediatric chronic graft-versus-host disease: primary results from the phase 1/2 iMAGINE study. Transplant Cell Ther 28(11):771.e1–771.e10. https://doi.org/10.1016/j.jtct.2022.08.021
Carpenter PA, Gooley TA, Boiko J et al (2024) Decreasing chronic graft-versus-host disease rates in all populations. Blood Adv 8(22):5829–5837. https://doi.org/10.1182/bloodadvances.2024012722
Chandra G, Wang J, Siirtola P, Röning J (2024) Leveraging machine learning for predicting acute graft-versus-host disease grades in allogeneic hematopoietic cell transplantation for T-cell prolymphocytic leukaemia. BMC Med Res Methodol 24(1):112. (In eng). https://doi.org/10.1038/s41591-024-03016-4
Chen X, Mayne CG (2018) The role of micronutrients in graft-VS.-host disease: immunomodulatory effects of vitamins a and D. Front Immunol 9:2853. https://doi.org/10.3389/fimmu.2018.02853
Chen GL, Onstad L, Martin PJ et al (2023) Durable discontinuation of systemic therapy in patients affected by chronic graft-versus-host disease. Haematologica 108(2):483–489. https://doi.org/10.3324/haematol.2021.279814
Chen YB, Mohty M, Zeiser R et al (2024) Vedolizumab for the prevention of intestinal acute GVHD after allogeneic hematopoietic stem cell transplantation: a randomized phase 3 trial. Nat Med 30(8):2277–2287. (In eng). https://doi.org/10.1080/13543784.2024.2377322
Chin MM, Tamaresis JS, Johnston LJ et al (2024) Belumosudil combination therapy for chronic graft-versus-host-disease in real-world clinical practice. Bone Marrow Transplant. https://doi.org/10.1038/s41409-024-02476-z
Couriel DR, Beguelin GZ, Giralt S et al (2002) Chronic graft-versus-host disease manifesting as polymyositis: an uncommon presentation. Bone Marrow Transplant 30(8):543–546. https://doi.org/10.1038/sj.bmt.1703711
Couriel D, Saliba R, Hicks K et al (2004) Tumor necrosis factor-alpha blockade for the treatment of acute GVHD. Blood 104(3):649–654. https://doi.org/10.1182/blood-2003-12-4241
Couriel DR, Saliba R, Escalon MP et al (2005) Sirolimus in combination with tacrolimus and corticosteroids for the treatment of resistant chronic graft-versus-host disease. Br J Haematol 130(3):409–417. https://doi.org/10.1111/j.1365-2141.2005.05616.x
Cutler C, Lee SJ, Arai S et al (2021) Belumosudil for chronic graft-versus-host disease after 2 or more prior lines of therapy: the ROCKstar study. Blood 138(22):2278–2289. (In eng). https://doi.org/10.1182/hematology.2021000301
Cuvelier GDE, Schoettler M, Buxbaum NP et al (2022) Toward a better understanding of the atypical features of chronic graft-versus-host disease: a report from the 2020 National Institutes of Health Consensus Project Task Force. Transplant Cell Ther 28(8):426–445. (In eng). https://doi.org/10.1111/bjh.18300
Deeg HJ (2007) How I treat refractory acute GVHD. Blood 109(10):4119–4126. https://doi.org/10.1182/blood-2006-12-041889
Deeg HJ, Seidel K, Bruemmer B, Pepe MS, Appelbaum FR (1995) Impact of patient weight on non-relapse mortality after marrow transplantation. Bone Marrow Transplant 15(3):461–468. https://www.ncbi.nlm.nih.gov/pubmed/7599573
DeFilipp Z, Alousi AM, Pidala JA et al (2021) Nonrelapse mortality among patients diagnosed with chronic GVHD: an updated analysis from the chronic GVHD consortium. Blood Adv 5(20):4278–4284. https://doi.org/10.1182/bloodadvances.2021004941
Drexler B, Buser A, Infanti L, Stehle G, Halter J, Holbro A (2020) Extracorporeal Photopheresis in graft-versus-host disease. Transfus Med Hemother 47(3):214–225. https://doi.org/10.1159/000508169
Drobyski WR, Pasquini M, Kovatovic K et al (2011) Tocilizumab for the treatment of steroid refractory graft-versus-host disease. Biol Blood Marrow Transplant 17(12):1862–1868. https://doi.org/10.1016/j.bbmt.2011.07.001
El Jurdi N, Rayes A, MacMillan ML et al (2021) Steroid-dependent acute GVHD after allogeneic hematopoietic cell transplantation: risk factors and clinical outcomes. Blood Adv 5(5):1352–1359. https://doi.org/10.1182/bloodadvances.2020003937
Esteban NV, Loughlin T, Yergey AL et al (1991) Daily cortisol production rate in man determined by stable isotope dilution/mass spectrometry. J Clin Endocrinol Metab 72(1):39–45. https://doi.org/10.1210/jcem-72-1-39
Eurofins-Viracor L.. Diagnostics for acute graft vs host disease. https://www.eurofins-viracor.com/our-testing/agvhd/
Ferreira AM, Szor RS, Molla VC et al (2021) Long-term follow-up of Ruxolitinib in the treatment of steroid-refractory chronic graft-versus-host disease. Transplant Cell Ther 27(9):777.e1–777.e6. https://doi.org/10.1016/j.jtct.2021.06.002
Filipovich AH, Weisdorf D, Pavletic S et al (2005) National Institutes of Health consensus development project on criteria for clinical trials in chronic graft-versus-host disease: I. Diagnosis and staging working group report. Biol Blood Marrow Transplant 11(12):945–956. https://doi.org/10.1016/j.bbmt.2005.09.004
Floisand Y, Lazarevic VL, Maertens J et al (2019) Safety and effectiveness of Vedolizumab in patients with steroid-refractory gastrointestinal acute graft-versus-host disease: a retrospective record review. Biol

Blood Marrow Transplant 25(4):720–727. https://doi.org/10.1016/j.bbmt.2018.11.013

Foord AM, Cushing-Haugen KL, Boeckh MJ et al (2020) Late infectious complications in hematopoietic cell transplantation survivors: a population-based study. Blood Adv 4(7):1232–1241. https://doi.org/10.1182/bloodadvances.2020001470

Furlong T, Martin P, Flowers ME et al (2009) Therapy with mycophenolate mofetil for refractory acute and chronic GVHD. Bone Marrow Transplant 44(11):739–748. https://doi.org/10.1038/bmt.2009.76

Ganetsky A, Frey NV, Hexner EO et al (2019) Tocilizumab for the treatment of severe steroid-refractory acute graft-versus-host disease of the lower gastrointestinal tract. Bone Marrow Transplant 54(2):212–217. https://doi.org/10.1038/s41409-018-0236-z

Garcia-Cadenas I, Rivera I, Martino R et al (2017) Patterns of infection and infection-related mortality in patients with steroid-refractory acute graft versus host disease. Bone Marrow Transplant 52(1):107–113. https://doi.org/10.1038/bmt.2016.225

Giaccone L, Martin P, Carpenter P et al (2005) Safety and potential efficacy of low-dose methotrexate for treatment of chronic graft-versus-host disease. Bone Marrow Transplant 36(4):337–341. https://doi.org/10.1038/sj.bmt.1705022

Gilman AL, Schultz KR, Goldman FD et al (2012) Randomized trial of hydroxychloroquine for newly diagnosed chronic graft-versus-host disease in children: a Children's Oncology Group study. Biol Blood Marrow Transplant 18(1):84–91. https://doi.org/10.1016/j.bbmt.2011.05.016

Glucksberg H, Storb R, Fefer A et al (1974) Clinical manifestations of graft-versus-host disease in human recipients of marrow from HL-A-matched sibling donors. Transplantation 18(4):295–304. https://doi.org/10.1097/00007890-197410000-00001

Goldberg JD, Jacobsohn DA, Margolis J et al (2003) Pentostatin for the treatment of chronic graft-versus-host disease in children. J Pediatr Hematol Oncol 25(7):584–588. https://doi.org/10.1097/00043426-200307000-00020

Gomez-Almaguer D, Ruiz-Arguelles GJ, del Carmen Tarin-Arzaga L et al (2008) Alemtuzumab for the treatment of steroid-refractory acute graft-versus-host disease. Biol Blood Marrow Transplant 14(1):10–15. https://doi.org/10.1016/j.bbmt.2007.08.052

Hamilton BK, Onstad L, Carpenter PA et al (2024) Study protocol: predicting the quality of response to specific treatments (PQRST) in chronic graft-versus-host disease. Contemp Clin Trials 145:107637. https://doi.org/10.1016/j.cct.2024.107637

Harper K, McMillen KK (2021) Chapter 15: Medical nutrition therapy for hematopoietic cell transplantation. In: Voss AC, Williams V (eds) Oncology nutrition for clinical practice, 2nd edn. Academy of Nutrition and Dietetics, pp 305–329

Harris AC, Young R, Devine S et al (2016) International, multicenter standardization of acute graft-versus-host disease clinical data collection: a report from the Mount Sinai acute GVHD international consortium. Biol Blood Marrow Transplant 22(1):4–10. (In eng). https://doi.org/10.1016/j.bbmt.2015.10.018

Hartwell MJ, Özbek U, Holler E et al (2017) An early-biomarker algorithm predicts lethal graft-versus-host disease and survival. JCI Insight 2(3):e89798. (In eng). https://doi.org/10.1182/blood-2016-09-691394

Hashmi S, Carpenter P, Khera N, Tichelli A, Savani BN (2015) Lost in transition: the essential need for long-term follow-up clinic for blood and marrow transplantation survivors. Biol Blood Marrow Transplant 21(2):225–232. https://doi.org/10.1016/j.bbmt.2014.06.035

Hashmi SK, Lee SJ, Savani BN et al (2018) ASBMT practice guidelines committee survey on long-term follow-up clinics for hematopoietic cell transplant survivors. Biol Blood Marrow Transplant 24(6):1119–1124. https://doi.org/10.1016/j.bbmt.2018.03.023

Hills RD Jr, Pontefract BA, Mishcon HR, Black CA, Sutton SC, Theberge CR (2019) Gut microbiome: profound implications for diet and disease. Nutrients 11(7). https://doi.org/10.3390/nu11071613

Hockenbery DM, Cruickshank S, Rodell TC et al (2007) A randomized, placebo-controlled trial of oral beclomethasone dipropionate as a prednisone-sparing therapy for gastrointestinal graft-versus-host disease. Blood 109(10):4557–4563. https://doi.org/10.1182/blood-2006-05-021139

Hoda D, Pidala J, Salgado-Vila N et al (2010) Sirolimus for treatment of steroid-refractory acute graft-versus-host disease. Bone Marrow Transplant 45(8):1347–1351. https://doi.org/10.1038/bmt.2009.343

Holler E, Butzhammer P, Schmid K et al (2014) Metagenomic analysis of the stool microbiome in patients receiving allogeneic stem cell transplantation: loss of diversity is associated with use of systemic antibiotics and more pronounced in gastrointestinal graft-versus-host disease. Biol Blood Marrow Transplant 20(5):640–645. (In eng). https://doi.org/10.1182/blood-2014-01-514786

Holtan SG, Pasquini M, Weisdorf DJ (2014) Acute graft-versus-host disease: a bench-to-bedside update. Blood 124(3):363–373. (In eng). https://doi.org/10.1172/jci75935

Holtan SG, Khera N, Levine JE et al (2016) Late acute graft-versus-host disease: a prospective analysis of clinical outcomes and circulating angiogenic factors. Blood 128(19):2350–2358. https://doi.org/10.1182/blood-2015-09-669846

Holtan SG, Yu J, Paranagama D et al (2022) Disease progression, hospital readmissions, and clinical outcomes for patients with steroid-refractory acute graft-versus-host disease: a multicenter, retrospective study. Bone Marrow Transplant 57(9):1399–1404. (In eng). https://doi.org/10.7759/cureus.29195

Huang XJ, Jiang Q, Chen H et al (2005) Low-dose methotrexate for the treatment of graft-versus-host disease after allogeneic hematopoietic stem cell transplantation. Bone Marrow Transplant 36(4):343–348. https://doi.org/10.1038/sj.bmt.1705034

Imataki O, Nakatani S, Hasegawa T et al (2006) Nutritional support for patients suffering from intestinal graft-versus-host disease after allogeneic hematopoietic stem cell transplantation. Am J Hematol 81(10):747–752. https://doi.org/10.1002/ajh.20700

Imbruciva (ibrutinib) [package insert] (2024) Sunnyvale, CA: Pharmacyclics

Jacobsohn DA, Chen AR, Zahurak M et al (2007) Phase II study of pentostatin in patients with corticosteroid-refractory chronic graft-versus-host disease. J Clin Oncol 25(27):4255–4261. https://doi.org/10.1200/JCO.2007.10.8456

Jagasia MH, Greinix HT, Arora M et al (2015) National Institutes of Health consensus development project on criteria for clinical trials in chronic graft-versus-host disease: I. The 2014 diagnosis and staging working group report. Biol Blood Marrow Transplant 21(3):389–401.e1. (In eng). https://doi.org/10.1016/j.bbmt.2015.01.001

Jagasia M, Perales MA, Schroeder MA et al (2020) Ruxolitinib for the treatment of steroid-refractory acute GVHD (REACH1): a multicenter, open-label phase 2 trial. Blood 135(20):1739–1749. (In eng). https://doi.org/10.1172/jci133909

Jaglowski SM, Blazar BR (2018) How ibrutinib, a B-cell malignancy drug, became an FDA-approved second-line therapy for steroid-resistant chronic GVHD. Blood Adv 2(15):2012–2019. https://doi.org/10.1182/bloodadvances.2018013060

Jakafi (ruxolitinib) [package insert] (2021) Wilmington, DE: Incyte Corporation

Jardine L, Cytlak U, Gunawan M et al (2020) Donor monocyte-derived macrophages promote human acute graft-versus-host disease. J Clin Invest 130(9):4574–4586. (In eng). https://doi.org/10.1001/jamanetworkopen.2020.34750

Jenq RR, Taur Y, Devlin SM et al (2015) Intestinal Blautia is associated with reduced death from graft-versus-host disease. Biol Blood Marrow Transplant 21(8):1373–1383. (In eng). https://doi.org/10.1016/j.bbmt.2015.09.001

Ji R, Li Y, Huang R, Xiong J, Wang X, Zhang X (2023) Recent advances and research progress in biomarkers for chronic graft versus host disease. Crit Rev Oncol Hematol 186:103993. https://doi.org/10.1016/j.critrevonc.2023.103993

Jim HS, Sutton SK, Jacobsen PB, Martin PJ, Flowers ME, Lee SJ (2016) Risk factors for depression and fatigue among survivors of hematopoietic cell transplantation. Cancer 122(8):1290–1297. https://doi.org/10.1002/cncr.29877

Johnston LJ, Brown J, Shizuru JA et al (2005) Rapamycin (sirolimus) for treatment of chronic graft-versus-host disease. Biol Blood Marrow Transplant 11(1):47–55. https://doi.org/10.1016/j.bbmt.2004.10.004

Jurado M, Vallejo C, Perez-Simon JA et al (2007) Sirolimus as part of immunosuppressive therapy for refractory chronic graft-versus-host disease. Biol Blood Marrow Transplant 13(6):701–706. https://doi.org/10.1016/j.bbmt.2007.02.003

Kebriaei P, Hayes J, Daly A et al (2020) A phase 3 randomized study of Remestemcel-L versus placebo added to second-line therapy in patients with steroid-refractory acute graft-versus-host disease. Biol Blood Marrow Transplant 26(5):835–844. https://doi.org/10.1016/j.bbmt.2019.08.029

Khandelwal P, Langenberg L, Luebbering N et al (2024) A randomized phase 2 trial of oral vitamin A for graft-versus-host disease in children and young adults. Blood 143(12):1181–1192. https://doi.org/10.1182/blood.2023022865

Kharfan-Dabaja MA, Mhaskar AR, Djulbegovic B, Cutler C, Mohty M, Kumar A (2009) Efficacy of rituximab in the setting of steroid-refractory chronic graft-versus-host disease: a systematic review and meta-analysis. Biol Blood Marrow Transplant 15(9):1005–1013. https://doi.org/10.1016/j.bbmt.2009.04.003

Khera N, Chow EJ, Leisenring WM et al (2011) Factors associated with adherence to preventive care practices among hematopoietic cell transplantation survivors. Biol Blood Marrow Transplant 17(7):995–1003. https://doi.org/10.1016/j.bbmt.2010.10.023

Kitko CL, Pidala J, Schoemans HM et al (2021) National Institutes of Health consensus development project on criteria for clinical trials in chronic graft-versus-host disease: IIa. The 2020 clinical implementation and early diagnosis working group report. Transplant Cell Ther 27(7):545–557. https://doi.org/10.1016/j.jtct.2021.03.033

Kitko CL, Abdel-Azim H, Carpenter PA et al (2022) A prospective, multicenter study of closed-system extracorporeal Photopheresis for children with steroid-refractory acute graft-versus-host disease. Transplant Cell Ther 28(5):261.e1–261.e7. https://doi.org/10.1016/j.jtct.2022.01.025

Kitko CL, Arora M, DeFilipp Z et al (2023) Axatilimab for chronic graft-versus-host disease after failure of at least two prior systemic therapies: results of a phase I/II study. J Clin Oncol 41(10):1864–1875. https://doi.org/10.1200/JCO.22.00958

Knipe RS, Tager AM, Liao JK (2015) The Rho kinases: critical mediators of multiple profibrotic processes and rational targets for new therapies for pulmonary fibrosis. Pharmacol Rev 67(1):103–117. https://doi.org/10.1124/pr.114.009381

Koc S, Leisenring W, Flowers ME et al (2000) Thalidomide for treatment of patients with chronic graft-versus-host disease. Blood 96(12):3995–3996. https://www.ncbi.nlm.nih.gov/pubmed/11090092

Koc S, Leisenring W, Flowers ME et al (2002) Therapy for chronic graft-versus-host disease: a randomized trial comparing cyclosporine plus prednisone versus prednisone alone. Blood 100(1):48–51. https://doi.org/10.1182/blood.v100.1.48

Koreth J, Kim HT, Jones KT et al (2016) Efficacy, durability, and response predictors of low-dose interleukin-2 therapy for chronic graft-versus-host disease. Blood 128(1):130–137. https://doi.org/10.1182/blood-2016-02-702852

Kurtzberg J, Abdel-Azim H, Carpenter P et al (2020) A phase 3, single-arm, prospective study of Remestemcel-L, ex vivo culture-expanded adult human mesenchymal stromal cells for the treatment of pediatric patients who failed to respond to steroid treatment for acute graft-versus-host disease. Biol Blood Marrow Transplant 26(5):845–854. https://doi.org/10.1016/j.bbmt.2020.01.018

Lee SJ, Vogelsang G, Flowers ME (2003) Chronic graft-versus-host disease. Biol Blood Marrow Transplant 9(4):215–233. https://doi.org/10.1053/bbmt.2003.50026

Lee SJ, Nguyen TD, Onstad L et al (2018) Success of immunosuppressive treatments in patients with chronic graft-versus-host disease. Biol Blood Marrow Transplant 24(3):555–562. (In eng). https://doi.org/10.1016/j.bbmt.2018.01.010

Lee CJ, Wickline M, Flowers MED (2021) Chapter 4: telemedicine in patient care of long-term transplant survivors. In: Savani B, Tichelli A (eds) Blood and marrow transplantation long-term management: prevention and complications, 2nd edn. Wiley, pp 25–32

Levine JE, Logan B, Wu J et al (2010) Graft-versus-host disease treatment: predictors of survival. Biol Blood Marrow Transplant 16(12):1693–1699. (In eng). https://doi.org/10.1155/2011/601953

Levine JE, Braun TM, Harris AC et al (2015) A prognostic score for acute graft-versus-host disease based on biomarkers: a multicentre study. Lancet Haematol 2(1):e21–e29. (In eng). https://doi.org/10.2147/ppa.S76563

Li AC, Dong C, Tay ST, Ananthakrishnan A, Ma KS (2022) Vedolizumab for acute gastrointestinal graft-versus-host disease: a systematic review and meta-analysis. Front Immunol 13:1025350. https://doi.org/10.3389/fimmu.2022.1025350

Linder BL, Esteban NV, Yergey AL, Winterer JC, Loriaux DL, Cassorla F (1990) Cortisol production rate in childhood and adolescence. J Pediatr 117(6):892–896. https://doi.org/10.1016/s0022-3476(05)80128-3

Lopez F, Parker P, Nademanee A et al (2005) Efficacy of mycophenolate mofetil in the treatment of chronic graft-versus-host disease. Biol Blood Marrow Transplant 11(4):307–313. https://doi.org/10.1016/j.bbmt.2005.01.011

Lounder DT, Khandelwal P, Dandoy CE et al (2017) Lower levels of vitamin A are associated with increased gastrointestinal graft-versus-host disease in children. Blood 129(20):2801–2807. https://doi.org/10.1182/blood-2017-02-765826

MacDonald KP, Hill GR, Blazar BR (2017a) Chronic graft-versus-host disease: biological insights from preclinical and clinical studies. Blood 129(1):13–21. https://doi.org/10.1182/blood-2016-06-686618

MacDonald KP, Blazar BR, Hill GR (2017b) Cytokine mediators of chronic graft-versus-host disease. J Clin Invest 127(7):2452–2463. https://doi.org/10.1172/JCI90593

MacMillan ML, Weisdorf DJ, Wagner JE et al (2002a) Response of 443 patients to steroids as primary therapy for acute graft-versus-host disease: comparison of grading systems. Biol Blood Marrow Transplant 8(7):387–394. https://doi.org/10.1053/bbmt.2002.v8.pm12171485

MacMillan ML, Weisdorf DJ, Davies SM et al (2002b) Early antithymocyte globulin therapy improves survival in patients with steroid-resistant acute graft-versus-host disease. Biol Blood Marrow Transplant 8(1):40–46. https://doi.org/10.1053/bbmt.2002.v8.pm11858189

MacMillan ML, DeFor TE, Weisdorf DJ (2010) The best endpoint for acute GVHD treatment trials. Blood 115(26):5412–5417. https://doi.org/10.1182/blood-2009-12-258442

MacMillan ML, Robin M, Harris AC et al (2015) A refined risk score for acute graft-versus-host disease that predicts response to initial therapy, survival, and transplant-related mortality. Biol Blood Marrow Transplant 21(4):761–767. (In eng). https://doi.org/10.1016/j.ajpath.2015.01.005

MacMillan ML, Holtan SG, Rashidi A, DeFor TE, Blazar BR, Weisdorf DJ (2020) Pediatric acute GVHD: clinical phenotype and response to upfront steroids. Bone Marrow Transplant 55(1):165–171. https://doi.org/10.1038/s41409-019-0651-9

Magenau JM, Goldstein SC, Peltier D et al (2018) alpha(1)-Antitrypsin infusion for treatment of steroid-resistant acute graft-versus-host disease. Blood 131(12):1372–1379. https://doi.org/10.1182/blood-2017-11-815746

Major-Monfried H, Renteria AS, Pawarode A et al (2018) MAGIC biomarkers predict long-term outcomes for steroid-resistant acute GVHD. Blood 131(25):2846–2855. https://doi.org/10.1182/blood-2018-01-822957

Malard F, Huang XJ, Sim JPY (2020) Treatment and unmet needs in steroid-refractory acute graft-versus-host disease. Leukemia 34(5):1229–1240. https://doi.org/10.1038/s41375-020-0804-2

Malard F, Holler E, Sandmaier BM, Huang H, Mohty M (2023a) Acute graft-versus-host disease. Nat Rev Dis Primers 9(1):27. https://doi.org/10.1038/s41572-023-00438-1

Malard F, Loschi M, Huynh A et al (2023b) Pooled allogeneic faecal microbiota MaaT013 for steroid-resistant gastrointestinal acute graft-versus-host disease: a single-arm, multicentre phase 2 trial. EClinicalMedicine 62:102111. https://doi.org/10.1016/j.eclinm.2023.102111

Martens MJ, Kou J, Logan BR, Paczesny S (2023) Machine learning validates risk biomarkers of chronic graft-versus-host disease in 936 patients from BMT CTN 0201 & 1202 cohorts. Blood 142(Supplement 1):479

Martin PJ (2020) How I treat steroid-refractory acute graft-versus-host disease. Blood 135(19):1630–1638. https://doi.org/10.1182/blood.2019000960

Martin PJ, Storer BE, Rowley SD et al (2009) Evaluation of mycophenolate mofetil for initial treatment of chronic graft-versus-host disease. Blood 113(21):5074–5082. https://doi.org/10.1182/blood-2009-02-202937

Martin PJ, Rizzo JD, Wingard JR et al (2012) First- and second-line systemic treatment of acute graft-versus-host disease: recommendations of the American Society of Blood and Marrow Transplantation. Biol Blood Marrow Transplant 18(8):1150–1163. https://doi.org/10.1016/j.bbmt.2012.04.005

McMillen KK, Coghlin-Dickson T, Adintori PA (2021) Optimization of nutrition support practices early after hematopoietic cell transplantation. Bone Marrow Transplant 56(2):314–326. https://doi.org/10.1038/s41409-020-01078-9

McNeil A, Parks K, Liu X et al (2022) Artificial intelligence recognition of cutaneous chronic graft-versus-host disease by a deep learning neural network. Br J Haematol 197(6):e69–e72. (In eng). https://doi.org/10.1182/bloodadvances.2021006220

Mielcarek M, Storer BE, Boeckh M et al (2009) Initial therapy of acute graft-versus-host disease with low-dose prednisone does not compromise patient outcomes. Blood 113(13):2888–2894. (In eng). https://doi.org/10.1182/blood-2009-02-202937

Mielcarek M, Furlong T, Storer BE et al (2015) Effectiveness and safety of lower dose prednisone for initial treatment of acute graft-versus-host disease: a randomized controlled trial. Haematologica 100(6):842–848. (In eng). https://doi.org/10.1016/j.bbmt.2015.02.025

Miklos D, Cutler CS, Arora M et al (2017) Ibrutinib for chronic graft-versus-host disease after failure of prior therapy. Blood 130(21):2243–2250. (In eng). https://doi.org/10.1056/NEJMra1609337

Miklos DB, Abu Zaid M, Cooney JP et al (2023) Ibrutinib for first-line treatment of chronic graft-versus-host disease: results from the randomized phase III iNTEGRATE study. J Clin Oncol 41(10):1876–1887. https://doi.org/10.1200/JCO.22.00509

Miyamura K, Yamashita T, Atsuta Y et al (2019) High probability of follow-up termination among AYA survivors after allogeneic hematopoietic cell transplantation. Blood Adv 3(3):397–405. https://doi.org/10.1182/bloodadvances.2018026039

Naymagon S, Naymagon L, Wong SY et al (2017) Acute graft-versus-host disease of the gut: considerations for the gastroenterologist. Nat Rev Gastroenterol Hepatol 14(12):711–726. https://doi.org/10.1038/nrgastro.2017.126

Nikiforow S, Wang T, Hemmer M et al (2018) Upper gastrointestinal acute graft-versus-host disease adds minimal prognostic value in isolation or with other graft-versus-host disease symptoms as currently diagnosed and treated. Haematologica 103(10):1708–1719. (In eng). https://doi.org/10.1038/s41409-018-0204-7

Niktimvo (axatilimab) [package insert] (2024) Wilmington, DE: Incyte Corporation

Norkin M, Shaw BE, Brazauskas R et al (2019) Characteristics of late fatal infections after allogeneic hematopoietic cell transplantation. Biol Blood Marrow Transplant 25(2):362–368. https://doi.org/10.1016/j.bbmt.2018.09.031

Ordentlich P, Wolfreys A, Da Costa A (2016) Targeting colony stimulating factor- 1 receptor (CSF-1R) with SNDX-6352, a novel anti-CSF-1R targeted antibody. J Immunother Cancer 4:P402

Pavletic SZ, Fowler DH (2012) Are we making progress in GVHD prophylaxis and treatment? Hematology Am Soc Hematol Educ Program 2012:251–264. https://doi.org/10.1182/asheducation-2012.1.251

Pavletic SZ, Martin PJ, Schultz KR, Lee SJ (2021) The future of chronic graft-versus-host disease: introduction to the 2020 National Institutes of Health consensus development project reports. Transplant Cell Ther 27(6):448–451. https://doi.org/10.1016/j.jtct.2021.02.034

Penack O, Marchetti M, Aljurf M et al (2024) Prophylaxis and management of graft-versus-host disease after stem-cell transplantation for haematological malignancies: updated consensus recommendations of the European Society for Blood and Marrow Transplantation. Lancet Haematol 11(2):e147–e159. https://doi.org/10.1016/S2352-3026(23)00342-3

Pidala J, Kim J, Anasetti C (2009) Sirolimus as primary treatment of acute graft-versus-host disease following allogeneic hematopoietic cell transplantation. Biol Blood Marrow Transplant 15(7):881–885. https://doi.org/10.1016/j.bbmt.2009.03.020

Pidala J, Kim J, Perkins J et al (2010) Mycophenolate mofetil for the management of steroid-refractory acute graft vs host disease. Bone Marrow Transplant 45(5):919–924. https://doi.org/10.1038/bmt.2009.252

Pidala J, Kurland B, Chai X et al (2011) Patient-reported quality of life is associated with severity of chronic graft-versus-host disease as measured by NIH criteria: report on baseline data from the chronic GVHD consortium. Blood 117(17):4651–4657. (In eng). https://doi.org/10.1200/jco.2010.33.7212

Przepiorka D, Chan KW, Champlin RE et al (1995) Prevention of graft-versus-host disease with anti-CD5 ricin A chain immunotoxin after CD3-depleted HLA-nonidentical marrow transplantation in pediatric leukemia patients. Bone Marrow Transplant 16(6):737–741. https://www.ncbi.nlm.nih.gov/pubmed/8750262

Pusic I, Lee C, Veeraputhiran M, Minor C, DiPersio JF (2024) Belumosudil and ruxolitinib combination for treatment of refractory chronic graft-versus-host disease. Bone Marrow Transplant 59(2):282–284. https://doi.org/10.1038/s41409-023-02165-3

Ragon BK, Mehta RS, Gulbis AM et al (2018) Pentostatin therapy for steroid-refractory acute graft versus host disease: identifying those who may benefit. Bone Marrow Transplant 53(3):315–325. https://doi.org/10.1038/s41409-017-0034-z

Redondo S, Esquirol A, Novelli S et al (2022) Efficacy and safety of Ruxolitinib in steroid-refractory/dependent chronic graft-versus-host disease: real-world data and challenges. Transplant Cell Ther 28(1):43.e1–43.e5. https://doi.org/10.1016/j.jtct.2021.10.015

Reichenbach DK, Schwarze V, Matta BM et al (2015) The IL-33/ST2 axis augments effector T-cell responses

during acute GVHD. Blood 125(20):3183–3192. (In eng). https://doi.org/10.3324/haematol.2014.118471
Reynolds G, Hall VG, Teh BW (2023) Vaccine schedule recommendations and updates for patients with hematologic malignancy post-hematopoietic cell transplant or CAR T-cell therapy. Transpl Infect Dis 25 Suppl 1(Suppl 1):e14109. https://doi.org/10.1111/tid.14109
Rezurock (belumosudil) [package insert] (2024) Paris, France: Sanofi
Riches DW, Backos DS, Redente EF (2015) ROCK and rho: promising therapeutic targets to ameliorate pulmonary fibrosis. Am J Pathol 185(4):909–912. (In eng). https://doi.org/10.1182/blood-2014-10-606830
Roddy JV, Haverkos BM, McBride A et al (2016) Tocilizumab for steroid refractory acute graft-versus-host disease. Leuk Lymphoma 57(1):81–85. https://doi.org/10.3109/10428194.2015.1045896
Rodriguez-Gil A, Carrillo-Cruz E, Marrero-Cepeda C, Rodriguez G, Perez-Simon JA (2022) Effect of vitamin D on graft-versus-host disease. Biomedicines 10(5). https://doi.org/10.3390/biomedicines10050987
Rowlings PA, Przepiorka D, Klein JP et al (1997) IBMTR severity index for grading acute graft-versus-host disease: retrospective comparison with Glucksberg grade. Br J Haematol 97(4):855–864. https://doi.org/10.1046/j.1365-2141.1997.1112925.x
Rubin LG, Levin MJ, Ljungman P et al (2014) 2013 IDSA clinical practice guideline for vaccination of the immunocompromised host. Clin Infect Dis 58(3):e44–e100. https://doi.org/10.1093/cid/cit684
Salit RB, Lee SJ, Burns LJ et al (2020) Return-to-work guidelines and programs for post-hematopoietic cell transplantation survivors: an initial survey. Biol Blood Marrow Transplant 26(8):1520–1526. https://doi.org/10.1016/j.bbmt.2020.04.022
Salit RB, Schoeppner K, De Biase C et al (2022) American Society for Transplantation and Cellular Therapy Return to Work Guidance Committee recommendations for health care providers who take care of hematopoietic cell transplantation patients. Transplant Cell Ther 28(12):822–828. https://doi.org/10.1016/j.jtct.2022.09.017
Schmidt-Hieber M, Fietz T, Knauf W et al (2005) Efficacy of the interleukin-2 receptor antagonist basiliximab in steroid-refractory acute graft-versus-host disease. Br J Haematol 130(4):568–574. https://doi.org/10.1111/j.1365-2141.2005.05631.x
Schnitzler M, Hasskarl J, Egger M, Bertz H, Finke J (2009) Successful treatment of severe acute intestinal graft-versus-host resistant to systemic and topical steroids with alemtuzumab. Biol Blood Marrow Transplant 15(8):910–918. https://doi.org/10.1016/j.bbmt.2009.04.002
Schoemans H, Goris K, Durm RV et al (2016) Development, preliminary usability and accuracy testing of the EBMT 'eGVHD App' to support GvHD assessment according to NIH criteria-a proof of concept. Bone Marrow Transplant 51(8):1062–1065. https://doi.org/10.1038/bmt.2016.26
Schoemans HM, Goris K, Van Durm R et al (2018a) The eGVHD app has the potential to improve the accuracy of graft-versus-host disease assessment: a multicenter randomized controlled trial. Haematologica 103(10):1698–1707. (In eng). https://doi.org/10.3324/haematol.2017.182550
Schoemans HM, Lee SJ, Ferrara JL et al (2018b) EBMT-NIH-CIBMTR task force position statement on standardized terminology & guidance for graft-versus-host disease assessment. Bone Marrow Transplant 53(11):1401–1415. (In eng). https://doi.org/10.3324/haematol.2018.195123
Schoemans HM, Goris K, Van Durm R et al (2018c) Accuracy and usability of the eGVHD app in assessing the severity of graft-versus-host disease at the 2017 EBMT annual congress. Bone Marrow Transplant 53(4):490–494. https://doi.org/10.1038/s41409-017-0017-0
Segon B, Lam L, Chan HY et al (2022) Vitamin requirements during stem cell transplantation: a systematic review. Support Care Cancer 30(12):10391–10405. https://doi.org/10.1007/s00520-022-07409-4
Seguy D, Duhamel A, Rejeb MB et al (2012) Better outcome of patients undergoing enteral tube feeding after myeloablative conditioning for allogeneic stem cell transplantation. Transplantation 94(3):287–294. https://doi.org/10.1097/TP.0b013e3182558f60
Shahzad M, Chaudhary SG, Basit A et al (2022) Chronic graft-versus-host disease presenting as acute polymyositis: a case series and systematic review. Transpl Immunol 70:101520. https://doi.org/10.1016/j.trim.2021.101520
Shulman HM, Sullivan KM, Weiden PL et al (1980) Chronic graft-versus-host syndrome in man. A long-term clinicopathologic study of 20 Seattle patients. Am J Med 69(2):204–217. https://doi.org/10.1016/0002-9343(80)90380-0
Sullivan KM, Witherspoon RP, Storb R et al (1988) Prednisone and azathioprine compared with prednisone and placebo for treatment of chronic graft-v-host disease: prognostic influence of prolonged thrombocytopenia after allogeneic marrow transplantation. Blood 72(2):546–554. https://www.ncbi.nlm.nih.gov/pubmed/3042041
van der Meij BS, de Graaf P, Wierdsma NJ et al (2013) Nutritional support in patients with GVHD of the digestive tract: state of the art. Bone Marrow Transplant 48(4):474–482. https://doi.org/10.1038/bmt.2012.124
Weisdorf D, Haake R, Blazar B et al (1990) Treatment of moderate/severe acute graft-versus-host disease after allogeneic bone marrow transplantation: an analysis of clinical risk features and outcome. Blood 75(4):1024–1030. https://www.ncbi.nlm.nih.gov/pubmed/2302454
Westin JR, Saliba RM, De Lima M et al (2011) Steroid-refractory acute GVHD: predictors and outcomes. Adv Hematol 2011:601953. (In eng). https://doi.org/10.1182/blood-2010-11-319509

Wolff D, Radojcic V, Lafyatis R et al (2021) National Institutes of Health consensus development project on criteria for clinical trials in chronic graft-versus-host disease: IV. The 2020 highly morbid forms report. Transplant Cell Ther 27(10):817–835. (In eng). https://doi.org/10.1182/blood.2021012021

Wolff D, Cutler C, Lee SJ et al (2024) Axatilimab in recurrent or refractory chronic graft-versus-host disease. N Engl J Med 391(11):1002–1014. https://doi.org/10.1056/NEJMoa2401537

Wu H, Shi J, Luo Y et al (2021) Evaluation of Ruxolitinib for steroid-refractory chronic graft-vs-host disease after allogeneic hematopoietic stem cell transplantation. JAMA Netw Open 4(1):e2034750. https://doi.org/10.1001/jamanetworkopen.2020.34750

Yu J, Judy JT, Parasuraman S, Sinha M, Weisdorf D (2020) Inpatient healthcare resource utilization, costs, and mortality in adult patients with acute graft-versus-host disease, including steroid-refractory or high-risk disease, following allogeneic hematopoietic cell transplantation. Biol Blood Marrow Transplant 26(3):600–605. https://doi.org/10.1016/j.bbmt.2019.10.028

Zama D, Gori D, Muratore E et al (2021) Enteral versus parenteral nutrition as nutritional support after allogeneic hematopoietic stem cell transplantation: a systematic review and meta-analysis. Transplant Cell Ther 27(2):180.e1–180.e8. https://doi.org/10.1016/j.jtct.2020.11.006

Zanin-Zhorov A, Weiss JM, Nyuydzefe MS et al (2014) Selective oral ROCK2 inhibitor down-regulates IL-21 and IL-17 secretion in human T cells via STAT3-dependent mechanism. Proc Natl Acad Sci USA 111(47):16814–16819. (In eng). https://doi.org/10.1124/pr.114.009381

Zeiser R, Blazar BR (2017a) Acute graft-versus-host disease – biologic process, prevention, and therapy. N Engl J Med 377(22):2167–2179. (In eng). https://doi.org/10.1038/nrgastro.2017.126

Zeiser R, Blazar BR (2017b) Pathophysiology of chronic graft-versus-host disease and therapeutic targets. N Engl J Med 377(26):2565–2579. https://doi.org/10.1056/NEJMra1703472

Zeiser R, Burchert A, Lengerke C et al (2015) Ruxolitinib in corticosteroid-refractory graft-versus-host disease after allogeneic stem cell transplantation: a multicenter survey. Leukemia 29(10):2062–2068. https://doi.org/10.1038/leu.2015.212

Zeiser R, von Bubnoff N, Butler J et al (2020) Ruxolitinib for glucocorticoid-refractory acute graft-versus-host disease. N Engl J Med 382(19):1800–1810. https://doi.org/10.1056/NEJMoa1917635

Zeiser R, Polverelli N, Ram R et al (2021) Ruxolitinib for glucocorticoid-refractory chronic graft-versus-host disease. N Engl J Med 385(3):228–238. https://doi.org/10.1056/NEJMoa2033122

Zeiser R, Russo D, Ram R et al (2023) Ruxolitinib in patients with chronic graft-versus-host disease: 3-year final analysis of efficacy and safety from the phase III REACH3 study. Blood 142(Supplement 1):654

Zhao D, Kim YH, Jeong S et al (2018) Survival signal REG3α prevents crypt apoptosis to control acute gastrointestinal graft-versus-host disease. J Clin Invest 128(11):4970–4979. (In eng). https://doi.org/10.1182/asheducation-2018.1.228

7 Acute Myeloid Leukemia-Use of Measurable Residual Disease to Improve Outcomes

Jesse M. Tettero and Christopher S. Hourigan

Abstract

Measurable residual disease (MRD) assessment following initial therapy attempts to detect low levels of leukemia cells that remain undetectable by conventional cytomorphological methods. MRD-negativity after induction therapy in patients with acute myeloid leukemia (AML) is strongly associated with improved disease-free and overall survival, making results of MRD testing a more reliable indicator of treatment response compared to traditional methods. Unfortunately, due in part to genetic and phenotypical heterogeneity, no perfect MRD assay for AML currently exists although multiple tests are currently in use. Understanding the performance characteristics of AML MRD tests is critical for their clinical use. After achieving first complete remission with chemotherapy, consolidation therapy aims to deepen remission and prolong response duration. For patients with difficult-to-treat AML, allogeneic stem cell transplantation (allo-SCT) is a key strategy for achieving durable remission. However, for patients with a relatively favorable risk profile, the benefits of allo-SCT may not outweigh its risks, and durable remission can often be achieved with chemotherapy alone. When the benefits of allo-SCT are less clear, MRD can refine patient selection. MRD assessment after induction chemotherapy using validated methods can signal the need for more aggressive or alternative treatments to mitigate relapse risk. MRD-negative patients eligible for transplantation might benefit from reduced treatment intensity or modified conditioning to avoid overtreatment. Additionally, MRD can guide the intensity of conditioning therapy in transplant-eligible patients. Post-transplant MRD monitoring can further identify patients who may benefit from preemptive treatments to maintain remission. Integrating MRD into treatment protocols allows clinicians to enhance therapeutic precision, optimize patient outcomes, and improve the likelihood of achieving long-term remission in AML. However, due to limitations associated with current AML MRD testing, careful consideration and validation are essential before incorporating MRD findings into routine clinical practice.

Keywords

Measurable Residual Disease (MRD) · Flow cytometry · Polymerase Chain Reaction (PCR) · Next-generation Sequencing · MRD-guidance · European LeukemiaNet

J. M. Tettero (✉) · C. S. Hourigan
Fralin Biomedical Research Institute, Virginia Tech FBRI Cancer Research Center, Washington, DC, USA
e-mail: jtettero@vt.edu

© The Author(s), under exclusive license to Springer Nature Switzerland AG 2025

K. Ballen (ed.), *Clinical Management of Challenges in Allogeneic Stem Cell Transplant*, Advances in Experimental Medicine and Biology 1475, https://doi.org/10.1007/978-3-031-84988-6_7

(ELN) risk classification · MRD-guided treatment · MRD-guided consolidation treatment · MRD-guided conditioning · Personalized medicine · Clinical decision-making

7.1 Introduction

7.1.1 Overview of Acute Myeloid Leukemia (AML) and Treatment Options

Acute Myeloid Leukemia (AML) is a heterogeneous hematologic malignancy characterized by the clonal expansion of myeloid progenitor cells in the bone marrow, leading to impaired hematopoiesis. Upon diagnosis, the first critical task is to assess whether the patient is eligible for intensive therapy, which is determined by factors such as age, overall fitness, and the presence of specific genetic aberrations (Döhner et al. 2022; Venditti et al. 2025). Younger patients with good performance status are typically candidates for intensive therapy, which involves a combination of cytarabine and an anthracycline. In contrast, non-intensive therapy is reserved for older patients or those with significant comorbidities who are deemed unfit for intensive chemotherapy. This often includes hypomethylating agents (HMAs) like azacitidine or decitabine, potentially in combination with other targeted therapies (Sorror et al. 2021).

The primary objective of induction therapy in AML is to achieve complete remission (CR), traditionally defined by the presence of less than 5% blasts in a bone marrow aspirate smear. This assessment, typically performed by morphological evaluation under a microscope, requires distinguishing between normal regenerating marrow blasts and leukemic blasts, a challenging task given the often-subtle morphological differences. Cytogenetic techniques, including karyotyping and fluorescence in situ hybridization (FISH), are employed to identify chromosomal abnormalities associated with myeloid neoplasms; however, these methods are limited by their sensitivity, detecting residual disease only when it is present at levels of 1–5% (Mancini et al. 2000).

Patients who achieve CR are then evaluated for consolidation therapy, which may involve allogeneic stem cell transplantation (allo-SCT) if deemed necessary, or additional cycles of chemotherapy, potentially followed by autologous stem cell transplantation (auto-SCT), depending on the specific chemotherapy regimen used and region where the patient is being treated (Cornelissen and Blaise 2016; Koreth et al. 2009; Ustun et al. 2019). Molecular classification and risk stratification at initial diagnosis remain the most important for choosing consolidation therapy. But despite achieving CR and risk stratified treatment, a significant percentage of patients relapse, underscoring the limitation of conventional techniques in detecting residual disease and highlighting the need for more sensitive methods to refine risk stratification, such as measurable residual disease (MRD; formerly referred to as minimal residual disease) assessment, to identify residual leukemia below the detection threshold of traditional methods (Hourigan and Karp 2013).

7.1.2 Significance of Measurable Residual Disease (MRD) in AML

MRD represents a critical advancement in the detection of residual leukemia, with the capability to identify disease at a detection limit of 0.1% or even lower (Heuser et al. 2021a). A meta-analysis of data from 81 studies demonstrated that achieving MRD negativity is strongly associated with superior disease-free survival (DFS) and overall survival (OS) in AML patients (Short et al. 2020). The consistent relationship between MRD status and clinical outcomes across various studies has led to the adoption of complete remission without MRD (CR_{MRD-}) as an official response criterion in AML (Döhner et al. 2022). Due to its significant prognostic value, MRD is increasingly being utilized to guide clinical decision-making at various stages of AML treatment, as will be explored in this chapter.

7.1.3 Purpose of the Chapter: Harnessing MRD to Improve AML Outcomes

The integration of MRD assessment into routine clinical practice is advocated by many experts, reflecting a paradigm shift in the management of AML (Ossenkoppele et al. 2019; Ravandi 2018; Luskin and Stone 2017; Ossenkoppele and Schuurhuis 2016; Freeman and Hourigan 2019). A survey conducted among hematologists in the United States revealed that 69% of respondents reported routinely using MRD in the management of AML patients, highlighting the growing importance of this practice (Epstein-Peterson et al. 2018). However, with the adoption of MRD as a clinical endpoint comes the responsibility to understand the strengths and limitations of the various MRD detection techniques. This chapter will delve into the methodologies for MRD assessment, addressing their challenges and limitations, and will review the current evidence from large clinical trials supporting the use of MRD in clinical practice, particularly in the context of allogeneic stem cell transplantation (allo-SCT).

7.2 Understanding Measurable Residual Disease (MRD)

7.2.1 Definition and Concept

Measurable residual disease (MRD), previously referred to as 'minimal residual disease,' refers to the presence of residual leukemia cells that remain after initial treatment, detectable using advanced methodologies with higher sensitivity than traditional morphological assessment. Given the heterogeneity of AML, various assays are employed at different clinical stages to detect MRD. These include multiparameter flow cytometry (MFC), polymerase chain reaction (PCR) for specific gene fusions or genomic alterations, and next-generation sequencing (NGS). Each method varies in its sensitivity, optimal tissue sources, and applicability to different patient populations, thus providing a complementary set of tools for MRD detection.

7.2.2 Multiparameter Flow Cytometry

The first major advancement in MRD assessment came with the introduction of multiparameter flow cytometry (MFC) in the 1980s and 1990s. MFC allowed for the detection of leukemic cells based on their expression of specific surface markers. Leukemic blast cells exhibit a combination of cell surface markers, referred to as leukemia-associated immunophenotypes (LAIPs), that are not present on healthy BM cells (Zeijlemaker et al. 2019; Wood et al. 2007; Feller et al. 2013). The LAIP combination on blast cells, characterized by CD45-dim expression, typically includes a myeloid marker (CD13 or CD33), a primitive marker (CD34, CD117, and/or CD133), and an aberrantly expressed cell surface marker. These antibody combinations can be cross-lineage expression, asynchronous expression, or lack/overexpression of antigens (Heuser et al. 2021a; Grimwade and Freeman 2014; Wood 2016). There are two primary methods for MFC-MRD assessment:

1. LAIP Method: In this approach, LAIPs identified at diagnosis are tracked over time. This method relies on the stability of these aberrant markers, which are critical for monitoring MRD at subsequent treatment stages.
2. Different from Normal (DfN) Method: When diagnostic LAIPs are unavailable, the DfN approach identifies emergent leukemic clones by comparing post-treatment samples to a reference population of normal cells. However, this approach may be less specific, as transient expression on regenerating nonleukemic progenitors can mimic leukemic blasts.

Ideally, the LAIP method is applied as it has shown more reliable results compared to the DfN approach (Hanekamp et al. 2020; Loghavi et al. 2021). Flow cytometry became widely adopted, in the past due to its ability to assess a large number of cells quickly and to provide information on multiple antigens simultaneously. However, the interpretation of flow cytometry results required significant expertise, and the method's sensitivity could vary depending on the quality of the sample and the specific markers used. For reliable MFC-MRD results, bone marrow (BM) samples are preferred, requiring rigorous sample processing to eliminate erythroblasts, doublets, and platelets (Cloos et al. 2018). In addition, settings and gating should adhere to international procedures and settings, such as the EuroFlow procedures or Harmonemia settings (Kalina et al. 2012; Lacombe et al. 2016; Zeijlemaker et al. 2019). The standard for determining MRD-negativity based on the guidelines for rare events in flow cytometry is the acquisition of >500.000 CD45-expressing cells and/or at least 200 viable cells in the blast compartment assessed for the best aberrancy(s) available (Tettero et al. 2022a). Different cut-offs are used to discriminate MRD-positive patients from MRD-negative patients, from anything measured is MRD-positive to the consensus cut-off of 0.1% LAIP of total WBC as recommended by the European LeukemiaNet (ELN)-MRD working party (Heuser et al. 2021a). This approach is applicable for over 90% of AML patients at diagnosis, making it well-suited for disease monitoring and assessing treatment response (Percival et al. 2017). However, its interpretation demands significant expertise, and sensitivity can fluctuate based on the assay's quality and the markers selected. The two clinical time points with the greatest evidence for risk stratification based on AML MRD testing results are following initial induction chemotherapy and before allogeneic transplantation. It is important to ensure that for clinical decision making, MRD assessment should be performed with a qualified assay (Tettero et al. 2023a).

7.2.3 Polymerase Chain Reaction (PCR)

Polymerase chain reaction (PCR) technology revolutionized MRD assessment by enabling the detection of specific genetic abnormalities in AML through the amplification of target DNA sequences. PCR techniques, such as real-time quantitative PCR (RT-qPCR) and digital droplet PCR (ddPCR), offer higher sensitivity than MFC, often detecting MRD at levels as low as 1 in 1,000,000 cells (Grimwade and Freeman 2014). Both peripheral blood (PB) and bone marrow (BM) can be utilized for PCR-based MRD assessment in AML. Although the sensitivity is typically one log lower in PB compared to BM, PB samples benefit from lower background levels, making them a valuable alternative in certain contexts (Ivey et al. 2016). To standardize results, most laboratories performing molecular MRD testing in AML use the ABL1 reference gene for normalization. The use of ABL1 as a reference gene is strongly encouraged to maintain consistency across assays (Scott et al. 2024). The ddPCR technique offers absolute quantification of target genes without requiring standard curves.

PCR is applicable primarily to AML patients harboring specific molecular targets (Hokland and Ommen 2011). However, as AML is a molecular heterogeneous disease, this means that there are many potential targets (Grimwade et al. 2016). In addition, markers are only suitable for molecular MRD analysis if the clonal malignant/driver alterations is demonstrating stability at relapse. These targets include:

- *NPM1* Mutation: The most common mutation in AML, providing a stable and specific MRD marker.

The most substantial evidence supports the use of PCR for detecting the *NPM1* mutation, the most prevalent mutation in AML, which serves as an ideal molecular target due to its specificity to AML and absence in pre-malignant conditions, making it an AML defining target (Ivey et al. 2016; Gorello et al. 2006). RT-PCR for *NPM1* MRD has been shown to be an independent prognostic factor (Ivey et al. 2016; Krönke et al. 2011; Shayegi et al. 2013; Tiong et al. 2021). MRD for *NPM1* is usually performed from RNA/cDNA because of the high expression of these genes and thus better sensitivity (Dillon et al. 2019). Furthermore, ddPCR has demonstrated utility in predicting relapse in *NPM1*-mutant AML patients who are MRD-positive before transplant (Bill et al. 2018).

- Core Binding Factor (CBF) Leukemias: Involving *RUNX1-RUNX1T1* and *CBFB-MYH11* fusion transcripts, both serve as reliable MRD markers.

RT-PCR MRD has been studied for Core Binding Factor (CBF) leukemias involving the fusion genes t(8;21) and inv. (Zeijlemaker et al. 2019) which result in the *RUNX1-RUNX1T1* and *CBFB-MYH11* transcripts, respectively. Both of these CBF fusions are reliable markers for MRD (Yin et al. 2012; Zhu et al. 2013; Jourdan et al. 2013). However, a broad consensus on the specific cutoff for MRD-negativity has not been established, partly due to variability in published studies (Rücker et al. 2019).

- Other Targets: Including *KMT2A* fusions, *DEK-NUP214*, and *WT1* expression, although these are less commonly used in routine practice.

Other targets for qPCR-MRD are less common and not routinely used in clinical practice. For example, MRD monitoring of the translocation t(9;11)/*KMT2A::MLLT3* is associated with long-term remission when MRD-negativity is achieved. Pre-transplant detection of MRD in these cases is strongly predictive of post-transplant relapse and poor survival outcomes (Scholl et al. 2005). Although data on other KMT2A::X fusions support MRD monitoring, there is currently insufficient evidence to influence clinical decisions (Heuser et al. 2021a; Liu et al. 2014). However, the relevance of KMT2A fusions is increasing with the development of novel agents use of menin inhibitors, which have shown potential in inducing both clinical and MRD responses (Issa et al. 2022).

The t(6;9)(p23;q34) translocation, generating the *DEK::NUP214* fusion protein, is a rare target, present in approximately 1% of AML cases (Slovak et al. 2006). This translocation is highly associated with relapse when assessed after initial therapy, reflecting the aggressive nature of this AML subtype (Ommen et al. 2015; Potluri et al. 2024).

Finally, the Wilms' Tumor Gene (*WT1*) expression has been correlated with relapse (Nomdedéu et al. 2013; Candoni et al. 2017a; Candoni et al. 2017b; Lambert et al. 2021). *WT1* has relatively high background expression in normal BM, which makes PB the preferred tissue for assessment (Cilloni et al. 2009).

Leukemia-specific PCR assays, such as those targeting *NPM1* or CBF AML, are preferred over these less specific markers like *KMT2A* or *WT1* expression. Overall, around 60% of AML patients have a validated qPCR target, underscoring the importance of PCR-based MRD assessment in the management of AML (Grimwade et al. 2016).

7.2.4 Next-Generation Sequencing (NGS)

The most recent and transformative advancement in MRD assessment is Next-Generation Sequencing (NGS). This technology offers ultra-deep sequencing capabilities, enabling the identification and quantification of leukemic mutations with remarkable sensitivity. NGS is particularly advantageous for its ability to detect multiple mutations simultaneously, making it an invaluable tool in MRD assessment, especially in cases with complex clonal evolution.

The potential of NGS is immense; it can identify a wide array of patient-specific gene mutations in a single, rapid assay. This technology has the capability to track multiple variants, in at least 40%–50% of AML patients, making it potentially one of the most comprehensive tools available for MRD monitoring (Papaemmanuil et al. 2016). Ideally, genetic markers should be identified at diagnosis and monitored throughout treatment, particularly after the patient achieves complete remission (CR). In situations where diagnostic material is unavailable, an agnostic gene panel may be used, although the optimal configuration of such a panel remains a topic of ongoing debate.

NGS has significantly deepened our understanding of the clonal architecture of AML, revealing subclonal mutations that could drive relapse. However, there are critical considerations when interpreting NGS-MRD results. The choice of the bioinformatics pipeline and its validation process are crucial for accurate analysis. A particular challenge is the persistence of certain mutations, such as those in *DNMT3A*, *TET2*, and *ASXL1*—commonly referred to as DTA mutations—following chemotherapy (Jongen-Lavrencic et al. 2018). These mutations are frequently found in individuals with age-related clonal hematopoiesis and may not represent residual leukemia, which complicates their interpretation. Consequently, DTA mutations should be excluded from MRD analysis pre-transplant to avoid false positives.

Excluding these DTA mutations enhances the clinical value of NGS-MRD, particularly when monitoring mutations like NPM1 and those in signaling pathway genes (e.g., *FLT3*-ITD, *FLT3*-TKD, *KIT*, *KRAS*, *NRAS*, and others), which are more likely to represent residual AML. There is high-quality evidence for the clinical utility of NGS MRD testing, particularly for persistence of *FLT3*-ITD and/or *NPM1* mutation in first remission prior to transplant, and outperforming flow cytometry as currently performed (Dillon et al. 2023). Despite its promise, NGS is still in the process of being refined for routine clinical use, with challenges including standardization, cost, and the complexity of data interpretation (Heuser et al. 2021a).

7.3 Challenges and Limitations of MRD Assessment in AML

7.3.1 Technical Challenges

Accurate MRD assessment in AML requires a high level of technical expertise and stringent procedural adherence. For BM samples, it is critical to collect only a small volume (<5 mL) to avoid hemodilution, a phenomenon where PB contaminates the sample, leading to false-negative results—this is particularly problematic in MFC (Tettero et al. 2024). Once collected, BM samples should be stored at room temperature and ideally analyzed within 72 h to maintain cell viability (Heuser et al. 2021a; Tettero et al. 2023a). When using PB for molecular MRD assessment, especially for targets where BM shows higher background levels (e.g., *WT1* expression), a minimum of 10 mL is typically required, depending on the white blood cell count and assay specifics.

For MFC-MRD, reliable results hinge on the precise processing of samples. Experts have provided guidelines on panel design and technical execution, which have led to consistent results across laboratories (Tettero et al. 2022a). Molecular MRD monitoring via RT-qPCR for fusion genes also involves several critical technical considerations. Valid testing necessitates specific criteria, including the use of negative and positive controls, control gene amplification, and target-specific standard curves (Larionov et al. 2005). Unlike RT-qPCR, digital droplet PCR (ddPCR) offers absolute quantification without the need for a standard curve, providing high precision and sensitivity but remaining target-specific.

NGS-based MRD detection is highly sensitive but requires careful differentiation between true leukemic variants and background errors introduced during library preparation and sequencing. Accurate data interpretation depends on understanding variant-specific false-positive rates and applying rigorous laboratory and bioinformatics approaches to mitigate errors (Ghannam et al. 2020). Although NGS can measure multiple targets simultaneously, challenges include long turnaround times, the need for specialized bioinformatics tools, and the high cost associated with sequencing.

Detection of MRD by NGS is contingent upon the ability to confidently distinguish variants due to true patient leukemic biology from those due to background error generated during library preparation and sequencing. The data should therefore be interpreted in the context of variant-specific false-positive rates, and laboratory and/or bioinformatics approaches to mitigate sources of error (Ghannam et al. 2020). For NGS, multiple targets can be measured simultaneously. The biggest challenges are the long-turnaround times which requires a bioinformatic tools and the cost that is associated with measuring.

Another significant challenge in MRD assessment is the inherent heterogeneity of AML itself. The disease's genetic and cellular diversity can make it difficult to capture the entire leukemic burden, as different subclones may show varying levels of resistance to treatment. Thus, even detected MRD might not represent the full extent of residual disease.

Moreover, the timing of MRD assessments is critical, as MRD levels can fluctuate during treatment (Tettero et al. 2022b; Tsai et al. 2021). A single MRD measurement may not be reliable; longitudinal monitoring may provide a more accurate reflection of disease status.

7.3.2 Standardization and Harmonization Issues

Standardization and harmonization are crucial for all MRD techniques, yet significant challenges remain. Ensuring consistent testing quality and reliable results is essential, particularly when MRD data informs clinical decisions. In MFC, while no single approach is universally superior or used, the European LeukemiaNet (ELN) provides specific recommendations for both general and technical aspects of MFC-MRD (Heuser et al. 2021a; Tettero et al. 2022a). For example, it is advised to report CD45-expressing cells (leukocytes) as the denominator for MRD percentage values (Tettero et al. 2022a). Although round-robin tests have shown comparable results across laboratories, the subjective nature of data interpretation and operator dependency introduces potential biases, making it difficult to achieve full harmonization.

In qPCR-MRD assays for targets like *NPM1* and Core Binding Factor (CBF) fusions, a commonly used quality indicator is the measurement of ≥10,000 ABL1 copies. Despite interlaboratory assessments in Europe, testing and interpretation errors persist, posing a risk of unreliable results with significant clinical consequences (Scott et al. 2023). An international expert advisory board has issued recommendations for molecular MRD testing, but these are only initial steps toward broader standardization (Scott et al. 2024).

For NGS-MRD, DNA is the preferred nucleic acid, but no uniform sequencing method or bioinformatics pipeline currently exists for detecting MRD above background noise (Ghannam et al. 2020). It is recommended to use at least 500 nanograms of DNA, with an optimal target of ≥10,000 unique molecule read families or ≥100,000 reads. The lack of uniformity complicates the comparison of NGS-MRD results across studies, limiting its adoption in routine clinical practice. Another layer of complexity in standardization for molecular MRD arises from the impact of clonal hematopoiesis, particularly in older patients. Mutations related to clonal hematopoiesis, such as DTA mutations, may persist post-treatment and confound MRD assessment, as they do not necessarily indicate residual leukemia. This underscores the need for careful interpretation to avoid overtreatment.

7.3.3 Considerations on MRD

MRD testing is still being refined for its predictive value in relapse, with approximately 30% of MRD-negative patients experiencing relapse and 30% of MRD-positive patients not relapsing (Othus et al. 2019). Several factors contribute to this variability, including AML's molecular heterogeneity, the lack of standardization, incomplete MRD characterization, suboptimal timing of tests, and inappropriate sample utilization (Blachly et al. 2022). Therefore, the absence of detectable MRD does not always equate to complete disease eradication, and MRD measurement alone is not a definitive predictor of relapse.

Lastly, cost and accessibility issues remain significant barriers to the widespread adoption of advanced MRD techniques, especially in resource-limited settings. The high cost of NGS and other advanced methods, coupled with the need for specialized bioinformatics, expert interpretation, and limited availability depending on geographic location, may contribute to disparities in patient outcomes. In the United States, insurance coverage for MRD testing is evolving, but highly variable. Currently, MRD assessment via flow cytometry and qPCR is widely available in the Western world, with ongoing efforts to expand access in developing countries.

7.4 Predicting Relapse and Prognosis

7.4.1 Prognostic Value of MRD Assessment

Over the past two decades, substantial evidence has established the prognostic significance of MRD clearance via MFC after intensive chemotherapy in patients with AML. Numerous studies have consistently demonstrated that MRD detection following induction chemotherapy and prior to allo-SCT is independently linked with an elevated risk of relapse and reduced survival (Zeijlemaker et al. 2019; Terwijn et al. 2013; Freeman et al. 2013, 2018; Inaba et al. 2012; Buccisano et al. 2010; Feller et al. 2004; Maurillo et al. 2007; Köhnke et al. 2015; Chen et al. 2015; Vidriales et al. 2016; Othus et al. 2016; Ravandi et al. 2017; Lacombe et al. 2018; Liu et al. 2021; Bazinet et al. 2023; Caballero-Velázquez et al. 2023; Löwenberg et al. 2021). Consequently, MRD assessment is crucial for refining relapse risk in patients achieving complete remission (CR). Although various cut-offs exist to distinguish MRD-negative from MRD-positive cases, the 0.1% level of LAIP cells within the total WBC is the most commonly used threshold, as endorsed by the ELN-MRD working group (Heuser et al. 2021a). This threshold classifies approximately 20%–25% of patients as MRD-positive. Data collected at other time-points, such as post-transplant have yielded similar prognostic outcomes (Zhou et al. 2016; Shah et al. 2018). For MFC, the optimal approach involves identifying the MRD marker at diagnosis and tracking it over time. If diagnostic material is unavailable, a 'different from normal' (DfN) approach may be employed.

Molecular MRD assessment, particularly through reverse transcription RT-qPCR for mutant *NPM1*, serves as a significant prognostic marker before allo-SCT (Kayser et al. 2016). MRD-positive patients exhibit a higher relapse risk and poorer survival outcomes compared to MRD-negative counterparts. Additionally, ddPCR has been evaluated in *NPM1*-mutated AML, where pre-transplant MRD positivity consistently predicts relapse (Bill et al. 2018). In such cases, MRD should be assessed primarily in PB after two chemotherapy cycles, in BM at the end of consolidation, and every 3 months in BM for 2 years following consolidation therapy. The *NPM1* copy number at the end of treatment is strongly associated with relapse risk (Tiong et al. 2021). Importantly, end-of-treatment samples that are MRD-positive for *NPM1* at low, stable levels below 2%—referred to as 'molecular persistence at low copy number' (MP-LCN)—are linked to a very low relapse risk and may not require treatment modifications (Heuser et al. 2021a; Scott et al. 2023). The defining feature of MP-LCN is the stability over time, with minimal fluctuations or increases, suggesting the disease may not be actively progressing despite detectable MRD. While ongoing research continues to explore the underlying mechanisms, including the possibility of false positives from other cell types, a definitive explanation for this phenomenon remains unclear (Cook et al. 2024).

For CBF-AML (e.g., *RUNX1-RUNX1T1* or *CBFB-MYH11*), MRD monitoring via qPCR post-induction therapy correlates with an increased relapse risk and worse survival (Yin et al. 2012; Zhu et al. 2013; Jourdan et al. 2013). Failing to achieve a 3- to 4-log reduction in MRD in the same tissue following consolidation therapy indicates a higher relapse risk and poorer prognosis. Additionally, a rise in MRD levels by ≥1 log between successive samples is prognostic for relapse (Schuurhuis et al. 2018; Puckrin et al. 2021). CBF-MRD should be monitored in PB after two chemotherapy cycles, in BM at the end of consolidation, and every 4 to 6 weeks in PB for up to 24 months post-consolidation, as the majority of relapses occur within this period (Rücker et al. 2019; Puckrin et al. 2021). Similar to *NPM1* MRD, end-of-treatment MP-LCN can occur in CBF-MRD, where an increase of <1-log is not associated with an elevated risk of relapse (Heuser et al. 2021a; Cook et al. 2024).

For AML cases outside the *NPM1* and CBF subgroups, MFC remains the recommended method for MRD monitoring, given the limited prognostic data for other targets. Emerging evidence supports the prognostic value of *FLT3*-ITD as an MRD marker, with several studies confirming its significance (Ivey et al. 2016; Shayegi et al. 2013; Chou et al. 2007; Schnittger et al. 2009; Krönke et al. 2011; Hubmann et al. 2014; Jain et al. 2014; Grob et al. 2023).

WT1 expression is another independent prognostic marker in allo-SCT candidates (Nomdedéu et al. 2013; Candoni et al. 2017a, b; Lambert et al. 2021). Detectable *WT1* post-induction is associated with shorter survival and a higher relapse risk, though it is less specific than other molecular markers and thus less preferred compared to other molecular markers.

NGS-based MRD assessment has shown prognostic relevance for certain mutations identified at diagnosis or in CR samples (Kim et al. 2018; Hourigan et al. 2020). Similar to MFC-MRD, NGS-MRD is most effective when a diagnostic sample is available to track mutations over time. In cases where a diagnostic sample is not available, an agnostic NGS gene panel may be utilized. While NGS-MRD adds prognostic value beyond MFC, current data are insufficient to support its routine use as a stand-alone technique (Jongen-Lavrencic et al. 2018). *FLT3*-ITD is a notable exception that has been validated as a NGS-MRD target. *FLT3*-ITD positive patients demonstrating particularly poor outcomes; if MRD is detectable pre-transplant, these patients face high relapse and mortality rates post-transplant (Dillon et al. 2023; Grob et al. 2023; Hourigan et al. 2020). Notably, patients with peri-transplant detectable *FLT3*-ITD MRD benefit from post-transplant gilteritinib therapy (Levis et al. 2024). Conversely, NGS-MRD measurement of TP53 lacks prognostic value, as both MRD-positive and MRD-negative groups exhibit similarly poor overall survival (Grob et al. 2022).

While some studies have suggested that combining multiple MRD assessment techniques, such as MFC with RT-qPCR or NGS, may enhance prognostic accuracy (Jongen-Lavrencic et al. 2018; Morita et al. 2018), other studies combining NGS with MFC have also shown no added value of MFC (Dillon et al. 2023; Grob et al. 2023). Discordances between MRD methods are common, and the underlying causes are not yet fully understood. The prevailing recommendation is to classify a patient as MRD-positive if any method detects MRD (Heuser et al. 2021a). Ongoing developments in biomarkers and MRD detection methodologies may further refine prognostic models, enabling more personalized treatment strategies.

7.4.2 Monitoring Response to Therapy with MRD

MRD status provides a continuous measure for risk assessment throughout the AML treatment continuum. As early as after one chemotherapy cycle, MRD can be evaluated to gauge treatment response, using both MFC and qPCR for *NPM1* (Tettero et al. 2022b; Freeman et al. 2018). Patients who convert from MRD-positive to MRD-negative between the first and second induction cycles generally experience improved prognosis, whereas those who revert to MRD-positive status face increased relapse risk and diminished overall survival, potentially necessitating immediate therapeutic adjustments. Sequential MRD assessments offer critical insights into the disease's dynamic response to treatment, affording a more comprehensive understanding of therapeutic efficacy. For instance, decreasing MRD levels over time typically indicate effective treatment, while stable or rising MRD levels may signal suboptimal therapy, prompting the need for intervention.

7.5 MRD in Treatment Decision-Making

7.5.1 Post-induction

In AML patients who are fit for intensive chemotherapy, allo-SCT in CR1 is typically recommended for those with adverse risk cytogenetics, as for instance classified by the ELN risk categories (Döhner et al. 2022). This is due to the high likelihood that chemotherapy alone will not be sufficient to overcome the intrinsic resistance of the leukemic clone. Conversely, patients with favorable risk AML often achieve long-term remission with chemotherapy alone, allowing them to avoid allo-SCT and the associated transplant-related mortality. However, for patients classified as intermediate risk, the optimal post-remission therapy remains less clear.

The emergence of MRD as a robust prognostic marker has prompted investigations into its potential role in guiding clinical decision-making, particularly in cases where there the choice of consolidation therapy is uncertain. Following induction therapy, MRD status is assessed to evaluate the depth of response. An MRD-negative status at this stage generally indicates a better prognosis, often reducing the need for an allo-SCT. On the other hand, patients who remain MRD-positive after induction therapy face a higher risk of relapse, which may prompt clinicians to recommend more intensive treatment, including allo-SCT, to increase the chances of eradicating the disease.

The utility of MRD in guiding treatment decisions for intermediate-risk AML patients has been prospectively evaluated using MFC-MRD in two large cohorts: the GIMEMA-1310 and HOVON-132 studies (Löwenberg et al. 2021; Venditti et al. 2019). These studies demonstrated that MRD-positive patients who underwent allo-SCT had comparable relapse rates and overall survival outcomes to MRD-negative patients who received chemotherapy-based consolidation therapy. This suggests that allo-SCT could be safely omitted in MRD-negative patients. However, it is important to note that these studies were not randomized. In addition, the consolidation chemotherapy was occasionally followed by autologous stem cell transplantation (auto-SCT), which is considered a valid therapeutic option for low-risk AML patients in CR1 in regions like Japan and Europe (Yanada et al. 2020). Retrospective analyses further revealed that patients who relapsed after chemotherapy consolidation could often achieve a second complete remission (CR2) and subsequently undergo allo-SCT without a significant impact on surivival (Tettero et al. 2023b). In contrast, for patients with adverse risk AML, both MRD-negative and MRD-positive statuses were associated with better outcomes when allo-SCT was performed, indicating that MFC-MRD might not be sufficient to guide consolidation therapy decisions in this subgroup, as all eligible patients should receive allo-SCT.

As *NPM1* is an AML defining marker, patients with an *NPM1* positive AML at diagnosis can be monitored through qPCR-based MRD assessment. Patients with *NPM1*-positive AML at diagnosis who remain MRD-positive after induction chemotherapy appear to particularly benefit from allo-SCT compared to those with a greater than 4-log reduction in NPM1 transcript levels (Balsat et al. 2017; Othman et al. 2024). MRD-negative patients with *NPM1* mutation do not seem to derive a significant overall survival benefit from allo-SCT, though their relapse risk may still be lower. The presence of an *FLT3*-ITD co-mutation alongside *NPM1* at diagnosis typically predicts a worse outcome, and MRD status can be instrumental in guiding consolidation therapy in these cases (Othman et al. 2024).

For Core-Binding Factor AML, MRD can be measured for prognostication purposes, but it is generally not recommended for guiding transplant decisions, as CBF-AML is considered highly responsive to chemotherapy alone. However, MRD monitoring can be done in AML with the t(9;11)/KMT2A::MLLT3 translocation, as these patients fall under the ELN-intermediate risk category. In t(8;21) AML, MRD-directed post-remission therapy has been implemented, with MRD-positive patients allocated to allo-SCT and MRD-negative patients to chemotherapy/auto-SCT (Zhu et al. 2013). Despite the limited size of treatment arms in studies, evidence suggests that patients with a positive MRD result derive significant benefit from allo-SCT, supporting the hypothesis that allo-SCT should be reserved for those with a higher risk of relapse. This approach underscores the potential role of MRD as a critical factor in tailoring post-remission therapies to optimize patient outcomes.

Although additional markers are used as MRD-target (e.g., *KIT, NRAS,* and *KRAS*), these mutations have not been assessed as stand-alone MRD marker. However, data on NGS-panels suggest that when these markers are detected post-induction, they are associated with an increased risk of relapse (Heuser et al. 2021a; Patkar et al. 2021).

Although the prognostic value of MRD assessed by NGS for markers such as *NPM1* and *FLT3*-ITD has been established, its application in clinical decision-making is currently limited by the need for further standardization and validation.

7.5.2 Pre-allo-SCT and Conditioning

Detectable MRD, whether assessed by MFC, qPCR, or NGS, before allo-SCT is an independent predictor of poorer outcomes, including higher relapse rates and reduced overall survival (Short et al. 2020; Hourigan et al. 2020; Othman et al. 2024; Walter et al. 2015; Paras et al. 2022; Craddock et al. 2021; Loo et al. 2022). For patients who are MRD-positive before allo-SCT, the dilemma arises whether to proceed with the transplant or administer salvage therapy first. Currently, there is no evidence supporting the benefit of additional intensive chemotherapy in MRD-positive CR1 patients before transplantation. Retrospective analyses of MFC-MRD and *NPM1* MRD indicate that allo-SCT can convert MRD-positive status to MRD-negative, which is associated with improved outcomes (Paras et al. 2022). Specifically, for *NPM1* MRD, transitioning from MRD-positive post-induction to MRD-negative post-transplant appears to increase survival probabilities (Othman et al. 2024). Despite advancements in MRD measurement techniques, randomized studies have yet to demonstrate that specific therapeutic interventions in response to MRD positivity during the post-induction period significantly impact long-term outcomes.

The choice of conditioning regimen, whether reduced-intensity conditioning (RIC) or myeloablative conditioning (MAC), can also be informed by MRD status, particularly with *FLT3*-ITD/*NPM1* qPCR-MRD or MFC-MRD. Studies have shown a significant difference in relapse and survival outcomes for patients with persistent *FLT3*-ITD or *NPM1* MRD when receiving RIC compared to MAC. (Dillon et al. 2023; Grob et al. 2023; Othman et al. 2024) Therefore, for MRD-positive patients who are fit enough, a MAC regimen should be considered, as they seem to derive significant benefit from this more intensive approach. Conversely, in MRD-negative patients, increased conditioning intensity does not appear to confer added benefit.

7.5.3 Post-transplant Monitoring and Pre-emptive Treatment

Post-transplant MRD positivity, as detected by various modalities, is also associated with an increased risk of relapse and poorer overall survival (Kim et al. 2018; Liu et al. 2019; Nagler et al. 2023; Loke et al. 2023; Heuser et al. 2021b). However, the timing of post-transplant MRD assessments is less standardized, and the evidence is less robust compared to pre-transplant scenarios. Nevertheless, MRD monitoring during remission can provide an early warning of impending relapse, allowing for pre-emptive therapeutic interventions that may prevent full hematologic relapse. This approach can significantly improve the likelihood of regaining remission and extending survival.

For example, in a retrospective analysis using NGS-MRD, at least one MRD marker became detectable in 10%, 38%, and 64% of patients at 6, 3, and 1 month before relapse, respectively (Wienecke et al. 2024). Regular MRD monitoring post-transplant can thus detect early signs of relapse before clinical symptoms manifest, enabling timely interventions. Donor lymphocyte infusions (DLI) administration in case of MRD-positivity or mixed chimerism post-transplant can be considered as it is able to prevent relapse (Dominietto et al. 2007; Tan et al. 2014). However, DLI administration carries an increased risk of graft-versus-host disease (GVHD).

Preemptive treatment with azacitidine monotherapy has also shown efficacy in delaying hematological relapse in MRD-positive patients, as assessed by qPCR for NPM1, leukemic-specific fusion genes, or MFC (Platzbecker et al. 2018). Besides MRD-positive patients, other high-risk AML patients might benefit from maintenance therapy with azacitidine (Oran et al. 2020). When relapse does occur, achieving MRD negativity before a second complete remission (CR2) is associated with a reduced risk of relapse post-transplant (Gilleece et al. 2021).

7.5.4 Maintenance Therapy

In addition to pre-emptive treatments, ongoing studies are exploring maintenance therapies post-transplant with the goal of preventing relapse in AML patients, especially in those patients at risk. Promising results have been published for sorafenib for *FLT3*-ITD mutated patients (Xuan et al. 2023). Many trials are still ongoing and the optimal duration of post-allo-HCT maintenance therapy remains unclear, with most studies arbitrarily selecting a 1 to 2-year duration. Gilteritinib has also been studied as a maintenance therapy for *FLT3*-ITD mutated patients, showing effectiveness in increasing relapse-free survival for peri-transplant MRD-positive AML patients (Levis et al. 2024). This is the first randomized study to demonstrate that additional therapy targeting MRD-positivity after an ablative allo transplant can improve outcome. Conversely, MRD-negative patients did not derive additional benefit from maintenance therapy, suggesting that MRD-negativity post-transplant could justify to refrain from maintenance therapy. MRD assessment during maintenance therapy could potentially guide the duration or intensity of treatment, with persistent MRD positivity indicating a need for prolonged or intensified therapy to eradicate residual disease. However, current evidence is insufficient to support this approach universally, and the heterogeneity in access to mutational testing, MRD evaluation, and specific agents complicates the establishment of a standardized approach.

7.6 Case Studies and Clinical Examples

7.6.1 Case A: MRD-Guided Therapy Optimization

Patient Background and Diagnosis A 60-year-old male presented with symptoms of extreme fatigue and dyspnea, leading to a diagnosis of pancytopenia upon blood examination. Subsequent bone marrow (BM) biopsy confirmed the presence of acute myeloid leukemia (AML) with 23% blasts. Flow cytometry analysis revealed a leukemia-associated immunophenotype (LAIP) positive for CD34, CD13, and CD7. Molecular testing identified both *FLT3*-ITD and *NPM1* mutations, classifying the patient as having intermediate-risk AML according to the European LeukemiaNet (ELN) classification. Given his fitness for intensive chemotherapy, the patient was initiated on a standard '7 + 3' induction regimen, combining cytarabine (200 mg/m^2 on days 1–7) and daunorubicin (60 mg/m^2 on days 1–3), with the addition of midostaurin (50 mg twice daily on days 8–21) targeting the *FLT3*-ITD mutation.

MRD Assessment and Clinical Decision-Making After two cycles of induction therapy, a follow-up BM aspiration confirmed complete remission (CR1). The BM sample was then analyzed for measurable residual disease (MRD) using both multiparametric flow cytometry (MFC) and real-time quantitative PCR (RT-qPCR) for *NPM1*. The results showed a discordance: MRD-negative by MFC but MRD-positive for *NPM1*. In cases of contradictory MRD results, clinical guidelines generally recommend considering the patient MRD-positive, especially when the molecular marker persists. Given the high risk of relapse associated with persistent *NPM1* positivity, an allogeneic stem cell transplant (allo-SCT) was recommended as the optimal therapeutic strategy to minimize relapse risk and improve long-term survival.

Outcome and Implications The decision to proceed with allo-SCT was based on the integration of MRD data, which indicated a residual disease burden despite morphological remission. This case underscores the importance of using MRD as a key prognostic tool in guiding post-remission therapy in AML, particularly when molecular markers like *NPM1* remain detectable. It highlights the need for a personalized treatment approach, leveraging MRD assessment to refine relapse risk and tailor therapy accordingly.

7.6.2 Case B: Therapy Regardless of MRD-Status

Patient Background and Diagnosis A 58-year-old female was diagnosed with AML characterized by the presence of the t(8;21) fusion gene (*RUNX1-RUNX1T1*) and an *IDH2* mutation, confirmed by fluorescence in situ hybridization (FISH). Flow cytometry failed to identify any aberrant immunophenotypic markers, leaving the patient without a LAIP target for MFC-MRD assessment. However, the t(8;21) fusion provided a molecular MRD target via RT-qPCR. The patient was classified as having core-binding factor (CBF)-AML, a subtype generally associated with a favorable prognosis.

MRD Monitoring and Clinical Strategy Following intensive induction chemotherapy, the patient achieved complete remission (CR), confirmed by BM biopsy. Despite the availability of molecular MRD monitoring for t(8;21), the clinical decision was made to forego allogeneic stem cell transplantation (allo-SCT). This decision was informed by the favorable prognosis typically associated with CBF-AML and the high likelihood of a positive response to chemotherapy alone, independent of MRD status. The absence of a LAIP for MFC-MRD further supported the decision to rely primarily on the patient's cytogenetic and molecular profile rather than MRD results for post-remission management.

Outcome and Implications The patient continued to be monitored with routine molecular assessments for t(8;21), but no further intervention was deemed necessary given the favorable cytogenetic profile. This case illustrates that while MRD is a powerful prognostic tool in AML, certain subtypes like CBF-AML may achieve excellent outcomes with standard chemotherapy alone, allowing for a less aggressive post-remission strategy. It emphasizes the importance of considering the broader clinical context, including cytogenetics and the specific AML subtype, when deciding on the role of MRD in treatment planning.

7.7 Conclusion

7.7.1 Summary of the Impact of MRD on AML Outcomes

Traditional methods for assessing treatment response in AML, primarily based on morphological evaluation, have significant limitations in sensitivity, often failing to detect residual leukemic cells that can lead to relapse. The advent of measurable residual disease (MRD) assessment has revolutionized this landscape, allowing for a better appreciation of relapse risk. By integrating technologies such as MFC, qPCR, and NGS, clinicians can now detect MRD at much lower thresholds, significantly enhancing the ability to predict patient outcomes and tailor treatment strategies.

MFC remains a widely used tool for monitoring AML due to its broad applicability, but it has lower sensitivity compared to molecular methods like qPCR and NGS. PCR techniques, including RT-qPCR and ddPCR, offer higher sensitivity for detecting specific genetic mutations such as *NPM1*, *RUNX1-RUNX1T1*, and *CBFB-MYH11*, but are only validated for selected targets. NGS, while currently primarily research-focused, provides ultra-deep sequencing capabilities that allow for the detection of leukemic mutations with remarkable sensitivity and specificity. However, challenges such as technical complexity, standardization issues, and variability in prognostic value across different mutations complicate the interpretation of MRD results. For instance, mutations in genes like *DNMT3A*, *TET2*, and *ASXL1* may not necessarily indicate residual leukemia and should be interpreted cautiously.

MRD assessment has been shown to be a robust predictor of relapse risk and overall survival. Achieving MRD-negativity after induction chemotherapy is strongly associated with decreased relapse risk and better overall survival in AML patients, allowing for the integration into clinical practice for relapse risk stratification. In addition, incorporating MRD detection into AML management has become essential, influencing clinical decisions such as whether to proceed with additional chemotherapy or allo-SCT. MRD-positive patients may require more aggressive or alternative therapies to mitigate relapse risk, while MRD-negative patients may avoid unnecessary overtreatment. Such approach has for instance been seen with personalized conditioning regimens based on MRD levels. Moreover, post-transplant maintenance therapy is often deployed in patients who are MRD-negative in complete remission, while pre-emptive treatment is initiated upon the detection of MRD, both aimed at preventing relapse. Importantly, even as MRD measurement approaches in AML are refined, no randomized studies have shown that specific therapeutic interventions deployed in response to MRD positivity in the post-induction period impact long-term outcomes. Additionally, MRD measurement methods are not infallible; some MRD-negative patients still relapse, and some MRD-positive patients do not, underscoring the need for ongoing refinement and standardization of MRD assessment techniques. Nevertheless, MRD remains a crucial component in the modern management of AML, particularly in determining the necessity for additional therapies and guiding post-transplant maintenance strategies.

In summary, the evolution of MRD assessment from MFC to advanced molecular techniques such as NGS reflects the broader trend towards precision medicine in oncology. The ultimate goal remains to enhance patient outcomes through early detection and targeted intervention, highlighting the transformative role of MRD in the management of AML.

7.7.2 Future Prospects and Recommendations for Clinical Practice

Continuous optimization of MRD detection techniques is essential for enhancing their clinical utility. As novel targeted therapies, including midostaurin, gilteritinib, and quizartinib for *FLT3*-ITD, become integrated into treatment regimens, patients may achieve deeper molecular remissions, which could alter the prognostic significance of MRD. For example, patients who achieve MRD-negativity through targeted therapies may have a better long-term prognosis compared to those achieving MRD-negativity with standard chemotherapy alone. However, MRD-negativity during targeted therapy could negatively impact MRD sensitivity, particularly in cases where MRD is a direct target these drugs. Ongoing assessment of how these therapies influence MRD is crucial for optimizing treatment strategies.

This approach could improve disease control and reduce the risk of relapse. Additionally, as more data becomes available on post-transplant MRD monitoring and preemptive treatments, MRD may gain a more significant role in managing AML. The incorporation of MRD as an endpoint in clinical trials could accelerate drug development and refine personalized treatment strategies. However, challenges remain, such as standardizing MRD measurement techniques, validating MRD as a reliable surrogate marker linked to long-term clinical outcomes, cost, and ensuring consistent prognostic value across different treatments and patient subgroups. Careful consideration of MRD's limitations is essential to ensure it reflects meaningful clinical benefits.

Future advancements in MRD assessment may include emerging technologies such as single-cell sequencing and machine learning-based analysis, which could complement or even surpass current methods. Well-designed studies

Table 7.1 Value of different Measurable Residual Disease (MRD) techniques at different time-points

<table>
<tr><th></th><th>Post-induction</th><th>Pre-allo</th><th>Post-transplant</th></tr>
<tr><td>MFC</td><td>MRD holds prognostic value for all patients and may guide consolidation therapy in patients with intermediate risk</td><td>When detected, MRD serves as a prognostic predictor of poorer outcomes after transplant, with no current evidence supporting additional therapy before transplantation</td><td rowspan="2">Prognostic predictor of increased risk of relapse. Pre-emptive treatment with azacitidine monotherapy or DLI might be considered when MRD-positivity is found</td></tr>
<tr><td>NPM1 (qPCR/dPCR/ NGS)</td><td>MRD holds prognostic value for all patients diagnosed with NPM1 mutated AML and help guide consolidation therapy</td><td>When detected, MRD predicts poorer outcomes after transplant, and can guide the choice of conditioning therapy based on MRD status</td></tr>
<tr><td>FLT3-ITD (NGS)</td><td>Prognostic value for patients who are FLT3-ITD positive at diagnosis</td><td colspan="2">Patients with peri-transplant detectable FLT3-ITD MRD benefit from post-transplant gilteritinib therapy</td></tr>
<tr><td>CBF-MRD (qPCR)</td><td>Prognostic value for patients diagnosed with CBF-mutations at diagnosis</td><td>When detected, MRD serves as a prognostic predictor of poorer outcomes after transplant, with no current evidence supporting additional therapy before transplantation. In patients with detectable MRD after chemotherapy can be upstaged to be transplant eligible</td><td>Prognostic predictor of increased risk of relapse. Pre-emptive treatment with azacitidine monotherapy or DLI might be considered when MRD-positivity is found</td></tr>
</table>

focused on MRD-driven clinical decision-making are needed to transition MRD from a prognostic to a predictive marker for individual patients. Until then, MRD will continue to be a valuable tool in specific scenarios, such as selecting candidates for allo-SCT and guiding post-transplant therapy. Decisions should always be made within the broader clinical context of each patient (Table 7.1).

References

Balsat M, Renneville A, Thomas X, Botton S, Caillot D, Marceau A et al (2017) Postinduction minimal residual disease predicts outcome and benefit from allogeneic stem cell transplantation in acute myeloid leukemia with NPM1 mutation: a study by the Acute Leukemia French Association Group. J Clin Oncol 35(2):185–193

Bazinet A, Kadia TM, Short NJ, Borthakur G, Wang SA, Wang W et al (2023) Undetectable measurable residual disease is associated with improved outcomes in AML irrespective of treatment intensity. Blood Adv 7:3284

Bill M, Grimm J, Jentzsch M, Kloss L, Goldmann K, Schulz J et al (2018) Digital droplet PCR-based absolute quantification of pre-transplant NPM1 mutation burden predicts relapse in acute myeloid leukemia patients. Ann Hematol 97(10):1757–1765

Blachly JS, Walter RB, Hourigan CS (2022) The present and future of measurable residual disease testing in acute myeloid leukemia. Haematologica 107(12):2810–2822

Buccisano F, Maurillo L, Spagnoli A, Del Principe MI, Fraboni D, Panetta P et al (2010) Cytogenetic and molecular diagnostic characterization combined to postconsolidation minimal residual disease assessment by flow cytometry improves risk stratification in adult acute myeloid leukemia. Blood 116(13):2295–2303

Caballero-Velázquez T, Pérez-López O, Yeguas Bermejo A, Rodríguez Arbolí E, Colado Varela E, Sempere Talens A et al (2023) Prognostic value of measurable residual disease in patients with AML undergoing HSCT: a multicenter study. Cancers (Basel) 15(5)

Candoni A, De Marchi F, Zanini F, Zannier ME, Simeone E, Toffoletti E et al (2017a) Predictive value of pretransplantation molecular minimal residual disease assessment by WT1 gene expression in FLT3-positive acute myeloid leukemia. Exp Hematol 49:25–33

Candoni A, De Marchi F, Zannier ME, Lazzarotto D, Filì C, Dubbini MV et al (2017b) High prognostic value of pre-allogeneic stem cell transplantation minimal residual disease detection by WT1 gene expression in AML transplanted in cytologic complete remission. Leuk Res 63:22–27

Chen X, Xie H, Wood BL, Walter RB, Pagel JM, Becker PS et al (2015) Relation of clinical response and minimal residual disease and their prognostic impact on outcome in acute myeloid leukemia. J Clin Oncol 33(11):1258–1264

Chou WC, Tang JL, Wu SJ, Tsay W, Yao M, Huang SY et al (2007) Clinical implications of minimal residual disease monitoring by quantitative polymerase chain reaction in acute myeloid leukemia patients bearing nucleophosmin (NPM1) mutations. Leukemia 21(5):998–1004

Cilloni D, Renneville A, Hermitte F, Hills RK, Daly S, Jovanovic JV et al (2009) Real-time quantitative polymerase chain reaction detection of minimal residual disease by standardized WT1 assay to enhance risk stratification in acute myeloid leukemia: a European LeukemiaNet study. J Clin Oncol 27(31):5195–5201

Cloos J, Harris JR, Janssen JJWM, Kelder A, Huang F, Sijm G et al (2018) Comprehensive protocol to sample and process bone marrow for measuring measurable residual disease and leukemic stem cells in acute myeloid leukemia. JoVE 133:56386

Cook JA, Lott L, Perry J, Eckel AM, Xu D, Hudson CA et al (2024) Fusion-harboring mast cells can explain molecular positivity in flow cytometric MRD-negative core-binding factor AML. Blood 144(5):581–585

Cornelissen JJ, Blaise D (2016) Hematopoietic stem cell transplantation for patients with AML in first complete remission. Blood 127(1):62–70

Craddock C, Jackson A, Loke J, Siddique S, Hodgkinson A, Mason J et al (2021) Augmented reduced-intensity regimen does not improve postallogeneic transplant outcomes in acute myeloid leukemia. J Clin Oncol 39(7):768–778

Dillon LW, Hayati S, Roloff GW, Tunc I, Pirooznia M, Mitrofanova A et al (2019) Targeted RNA-sequencing for the quantification of measurable residual disease in acute myeloid leukemia. Haematologica 104(2):297–304

Dillon LW, Gui G, Page KM, Ravindra N, Wong ZC, Andrew G et al (2023) DNA sequencing to detect residual disease in adults with acute myeloid leukemia prior to hematopoietic cell transplant. JAMA 329(9):745–755

Döhner H, Wei AH, Appelbaum FR, Craddock C, DiNardo CD, Dombret H et al (2022) Diagnosis and management of AML in adults: 2022 recommendations from an international expert panel on behalf of the ELN. Blood 140(12):1345–1377

Dominietto A, Pozzi S, Miglino M, Albarracin F, Piaggio G, Bertolotti F et al (2007) Donor lymphocyte infusions for the treatment of minimal residual disease in acute leukemia. Blood 109(11):5063–5064

Epstein-Peterson ZD, Devlin SM, Stein EM, Estey E, Tallman MS (2018) Widespread use of measurable residual disease in acute myeloid leukemia practice. Leuk Res 67:92–98

Feller N, van der Pol MA, van Stijn A, Weijers GWD, Westra AH, Evertse BW et al (2004) MRD parameters using immunophenotypic detection methods are highly reliable in predicting survival in acute myeloid leukaemia. Leukemia 18(8):1380–1390

Feller N, van der Velden VHJ, Brooimans RA, Boeckx N, Preijers F, Kelder A et al (2013) Defining consensus leukemia-associated immunophenotypes for detection of minimal residual disease in acute myeloid leukemia in a multicenter setting. Blood Cancer J 3(8):e129–e129

Freeman SD, Hourigan CS (2019) MRD evaluation of AML in clinical practice: are we there yet? Hematology Am Soc Hematol Educ Program 2019(1):557–569

Freeman SD, Virgo P, Couzens S, Grimwade D, Russell N, Hills RK et al (2013) Prognostic relevance of treatment response measured by flow cytometric residual disease detection in older patients with acute myeloid leukemia. J Clin Oncol 31(32):4123–4131

Freeman SD, Hills RK, Virgo P, Khan N, Couzens S, Dillon R et al (2018) Measurable residual disease at induction redefines partial response in acute myeloid leukemia and stratifies outcomes in patients at standard risk without NPM1 mutations. J Clin Oncol 36(15):1486–1497

Ghannam J, Dillon LW, Hourigan CS (2020) Next-generation sequencing for measurable residual disease detection in acute myeloid leukaemia. Br J Haematol 188(1):77–85

Gilleece MH, Shimoni A, Labopin M, Robinson S, Beelen D, Socié G et al (2021) Measurable residual disease status and outcome of transplant in acute myeloid leukemia in second complete remission: a study by the acute leukemia working party of the EBMT. Blood Cancer J 11(5):88

Gorello P, Cazzaniga G, Alberti F, Dell'Oro MG, Gottardi E, Specchia G et al (2006) Quantitative assessment of minimal residual disease in acute myeloid leukemia carrying nucleophosmin (NPM1) gene mutations. Leukemia 20(6):1103–1108

Grimwade D, Freeman SD (2014) Defining minimal residual disease in acute myeloid leukemia: which platforms are ready for "prime time"? Blood 124(23):3345–3355

Grimwade D, Ivey A, Huntly BJ (2016) Molecular landscape of acute myeloid leukemia in younger adults and its clinical relevance. Blood 127(1):29–41

Grob T, Al Hinai ASA, Sanders MA, Kavelaars FG, Rijken M, Gradowska PL et al (2022) Molecular characterization of mutant TP53 acute myeloid leukemia and high-risk myelodysplastic syndrome. Blood 139(15):2347–2354

Grob T, Sanders MA, Vonk CM, Kavelaars FG, Rijken M, Hanekamp DW et al (2023) Prognostic value of FLT3-internal tandem duplication residual disease in acute myeloid leukemia. J Clin Oncol 41(4):756–765

Hanekamp D, Bachas C, van de Loosdrecht A, Ossenkoppele G, Cloos J (2020) Re: Myeloblasts in normal bone marrows expressing leukaemia-associated immunophenotypes. Pathology 52(2):289–291

Heuser M, Freeman SD, Ossenkoppele GJ, Buccisano F, Hourigan CS, Ngai LL et al (2021a) 2021 Update on MRD in acute myeloid leukemia: a consensus document from the European LeukemiaNet MRD Working Party. Blood 138(26):2753–2767

Heuser M, Heida B, Büttner K, Wienecke CP, Teich K, Funke C et al (2021b) Posttransplantation MRD monitoring in patients with AML by next-generation sequencing using DTA and non-DTA mutations. Blood Adv 5(9):2294–2304

Hokland P, Ommen HB (2011) Towards individualized follow-up in adult acute myeloid leukemia in remission. Blood 117(9):2577–2584

Hourigan CS, Karp JE (2013) Minimal residual disease in acute myeloid leukaemia. Nat Rev Clin Oncol 10(8):460–471

Hourigan CS, Dillon LW, Gui G, Logan BR, Fei M, Ghannam J et al (2020) Impact of conditioning intensity of allogeneic transplantation for acute myeloid leukemia with genomic evidence of residual disease. J Clin Oncol 38(12):1273–1283

Hubmann M, Köhnke T, Hoster E, Schneider S, Dufour A, Zellmeier E et al (2014) Molecular response assessment by quantitative real-time polymerase chain reaction after induction therapy in NPM1-mutated patients identifies those at high risk of relapse. Haematologica 99(8):1317–1325

Inaba H, Coustan-Smith E, Cao X, Pounds SB, Shurtleff SA, Wang KY et al (2012) Comparative analysis of different approaches to measure treatment response in acute myeloid leukemia. J Clin Oncol 30(29):3625–3632

Issa GC, Aldoss I, DiPersio JF, Cuglievan B, Stone RM, Arellano ML et al (2022) The menin inhibitor SNDX-5613 (revumenib) leads to durable responses in patients (Pts) with KMT2A-rearranged or NPM1 mutant AML: updated results of a phase (Ph) 1 study. Blood 140(Supplement 1):150–152

Ivey A, Hills RK, Simpson MA, Jovanovic JV, Gilkes A, Grech A et al (2016) Assessment of minimal residual disease in standard-risk AML. N Engl J Med 374(5):422–433

Jain P, Kantarjian H, Patel K, Faderl S, Garcia-Manero G, Benjamini O et al (2014) Mutated NPM1 in patients with acute myeloid leukemia in remission and relapse. Leuk Lymphoma 55(6):1337–1344

Jongen-Lavrencic M, Grob T, Hanekamp D, Kavelaars FG, Aa H, Zeilemaker A et al (2018) Molecular minimal residual disease in acute myeloid leukemia. N Engl J Med 378(13):1189–1199

Jourdan E, Boissel N, Chevret S, Delabesse E, Renneville A, Cornillet P et al (2013) Prospective evaluation of gene mutations and minimal residual disease in patients with core binding factor acute myeloid leukemia. Blood 121(12):2213–2223

Kalina T, Flores-Montero J, van der Velden VHJ, Martin-Ayuso M, Böttcher S, Ritgen M et al (2012) EuroFlow standardization of flow cytometer instrument settings and immunophenotyping protocols. Leukemia 26(9):1986–2010

Kayser S, Benner A, Thiede C, Martens U, Huber J, Stadtherr P et al (2016) Pretransplant NPM1 MRD levels predict outcome after allogeneic hematopoietic stem cell transplantation in patients with acute myeloid leukemia. Blood Cancer J 6(7):e449–e449

Kim T, Moon JH, Ahn J-S, Kim Y-K, Lee S-S, Ahn S-Y et al (2018) Next-generation sequencing–based posttransplant monitoring of acute myeloid leukemia identifies patients at high risk of relapse. Blood 132(15):1604–1613

Köhnke T, Sauter D, Ringel K, Hoster E, Laubender RP, Hubmann M et al (2015) Early assessment of minimal residual disease in AML by flow cytometry during aplasia identifies patients at increased risk of relapse. Leukemia 29(2):377–386

Koreth J, Schlenk R, Kopecky KJ, Honda S, Sierra J, Djulbegovic BJ et al (2009) Allogeneic stem cell transplantation for acute myeloid leukemia in first complete remission: systematic review and meta-analysis of prospective clinical trials. JAMA 301(22):2349–2361

Krönke J, Schlenk RF, Jensen KO, Tschürtz F, Corbacioglu A, Gaidzik VI et al (2011) Monitoring of minimal residual disease in NPM1-mutated acute myeloid leukemia: a study from the German-Austrian Acute Myeloid Leukemia Study Group. J Clin Oncol 29(19):2709–2716

Lacombe F, Bernal E, Bloxham D, Couzens S, Porta MGD, Johansson U et al (2016) Harmonemia: a universal strategy for flow cytometry immunophenotyping – a European LeukemiaNet WP10 study. Leukemia 30(8):1769–1772

Lacombe F, Campos L, Allou K, Arnoulet C, Delabarthe A, Dumezy F et al (2018) Prognostic value of multicenter flow cytometry harmonized assessment of minimal residual disease in acute myeloblastic leukemia. Hematol Oncol 36(2):422–428

Lambert J, Lambert J, Thomas X, Marceau-Renaut A, Micol JB, Renneville A et al (2021) Early detection of WT1 measurable residual disease identifies high-risk patients, independent of transplantation in AML. Blood Adv 5(23):5258–5268

Larionov A, Krause A, Miller W (2005) A standard curve based method for relative real time PCR data processing. BMC Bioinformatics 6:62

Levis MJ, Hamadani M, Logan B, Jones RJ, Singh AK, Litzow M et al (2024) Gilteritinib as post-transplant maintenance for AML with internal tandem duplication mutation of FLT3. J Clin Oncol 42(15):1766–1775

Liu J, Wang Y, Xu L-P, Liu D-H, Qin Y-Z, Chang Y-J et al (2014) Monitoring mixed lineage leukemia expression may help identify patients with mixed lineage leukemia–rearranged acute leukemia who are at high risk of relapse after allogeneic hematopoietic stem cell transplantation. Biol Blood Marrow Transplant 20(7):929–936

Liu J, Ma R, Liu Y-R, Xu L-P, Zhang X-H, Chen H et al (2019) The significance of peri-transplantation minimal residual disease assessed by multiparameter flow cytometry on outcomes for adult AML patients receiving haploidentical allografts. Bone Marrow Transplant 54(4):567–577

Liu FJ, Cheng WY, Lin XJ, Wang SY, Jiang TY, Ma TT et al (2021) Measurable residual disease detected by multiparameter flow cytometry and sequencing improves prediction of relapse and survival in acute myeloid leukemia. Front Oncol 11:677833

Loghavi S, DiNardo CD, Furudate K, Takahashi K, Tanaka T, Short NJ et al (2021) Flow cytometric immunophenotypic alterations of persistent clonal haematopoiesis in remission bone marrows of patients with NPM1-mutated acute myeloid leukaemia. Br J Haematol 192(6):1054–1063

Loke J, McCarthy N, Jackson A, Siddique S, Hodgkinson A, Mason J et al (2023) Posttransplant MRD and T-cell chimerism status predict outcomes in patients who received allografts for AML/MDS. Blood Adv 7(14):3666–3676

Loo S, Dillon R, Ivey A, Anstee NS, Othman J, Tiong IS et al (2022) Pretransplant FLT3-ITD MRD assessed by high-sensitivity PCR-NGS determines posttransplant clinical outcome. Blood 140(22):2407–2411

Löwenberg B, Pabst T, Maertens J, Gradowska P, Biemond BJ, Spertini O et al (2021) Addition of lenalidomide to intensive treatment in younger and middle-aged adults with newly diagnosed AML: the HOVON-SAKK-132 trial. Blood Adv 5(4):1110–1121

Luskin MR, Stone RM (2017) Can minimal residual disease determination in acute myeloid leukemia be used in clinical practice? J Oncol Pract 13(8):471–480

Mancini M, Cedrone M, Diverio D, Emanuel B, Stul M, Vranckx H et al (2000) Use of dual-color interphase FISH for the detection of inv(16) in acute myeloid leukemia at diagnosis, relapse and during follow-up: a study of 23 patients. Leukemia 14(3):364–368

Maurillo L, Buccisano F, Spagnoli A, Del Poeta G, Panetta P, Neri B et al (2007) Monitoring of minimal residual disease in adult acute myeloid leukemia using peripheral blood as an alternative source to bone marrow. Haematologica 92(5):605–611

Morita K, Kantarjian HM, Wang F, Yan Y, Bueso-Ramos C, Sasaki K et al (2018) Clearance of Somatic Mutations at Remission and the Risk of Relapse in Acute Myeloid Leukemia. J Clin Oncol 36(18):1788–1797

Nagler A, Labopin M, Dholaria B, Blaise D, Bondarenko S, Vydra J et al (2023) Impact of measurable residual disease on outcomes of unrelated donor haematopoietic cell transplantation with post-transplant cyclophosphamide in AML in first complete remission. Br J Haematol 201(6):1169–1178

Nomdedéu JF, Hoyos M, Carricondo M, Bussaglia E, Estivill C, Esteve J et al (2013) Bone marrow WT1 levels at diagnosis, post-induction and post-intensification in adult de novo AML. Leukemia 27(11):2157–2164

Ommen HB, Touzart A, MacIntyre E, Kern W, Haferlach T, Haferlach C et al (2015) The kinetics of relapse in −214-positive acute myeloid leukemia patients. Eur J Haematol 95(5):436–441

Oran B, de Lima M, Garcia-Manero G, Thall PF, Lin R, Popat U et al (2020) A phase 3 randomized study of 5-azacitidine maintenance vs observation after transplant in high-risk AML and MDS patients. Blood Adv 4(21):5580–5588

Ossenkoppele G, Schuurhuis GJ (2016) MRD in AML: does it already guide therapy decision-making? Hematology Am Soc Hematol Educ Program 2016(1):356–365

Ossenkoppele G, Schuurhuis GJ, van de Loosdrecht A, Cloos J (2019) Can we incorporate MRD assessment into clinical practice in AML? Best Pract Res Clin Haematol 32(2):186–191

Othman J, Potter N, Ivey A, Jovanovic J, Runglall M, Freeman SD et al (2024) Postinduction molecular MRD identifies patients with NPM1 AML who benefit from allogeneic transplant in first remission. Blood 143(19):1931–1936

Othus M, Wood BL, Stirewalt DL, Estey EH, Petersdorf SH, Appelbaum FR et al (2016) Effect of measurable ('minimal') residual disease (MRD) information on prediction of relapse and survival in adult acute myeloid leukemia. Leukemia 30(10):2080–2083

Othus M, Gale RP, Hourigan CS, Walter RB (2019) Statistics and measurable residual disease (MRD) testing: uses and abuses in hematopoietic cell transplantation. Bone Marrow Transplant 55:843

Papaemmanuil E, Gerstung M, Bullinger L, Gaidzik VI, Paschka P, Roberts ND et al (2016) Genomic classification and prognosis in acute myeloid leukemia. N Engl J Med 374(23):2209–2221

Paras G, Morsink LM, Othus M, Milano F, Sandmaier BM, Zarling LC et al (2022) Conditioning intensity and peritransplant flow cytometric MRD dynamics in adult AML. Blood 139(11):1694–1706

Patkar N, Kakirde C, Shaikh AF, Salve R, Bhanshe P, Chatterjee G et al (2021) Clinical impact of panel-based error-corrected next generation sequencing versus flow cytometry to detect measurable residual disease (MRD) in acute myeloid leukemia (AML). Leukemia 35(5):1392–1404

Percival ME, Lai C, Estey E, Hourigan CS (2017) Bone marrow evaluation for diagnosis and monitoring of acute myeloid leukemia. Blood Rev 31(4):185–192

Platzbecker U, Middeke JM, Sockel K, Herbst R, Wolf D, Baldus CD et al (2018) Measurable residual disease-guided treatment with azacitidine to prevent haematological relapse in patients with myelodysplastic syndrome and acute myeloid leukaemia (RELAZA2): an open-label, multicentre, phase 2 trial. Lancet Oncol 19(12):1668–1679

Potluri S, Kellaway SG, Coleman DJL, Keane P, Imperato MR, Assi SA et al (2024) Gene regulation in t(6;9) DEK::NUP214 Acute Myeloid Leukemia resembles that of FLT3-ITD/NPM1 Acute Myeloid Leukemia but with an altered HOX/MEIS axis. Leukemia 38(2):403–407

Puckrin R, Atenafu EG, Claudio JO, Chan S, Gupta V, Maze D et al (2021) Measurable residual disease monitoring provides insufficient lead-time to prevent morphologic relapse in the majority of patients with core-binding factor acute myeloid leukemia. Haematologica 106(1):56–63

Ravandi F (2018) Is it time to routinely incorporate MRD into practice? Best Pract Res Clin Haematol 31(4):396–400

Ravandi F, Jorgensen J, Borthakur G, Jabbour E, Kadia T, Pierce S et al (2017) Persistence of minimal residual disease assessed by multiparameter flow cytometry is highly prognostic in younger patients with acute myeloid leukemia. Cancer 123(3):426–435

Rücker FG, Agrawal M, Corbacioglu A, Weber D, Kapp-Schwoerer S, Gaidzik VI et al (2019) Measurable residual disease monitoring in acute myeloid leukemia with t(8;21)(q22;q22.1): results from the AML Study Group. Blood 134(19):1608–1618

Schnittger S, Kern W, Tschulik C, Weiss T, Dicker F, Falini B et al (2009) Minimal residual disease levels assessed by NPM1 mutation-specific RQ-PCR provide important prognostic information in AML. Blood 114(11):2220–2231

Scholl C, Schlenk RF, Eiwen K, Döhner H, Fröhling S, Döhner K (2005) The prognostic value of MLL-AF9 detection in patients with t(9;11)(p22;q23)-positive acute myeloid leukemia. Haematologica 90(12):1626–1634

Schuurhuis GJ, Heuser M, Freeman S, Béné M-C, Buccisano F, Cloos J et al (2018) Minimal/measurable residual disease in AML: a consensus document from the European LeukemiaNet MRD Working Party. Blood 131(12):1275–1291

Scott S, Dillon R, Thiede C, Sadiq S, Cartwright A, Clouston HJ et al (2023) Assessment of acute myeloid leukemia molecular measurable residual disease testing in an interlaboratory study. Blood Adv 7(14):3686–3694

Scott S, Devonshire A, Dillon R, Thiede C, Cross NCP, White HE et al (2024) Recommendations from the AML molecular MRD expert advisory board. Leukemia 38(7):1638–1641

Shah MV, Jorgensen JL, Saliba RM, Wang SA, Alousi AM, Andersson BS et al (2018) Early post-transplant minimal residual disease assessment improves risk stratification in acute myeloid leukemia. Biol Blood Marrow Transplant 24(7):1514–1520

Shayegi N, Kramer M, Bornhäuser M, Schaich M, Schetelig J, Platzbecker U et al (2013) The level of residual disease based on mutant NPM1 is an independent prognostic factor for relapse and survival in AML. Blood 122(1):83–92

Short NJ, Zhou S, Fu C, Berry DA, Walter RB, Freeman SD et al (2020) Association of measurable residual disease with survival outcomes in patients with acute myeloid leukemia: a systematic review and meta-analysis. JAMA Oncol 6(12):1890–1899

Slovak ML, Gundacker H, Bloomfield CD, Dewald G, Appelbaum FR, Larson RA et al (2006) A retrospective study of 69 patients with t(6;9)(p23;q34) AML emphasizes the need for a prospective, multicenter initiative for rare 'poor prognosis' myeloid malignancies. Leukemia 20(7):1295–1297

Sorror ML, Storer BE, Fathi AT, Brunner A, Gerds AT, Sekeres MA et al (2021) Multisite 11-year experience of less-intensive vs intensive therapies in acute myeloid leukemia. Blood 138(5):387–400

Tan Y, Du K, Luo Y, Shi J, Cao L, Zheng Y et al (2014) Superiority of preemptive donor lymphocyte infusion based on minimal residual disease in acute leukemia patients after allogeneic hematopoietic stem cell transplantation. Transfusion 54(6):1493–1500

Terwijn M, van Putten WL, Kelder A, van der Velden VH, Brooimans RA, Pabst T et al (2013) High prognostic impact of flow cytometric minimal residual disease detection in acute myeloid leukemia: data from the HOVON/SAKK AML 42A study. J Clin Oncol 31(31):3889–3897

Tettero JM, Freeman S, Buecklein V, Venditti A, Maurillo L, Kern W et al (2022a) Technical aspects of flow cytometry-based measurable residual disease quantification in acute myeloid leukemia: experience of the European LeukemiaNet MRD Working Party. Hema 6(1):e676

Tettero JM, Al-Badri WKW, Ngai LL, Bachas C, Breems DA, van Elssen CHMJ et al (2022b) Concordance in measurable residual disease result after first and second induction cycle in acute myeloid leukemia: an outcome- and cost-analysis. Front Oncol 12

Tettero JM, Dakappagari N, Heidinga ME, Oussoren-Brockhoff Y, Hanekamp D, Pahuja A et al (2023a) Analytical assay validation for acute myeloid leukemia measurable residual disease assessment by multiparametric flow cytometry. Cytometry B Clin Cytom 104(6):426–439

Tettero JM, Ngai LL, Bachas C, Breems DA, Fischer T, Gjertsen BT et al (2023b) Measurable residual disease-guided therapy in intermediate-risk acute myeloid leukemia patients is a valuable strategy in reducing allogeneic transplantation without negatively affecting survival. Haematologica 108(10):2794–2798

Tettero JM, Heidinga ME, Mocking TR, Fransen G, Kelder A, Scholten WJ et al (2024) Impact of hemodilution on flow cytometry based measurable residual disease assessment in acute myeloid leukemia. Leukemia 38(3):630–639

Tiong IS, Dillon R, Ivey A, Kuzich JA, Thiagarajah N, Sharplin KM et al (2021) Clinical impact of NPM1-mutant molecular persistence after chemotherapy for acute myeloid leukemia. Blood Adv 5(23):5107–5111

Tsai CH, Tang JL, Tien FM, Kuo YY, Wu DC, Lin CC et al (2021) Clinical implications of sequential MRD monitoring by NGS at 2 time points after chemotherapy in patients with AML. Blood Adv 5(10):2456–2466

Ustun C, Le-Rademacher J, Wang H-L, Othus M, Sun Z, Major B et al (2019) Allogeneic hematopoietic cell transplantation compared to chemotherapy consolidation in older acute myeloid leukemia (AML) patients 60–75 years in first complete remission (CR1): an alliance (A151509), SWOG, ECOG-ACRIN, and CIBMTR study. Leukemia 33(11):2599–2609

Venditti (2025) Blood Adv:https://ashpublications.org/bloodadvances/article/doi/10.1182/bloodadvances.2024013744/535487/FITNESS-ASSESSMENTIN-ACUTE-MYELOID-LEUKEMIA

Venditti A, Piciocchi A, Candoni A, Melillo L, Calafiore V, Cairoli R et al (2019) GIMEMA AML1310 trial of risk-adapted, MRD-directed therapy for young adults with newly diagnosed acute myeloid leukemia. Blood 134(12):935–945

Vidriales MB, Pérez-López E, Pegenaute C, Castellanos M, Pérez JJ, Chandía M et al (2016) Minimal residual disease evaluation by flow cytometry is a complementary tool to cytogenetics for treatment decisions in acute myeloid leukaemia. Leuk Res 40:1–9

Walter RB, Gyurkocza B, Storer BE, Godwin CD, Pagel JM, Buckley SA et al (2015 Jan) Comparison of minimal residual disease as outcome predictor for AML patients in first complete remission undergoing myeloablative or nonmyeloablative allogeneic hematopoietic cell transplantation. Leukemia 29(1):137–144

Wienecke CP, Heida B, Venturini L, Gabdoulline R, Krüger K, Teich K et al (2024) Clonal relapse dynamics in acute myeloid leukemia following allogeneic hematopoietic cell transplantation. Blood 144(3):296–307

Wood BL (2016) Principles of minimal residual disease detection for hematopoietic neoplasms by flow cytometry. Cytometry B Clin Cytom 90(1):47–53

Wood BL, Arroz M, Barnett D, DiGiuseppe J, Greig B, Kussick SJ et al (2007) 2006 Bethesda International Consensus recommendations on the immunophenotypic analysis of hematolymphoid neoplasia by flow cytometry: Optimal reagents and reporting for the flow cytometric diagnosis of hematopoietic neoplasia. Cytometry B Clin Cytom 72B(S1):S14–S22

Xuan L, Wang Y, Yang K, Shao R, Huang F, Fan Z et al (2023) Sorafenib maintenance after allogeneic haemopoietic stem-cell transplantation in patients with *FLT3*-ITD acute myeloid leukaemia: long-term follow-up of an open-label, multicentre, randomised, phase 3 trial. Lancet Haematol 10(8):e600–e611

Yanada M, Takami A, Mizuno S, Mori J, Chou T, Usuki K et al (2020) Autologous hematopoietic cell transplantation for acute myeloid leukemia in adults: 25 years of experience in Japan. Int J Hematol 111(1):93–102

Yin JAL, O'Brien MA, Hills RK, Daly SB, Wheatley K, Burnett AK (2012) Minimal residual disease monitoring by quantitative RT-PCR in core binding factor AML allows risk stratification and predicts relapse: results of the United Kingdom MRC AML-15 trial. Blood 120(14):2826–2835

Zeijlemaker W, Kelder A, Cloos J, Schuurhuis GJ (2019) Immunophenotypic detection of measurable residual (stem cell) disease using LAIP approach in acute myeloid leukemia. Curr Protoc Cytom 91(1):e66–e66

Zhou Y, Othus M, Araki D, Wood BL, Radich JP, Halpern AB et al (2016) Pre- and post-transplant quantification of measurable ('minimal') residual disease via multiparameter flow cytometry in adult acute myeloid leukemia. Leukemia 30(7):1456–1464

Zhu HH, Zhang XH, Qin YZ, Liu DH, Jiang H, Chen H et al (2013) MRD-directed risk stratification treatment may improve outcomes of t(8;21) AML in the first complete remission: results from the AML05 multicenter trial. Blood 121(20):4056–4062

8 Acute Lymphoid Leukemia: Is Transplant Indicated for Philadelphia Chromosome Positive ALL?

Ryszard Swoboda and Arnon Nagler

Abstract

Philadelphia chromosome-positive acute lymphoblastic leukemia (Ph+ ALL) is a distinct subtype of ALL that has historically been associated with a very poor prognosis. Allogeneic hematopoietic stem cell transplantation (alloHSCT) has improved the 5-year overall survival rate to approximately 44% and has been considered the only potentially curative treatment option for these patients. The introduction of tyrosine kinase inhibitors (TKIs) represented a breakthrough in the management of Ph+ ALL, significantly improving patient outcomes. Imatinib, the first-generation TKI, led to high complete remission (CR) rates when added to induction and consolidation chemotherapy. Next-generation TKIs, such as dasatinib and ponatinib, have also demonstrated remarkably high CR rates, even when combined with reduced-intensity chemotherapy or steroids. The advent of immunotherapies, including inotuzumab ozogamicin, blinatumomab, and chimeric antigen receptor T-cell (CAR-T) therapy, was another major milestone in Ph+ ALL treatment. The combination of TKIs with immunotherapy appears to be an optimal therapeutic strategy. Given the efficacy of these chemotherapy-free approaches, the role of alloHSCT in Ph+ ALL needs to be reassessed. Patients treated with next-generation TKIs, particularly in combination with novel immunotherapies, may achieve durable remissions without the need for alloHSCT, provided they do not harbour additional adverse cytogenetic or molecular abnormalities and achieve a deep response following induction and consolidation therapy.

Keywords

Allogeneic hematopoietic stem cell transplantation (alloHSCT) · Philadelphia chromosome-positive acute lymphoblastic leukemia (Ph+ ALL) · Tyrosine kinase inhibitor (TKI) · Immunotherapy · Blinatumomab · Inotuzumab ozogamicin · Chimeric antigen receptor T cell therapy (CAR-T) · Chemotherapy-free treatment

R. Swoboda
Department of Bone Marrow Transplantation and Onco-Hematology, Maria Sklodowska-Curie National Research Institute of Oncology, Gliwice Branch, Gliwice, Poland
e-mail: ryszard.swoboda@gliwice.nio.gov.pl

A. Nagler (✉)
Division of Hematology, Chaim Sheba Medical Center, Tel Hashomer, Israel
e-mail: arnon.nagler@sheba.health.gov.il

© The Author(s), under exclusive license to Springer Nature Switzerland AG 2025

K. Ballen (ed.), *Clinical Management of Challenges in Allogeneic Stem Cell Transplant*, Advances in Experimental Medicine and Biology 1475, https://doi.org/10.1007/978-3-031-84988-6_8

8.1 Introduction

Acute lymphoblastic leukemia (ALL) with the presence of Philadelphia chromosome (Ph+) is the only well-defined subtype of ALL, for which a distinct treatment algorithm is commonly recommended. The Philadelphia chromosome, resulting from a reciprocal translocation between the long arms of chromosome 9 (q34.1) and 22 (q11.2), is the most common cytogenetic abnormality diagnosed in about 10–40% of adult ALL patients with frequency increasing with patient age (Burmeister et al. 2008; Thomas et al. 2004; Faderl et al. 1998). As a consequence of this translocation, a fusion gene BCR-ABL1 arises, which encodes the oncoprotein with constitutive tyrosine kinase activity. Due to different breakpoint cluster regions in the BCR gene on chromosome 22, the hybrid gene BCR-ABL1 encodes for a 190-Kd protein (p190) or a 210-Kd protein (p210), respectively in about 70% and 30% of ALL patients (Shi et al. 2022; Bergfelt Lennmyr et al. 2021). Moreover, in about 70% of patients diagnosed with Ph+ ALL, additional chromosomal abnormalities (ACAs) can be detected, some of them, such as −7, +8, hypodiploidy (<44 chromosomes), or complex karyotype (≥5 chromosome abnormalities) are associated with poor prognosis (Aldoss et al. 2015).

Historically, the outcome of Ph+ ALL was considered poor with 5-year overall survival (OS) from 20% for patients treated with chemotherapy only to about 44% for those who were consolidated with allogeneic hematopoietic stem cell transplantation (alloHSCT) (Fielding et al. 2009). The advent of tyrosine kinase inhibitors (TKIs) in the 2000s was a breakthrough in the management of patients with Ph+ ALL. Imatinib, the first TKI added to induction and consolidation therapy in Ph+ ALL resulted in very high complete remission (CR) rates with 2-year OS at about 60% (Thomas et al. 2004; Lee et al. 2005; de Labarthe et al. 2007; Towatari et al. 2004; Wassmann et al. 2006). In the following years, second and third-generation TKIs have been evaluated in patients with Ph+ ALL resulting in improved outcomes in this setting. Another breakthrough in the management of ALL patients was the introduction of novel immunotherapies, such as inotuzumab ozogamicin, blinatumomab, or chimeric antigen receptor T cell therapy (CAR-T) (Kantarjian et al. 2016; Martinelli et al. 2021; Maude et al. 2018; Shah et al. 2021). TKIs and new immunotherapies revolutionized the treatment paradigm of Ph+ ALL and patients' prognosis. These unprecedented developments questioned the role of chemotherapy as well as its intensity, the choice of particular TKI, the recommended length of their use, and lastly the role of alloHSCT in the management of Ph+ ALL. In this chapter, we will focus on the role of alloHSCT in the context of particular generations of TKIs and new immunotherapies available for Ph+ ALL patients. On the basis of the literature, we will try to determine the group of Ph+ ALL patients who still benefit from alloHSCT and those who can do without transplantation in the era of new treatment options.

8.2 The Role of Tyrosine Kinase Inhibitors (TKIs) in the Treatment of Ph+ ALL

Imatinib, the first-generation TKI, inhibits selectively the BCR-ABL1 tyrosine kinase preventing the activation of downstream proliferative and survival pathways. Moreover, it inhibits *c-kit*, platelet-derived growth factor (PDGF), and stem cell factor (SCF) (Druker et al. 1996; Platanias 2011). In the first clinical trial, imatinib combined with hyper-CVAD (cyclophosphamide, vincristine, adriamycin, and dexamethasone) resulted in CR rate of 100% and a 2-year disease-free survival (DFS) of 85% in de novo ALL Ph+ (Thomas et al. 2004). The final results of that study, in which 54 Ph+ ALL patients (39 with de novo disease, 6 with primary refractory after induction without a TKI, and 9 in CR after induction therapy without TKI) with a median age of 51 years were enrolled, revealed CR rate of 93% with 5-year OS of 43%, although only 30% of patients underwent subsequent alloHSCT (Daver et al. 2015). In that relatively small studied group, there was no significant difference in 5-year DFS between patients being transplanted and those without transplantation (63% vs. 43%, respectively, $p = 0.52$). The only factor revealed to influence OS in the multivariate analysis was the patient's age over 60 years. Severe toxicities as well as hematologic recovery were not significantly different from those observed in patients treated with chemotherapy alone. In a larger prospective UKALLXII/ECOG2993 study the authors showed significantly better CR rate in Ph+ ALL patients treated with imatinib combined with chemotherapy when compared to those from the "pre-imatinib" cohort (92% vs. 82%, respectively, $p = 0.004$) (Fielding et al. 2014). Similarly, the 4-year OS was better in patients treated with imatinib (38% vs. 22%, $p = 0.003$), which may be partially due to the higher proportion of patients eligible for alloHSCT in this group (46% vs. 31%). The positive effect of imatinib was also shown with event-free survival (EFS, 33% vs. 18%, $p = 0.0001$) and relapse-free survival (RFS, 50% vs. 33%, $p = 0.0003$). To reduce treatment-related mortality (TRM), the authors of the GRAAPH-2005 trial compared prospectively imatinib (400 mg twice daily from day 1–28) in combination with reduced intensity chemotherapy (RICH; vincristine and dexamethasone) vs. imatinib (400 mg twice daily from days 1–14) with hyper-CVAD chemotherapy during induction course (Chalandon et al. 2015). There was a significant difference in CR rate in favor of RICH (98% vs. 91%, $p = 0.006$), which was mainly due to fewer induction deaths in this group. However, the study revealed no statistically significant differences in the main transplant outcomes between the studied groups with the median EFS and OS of 2.5 and 4.1 years in the RICH group and 1.8 and 3.3 years in the standard chemotherapy group, respectively ($p = 0.13$ for EFS and $p = 0.37$ for OS). Moreover, the authors showed significant benefit from alloHSCT in the first CR (CR1) with RFS (hazard ratio (HR) 0.69; 95% confidence interval (CI), 0.49–0.98, $p = 0.036$) and OS (HR 0.64; 95% CI, 0.44–0.93, $p = 0.02$) (Chalandon et al. 2015).

Dasatinib, a second-generation BCR-ABL1 inhibitor, is less dependent on P-loop interactions for drug binding, overcoming most of the imatinib-resistant kinase domain mutations, except for the T315I mutation (Shah et al. 2004; Tokarski et al. 2006). Dasatinib is the only TKI capable of crossing the blood-brain barrier. In a phase 2 trial dasatinib (140 mg a day) as a single agent resulted in 58% complete cytogenetic response (CCyR) in ALL Ph+ patients ($n = 36$) with resistance or intolerance to imatinib. It is of note that the response was rapid with a median time to major hematologic response of 1.8 months (Ottmann et al. 2007). The treatment was well-tolerated with only two patients (6%) who had to discontinue the treatment due to its toxicity. In another study from the GIMEMA group dasatinib was used only with steroids in the group of 53 Ph+ ALL patients (median age 54 years) resulting in 100% CR and the rate of major molecular remission increasing with time from 22% at day 22 to 52% at the end of an induction course. The OS rate at 20 months was 69.2% and the DFS rate was 51.1% (Foà et al. 2011). The study focused on the induction phase and the protocol did not include any planned post-remission treatment, therefore no conclusions on the importance of alloHSCT can be drawn. The final results from the GIMEMA LAL1509 study evaluating dasatinib in combination with steroids and central nervous system prophylaxis in 60 de novo Ph+ ALL patients revealed 56.3% OS and 47.2% DFS with a median follow-up of 57.4 months. Almost all patients (97%) achieved complete hematologic remission, however only 18.3% of patients obtained complete molecular response (CMR). Only 22 out of 60 patients underwent an alloHSCT and according to the study investigators, transplantation did not seem to have an impact on survival; however, the study was not powered to define its role (Chiaretti et al. 2021). In contrast, the result of a Chinese phase 2 study on dasatinib combined with prednisone as induction and consolidation therapy in newly diagnosed Ph+ ALL patients (median patient age was 53) revealed a definite survival advantage in patients transplanted in CR 1 (the 2-year DFS and OS were 100% for both outcomes) when compared to non-HSCT group (2-year DFS and OS were 33% and 45%, respectively) (Xie et al. 2023a). Dasatinib combined with low-dose chemotherapy (intrathecal chemotherapy, vincristine, and dexamethasone during induction therapy) was evaluated in 71 patients at the age above 55 years. Patients in CR continued consolidation with dasatinib, cytarabine, asparaginase, and methotrexate for 6 months, and then maintenance therapy with dasatinib/vincristine/dexamethasone was implemented for 18 months followed by dasatinib until relapse or death. Complete remission was achieved in 96% of patients and the OS rate was 36% at 5 years. Only seven patients were treated with alloHSCT, which makes it impossible to assess the role of transplantation in this setting (Rousselot et al. 2016). In another phase 2 trial dasatinib was combined with eight cycles of alternating hyper-CVAD and high-dose cytarabine and methotrexate in the frontline setting. Complete remission was achieved in 96% of 72 patients, among whom measurable residual disease (MRD) was negative in 94% at a median of 3 weeks (Ravandi et al. 2015). The 5-year OS and RFS were 46% and 43%, respectively, although only 12 patients (16.6%) were transplanted from allogeneic donors in CR1. Among those who were transplanted (median age 52.5 years) seven patients died after transplant in CR1 due to pneumonia ($n = 4$), graft-versus-host disease (GVHD) ($n = 1$), myocardial infarction ($n = 1$), and unknown cause ($n = 1$). The high transplant-related mortality is likely due to the relatively advanced age of the cohort (median patient age was 55 years). Despite the small number of transplanted subjects, the authors of the study revealed that patients below the age of 40, but not older may benefit from alloHSCT in CR1 with OS. However, the difference was not statistically significant (Ravandi et al. 2015). In another study by Ravandi F et al., dasatinib in combination with hyper-CVAD resulted in CR or CR with incomplete count recovery (CRi) in 88% of patients, OS in 69%, EFS in 55%, and RFS in 62% at 3 years (Ravandi et al. 2016). Out of 94 patients evaluable (median age was 44 years), 44% underwent alloHSCT with subsequent maintenance therapy with dasatinib from day 100 posttransplant for up to 5 years. In that group, the 12-month RFS and OS after transplant were 71% and 87%, respectively. Moreover, landmark analysis showed statistically superior advantages for RFS and OS ($p = 0.038$ and $p = 0.037$, respectively) for the transplanted patients. Another induction protocol including dasatinib with prednisolone in the first step and dasatinib plus intensive chemotherapy in the second step followed by alloHSCT was evaluated in Japanese centers (Sugiura et al. 2022). The authors revealed a 100% CR rate with 53% MRD-negativity after such induction therapy, which resulted in a high rate of patients proceeding to alloHSCT (74.4%) and very encouraging survival outcomes in that group (the 3-year EFS and OS rates were 72.1% and 87.9%, respectively). There is limited data in the literature comparing dasatinib with imatinib in Ph+ ALL patients. In a retrospective study by the Acute Leukemia Working Party (ALWP) of the European Society for Blood and Marrow Transplantation (EBMT), the authors showed no differences between imatinib and dasatinib used before alloHSCT in terms of LFS and OS (Giebel et al. 2024). However, dasatinib was associated with an increased risk of severe GVHD and decreased GVHD and relapse-free survival (GRFS).

Nilotinib, another second-generation TKI with a higher affinity for BCR-ABL1 than imatinib, was assessed in a phase 2 trial in 90 newly diagnosed Ph+ ALL patients (Kim et al. 2015). Combined with multiagent chemotherapy, nilotinib resulted in a high CR rate of 91%, of which 94% achieved CMR and 70% proceeded to alloHSCT in CR1. The 2-year RFS and OS were 72% for both. The authors revealed better RFS in transplanted patients with no difference in non-relapse mortality (NRM) and OS when compared to those without transplantation. In another study, nilotinib in combination with RICH was evaluated in a relatively small ($n = 47$) group of older patients (median age 65 years) showing high CR of 97% with 30% of CMR and only 7 patients eligible for alloHSCT (Oliver et al. 2014).

Ponatinib is a third-generation TKI capable of overcoming most of the kinase domain mutations, including T315I, which is a common cause of first and second-TKI resistance (O'Hare et al. 2009). In a single-center phase 2 study including 76 patients (median age was 47 years) ponatinib was combined with hyper-CVAD as a frontline therapy for Ph+ ALL. All patients with active disease at enrollment achieved CR, and CMR was obtained in 63 (83%) patients. The 5-year PFS, EFS, and OS were 83%, 67%, and 71%, respectively. In a post-hoc 6-month landmark analysis, the authors did not show a difference in OS between patients who underwent alloHSCT ($n = 15$) and those without transplantation (3-year OS rate was 70% vs. 87%, respectively, $p = 0.32$) (Jabbour et al. 2018a). The last update of the study confirmed those excellent results in a total of 86 patients (median age 46 years) showing the cumulative CMR rate at 86% (all patients achieved CR), the 6-year EFS and OS rate at 65% and 75%, respectively (Kantarjian et al. 2023). Similarly, there was no difference in 6-year OS in a landmark analysis between transplanted ($n = 20$ patients) and not transplanted patients (70% and 87%, respectively, $p = 0.13$) (Kantarjian et al. 2023). Ponatinib combined with prednisone was investigated in elderly patients (median age 66.5 years) unfit for intensive chemotherapy and alloHSCT (Martinelli et al. 2022). Complete hematologic response (CHR) was achieved in 95.5% of patients, while CMR at 24 weeks was obtained in 40.9%. With a median follow-up of 34.9 months, a median OS and duration of CHR were not reached. The excellent results of ponatinib combined with standard induction and consolidation chemotherapy followed by alloHSCT were displayed in the Spanish phase 2 PONALFIL trial including 30 patients (median age was 49 years) (Ribera et al. 2022). At the end of consolidation, CR and CMR were achieved in 100% and 71% of patients, respectively. With 26 (87%) patients being transplanted from allogeneic donors, EFS and OS at 3 years were 70% and 96%, respectively (Ribera et al. 2022). These results were better than those presented in similar designed and scheduled trials using imatinib instead of dasatinib (Ribera et al. 2012). Moreover, the authors of the PONALFIL trial updated the results showing EFS and OS at 4 years at 66% and 92%, respectively (Ribera et al. 2024). In another study, ponatinib combined with hyper-CVAD was compared to dasatinib plus hyper-CVAD using a propensity score analysis (Sasaki et al. 2016). The authors of that analysis revealed significantly higher rates of CCyR and CMR at 3 months in the group treated with ponatinib. Moreover, the 3-year EFS and OS were also in favor of ponatinib (69% vs. 46%, $p = 0.04$ and 83% vs. 56%, $p = 0.03$, respectively). Similarly, in a matching adjusted indirect comparison, ponatinib-based therapy prolonged OS and gave a higher CMR rate when compared to patients treated with imatinib-based protocols (Ribera et al. 2023). These results were also confirmed in a global registrational phase 3 study, in which ponatinib combined with RICH demonstrated a superior rate of MRD-negative CR (34.4% vs. 16.7%, $p = 0.002$) and longer median EFS (not reached in the ponatinib group vs. 29 months in the imatinib group) over imatinib combined with RICH (Jabbour et al. 2024a). The superiority of ponatinib over earlier generation TKIs in achieving CMR and survival rate was also shown in a meta-analysis of 26 studies (CMR 79% vs. 34% and 3-year OS 79% vs. 50%, respectively) (Jabbour et al. 2018b).

Another issue requiring consideration is the use of TKIs as a maintenance therapy in order to reduce relapse incidence after alloHSCT. In a study by Chen H et al., imatinib initiated at a median time of 70 days post alloHSCT and used for a median of 90 days resulted in lower estimated probability of relapse (10.2% vs. 33.1%, $p = 0.016$) and better estimated DFS and OS (81.5% and 86.7% vs. 33.5% and 34.3%, $p = 0.000$ for both, respectively) when compared to the non-imatinib group (Chen et al. 2012). In general, two TKI treatment strategies after alloHSCT exist. In the prophylactic strategy, TKI is used before MRD detection, while in the preemptive strategy, TKI therapy is initiated only when MRD is detected. In a randomized study by the GMALL study group, imatinib therapy after alloHSCT resulted in a low relapse rate and durable remissions regardless of its use in prophylactic or MRD-triggered strategy (Pfeifer et al. 2013). In the prophylactic and MRD-triggered groups, the median time from alloHSCT to the start of imatinib maintenance therapy was 48 and 70 days, respectively. About 50% of patients in the MRD-triggered group did not require imatinib therapy and nearly 70% of patients receiving imatinib therapy stopped this therapy prematurely (Pfeifer et al. 2013). In another prospective study, patients treated with preemptive TKI strategy were compared to historical group that used prophylactic TKI strategy. The authors of that analysis demonstrated superior OS (HR 0.353, $p = 0.005$) and DFS (HR 0.381, $p = 0.003$) in favor of the preemptive group; however, these findings should be treated with caution because more patients in the preemptive cohort were treated with second and third generation TKIs. The initiation of TKI therapy was at 45 day and 95 day after alloHSCT, respectively in the prophylactic and preemptive group. Interestingly, above 70% of patients in the preemptive group did not need TKI therapy after alloHSCT (Liu et al. 2021). In another multicenter retrospective analysis, imatinib at a dose of 400 mg per day was compared to dasatinib at a dose of 100 mg per day as a maintenance therapy after alloHSCT (Guan et al. 2024). The initiation of TKI maintenance therapy was at 98 and 95.5 days after transplantation for imatinib and dasatinib respectively and lasted about a year in both groups. In that study from China, the authors did not show significant differences in long-term outcomes between those TKIs, however the 5-year mild chronic GVHD and gastrointestinal bleeding was higher in the dasatinib group (37.3% vs. 14.9%, $p = 0.03$ and 25.5% vs. 2.3%, $p < 0.001$) (Guan et al. 2024). In another observational study on a relatively small group of Ph+ ALL patients ($n = 48$), ponatinib used prophylactically after alloHSCT showed better 5-year RFS when compared to its pre-emptive use (95% vs. 57%, respectively, $p = 0.02$) (Candoni et al. 2024). Other studies on even smaller groups confirmed the positive role of ponatinib maintenance used after alloHSCT; however, the questions of the treatment duration and dosage remain unanswered (Leotta et al. 2021; Chen et al. 2022; Nanno et al. 2020).

Undoubtedly, TKIs are the keystone of treatment for patients diagnosed with Ph+ ALL. This targeted therapy should be implemented into the treatment protocol at a very early stage. The choice of TKI should provide for their effectiveness, contraindications, and possible side effects. As we have discussed above, the effectiveness is increasing with the newer generation of TKIs. The maintenance therapy with TKIs should be used for all patients after alloHSCT; however, its type (prophylactic or preemptive), initiation time and duration have not been established so far. In the case of alloHSCT in CR1, the experts from the ALWP of the EBMT recommend using TKI prophylactically over 12 months. A preemptive strategy using TKI in case of MRD positivity is also possible, providing rigorous MRD monitoring every 6–8 weeks and starting on TKI treatment as soon as MRD is detectable (Giebel et al. 2016). Based on the cited literature, the Ph+ ALL patients treated with first or second generation TKI benefit from alloHSCT and all eligible patients should be referred to this procedure. In case of third generation TKI, the decision on which patients should proceed to alloHSCT is more complicated and it seems that some patients, especially older ones who achieved CR1 with MRD negativity may do well without transplantation. However, prospective trials are needed to confirm this assumption.

8.3 The Role of New Immunotherapies in the Treatment of Ph+ ALL

During the last decade, immunotherapy revolutionized the treatment paradigm in adult patients with ALL, including Ph+ ALL (Martinelli et al. 2021; Nagel et al. 2017; Assi et al. 2017; Hanif et al. 2018). Blinatumomab is the prototype first in human bispecific monoclonal antibody targeting the CD3 on T lymphocytes and a relevant protein/antigen on the malignant cells; in the case of Ph+ ALL, it targets the CD19 on the B-lineage blasts (Zimmerman et al. 2015). Blinatumomab exerts its antileukemic effect by activating the host T cells against the leukemic blasts leading to their lysis (Zimmerman et al. 2015). In a phase 2 single-arm ALCANTARA study, blinatumomab was evaluated in 45 Ph+ ALL patients who relapsed or were refractory or intolerant to second-generation or more advanced TKIs

(Martinelli et al. 2017). During the first two cycles of blinatumomab, CR or CRi was achieved in 16 (36%) patients, most of them (88%) obtained MRD negativity, and 7 (44%) patients proceeded to alloHSCT. Median RFS and OS were 6.7 and 7.1 months, respectively (Martinelli et al. 2017). The final analysis of that study confirmed the initial findings showing a median OS of 9.0 months with a median follow-up of 25.1 months (Martinelli et al. 2021). The rationale for combining TKI with blinatumomab in relapsed/refractory (R/R) Ph+ ALL is overcoming a potential mechanism of resistance to blinatumomab, resulting from the selection of CD19-negative blasts in the BCR-ABL1-positive clones (Nagel et al. 2017). In two retrospective analyses, blinatumomab combined with TKI showed encouraging results in achieving CMR (75% and 60%, respectively) in R/R Ph+ ALL (Assi et al. 2017; Hanif et al. 2018). However, the number of patients with Ph+ ALL included in these analyses was very small ($n = 9$ and $n = 5$, respectively). Another study evaluating retrospectively blinatumomab in 239 patients with ALL was conducted by Badar et al. (2020). In a subgroup analysis of R/R Ph+ ALL patients, the CR/CRi rate was 74% (68% with negative MRD) (Badar et al. 2020). Among those who were treated with blinatumomab in combination with TKI ($n = 23$), the CR/CRi was 83%, while in patients treated with blinatumomab as monotherapy ($n = 32$), the CR/CRi rate was 69%. Remarkably, 66% of MRD-positive Ph+ ALL patients achieved MRD negativity with blinatumomab. With a median follow-up of 14 months, median RFS and OS in R/R Ph+ ALL patients who received blinatumomab in combination with TKI was not reached while for those treated with blinatumomab alone, it was 32 months ($p = 0.28$) and 13.1 months ($p = 0.03$), respectively (Badar et al. 2020). Moreover, in the multivariate analysis, diagnosis of Ph+ ALL and alloHSCT consolidation after blinatumomab were favorable prognostic factors for OS. Although alloHSCT significantly improved survival outcomes, the results were not restricted only to ALL Ph+ patients. Given the encouraging results of blinatumomab in R/R Ph+ ALL, it was reasonable to evaluate its potency and potential role in frontline therapy. In phase 2, single-arm GIMEMA LAL2116 D-ALBA trial, dasatinib followed by blinatumomab was evaluated in 63 patients (median age was 54 years) with newly diagnosed Ph+ ALL (Foà et al. 2020). After 7 days of the pre-phase therapy with glucocorticoid and 85 days of the induction phase with dasatinib, CR and CMR were obtained in 98% and 29% of patients, respectively. CMR rate increased to 60% and 81%, respectively after two and four cycles of post-induction therapy with blinatumomab. With a median follow-up of 18 months, DFS and OS were 88% and 95%, respectively. The updated analysis with a median follow-up of 53 months showed an impressive DFS and OS of 75.8% and 80.7%, respectively (Foà et al. 2024). Out of 24 (38%) patients (median age was 52 years) who underwent alloHSCT, 20 (83.3%) patients were alive in CR at a median follow-up of 49 months (Foà et al. 2024). In a subsequent study, dasatinib combined with prednisone followed by blinatumomab as a frontline therapy was assessed in older (median age was 73 years) Ph+ ALL patients (Advani et al. 2023). This chemotherapy-free regimen resulted in a 92% CR rate during the induction phase of dasatinib and prednisone. Complete molecular remission was achieved in 89% of analyzed patients. At a median follow-up of 2.7 years, the estimated 3-year OS and DFS were 87% and 77%, respectively. Similar and even somewhat better outcomes were reported in another phase 2 trial, in which blinatumomab was combined with third-generation TKI, ponatinib. This drug combination was given for up to five 42-day cycles to patients with Ph+ ALL (median age was 51 years) achieving CMR in 87% and 92% of newly diagnosed and R/R ALL Ph+ patients, respectively (Jabbour et al. 2023). The estimated 2-year EFS and OS were 93% for both outcomes. The recently reported updated study that includes 60 patients confirmed the excellent outcomes previously reported with blinatumomab in combination with ponatinib (CR was 97% and CMR was 83%) (Kantarjian et al. 2024). Although only two patients (3%) underwent alloHSCT, the 3-year OS and EFS rates were 91% and 77%, respectively. In an additional study, blinatumomab in combination with ponatinib was evaluated in a group of 26 patients with R/R Ph+ ALL (Couturier et al. 2021). Complete morphological and molecular remission was obtained in 96% and 88%, respectively. The median EFS and OS were 15.3 and 20 months, respectively. Moreover, the investigators demonstrated a significant difference in relapse rate, but not in OS and EFS, between patients who underwent alloHSCT and those who did not (12.5% vs. 82.3%, respectively, $p = 0.003$). Nevertheless, these results should be taken with caution due to the relatively small number of analyzed subjects ($n = 8$ in the alloHSCT and $n = 18$ in the non-alloHSCT group, respectively).

Another monoclonal antibody, a drug-conjugated antibody, incorporated in the treatment of Ph+ ALL patients is inotuzumab ozogamicin (InO), a CD22 monoclonal antibody conjugated with the cytotoxic antibiotic calicheamicin (Yurkiewicz et al. 2018). In a phase 3 INO-VATE study, Kantarjian HM and colleagues demonstrated significantly higher CR and MRD-negativity rates, as well as prolonged PFS and OS in R/R ALL patients treated with InO when compared to standard chemotherapy (Kantarjian et al. 2016). In the analysis of the subgroup of patients with Ph+ ALL ($n = 22$), there was a trend toward a higher CR/CRi rate in patients receiving InO compared to the standard chemotherapy-controlled study arm (73% vs. 56%, respectively $p = 0.1$) (Stock et al. 2021). Among patients achieving CR/CRi, the rate of MRD negativity was significantly higher in the InO group (81% vs. 33%, respectively, $p = 0.009$). There were no differences between the studied groups as for OS (median OS was 8.7 and 8.4 months in the InO and chemotherapy group, respectively, $p = 0.7$) and PFS (median PFS was 3.9 and 3.1 months, respectively, $p = 0.1$); however, the probability of being alive and event-free at 12 months was higher for patients treated with InO compared to those treated with chemotherapy (20.1% vs. 4.8%, respectively) (Stock et al. 2021). More patients in the InO group proceeded to alloHSCT (41% vs. 19%, respectively). Notably, transplantation prolonged survival in both the InO group (median PFS was 9.2 vs. 2.4 months for non-transplanted patients; median OS was 16.5 vs. 4.4 months, respectively) and the chemotherapy group (median PFS was 6.5 vs. 2.4 months, while median OS was 16.4 vs. 6.9 months, respectively) (Stock et al. 2021). The final results of the INO-VATE study were in line with the interim results, demonstrating a lessened effect of InO on achieving CR/CRi and OS in the group of Ph+ ALL patients in comparison to those achieved in the Philadelphia negative (Ph-) ALL counterparts (Kantarjian et al. 2019). However, the study analysis does not provide a subgroup analysis for the Ph+ ALL patients (Kantarjian et al. 2019). The encouraging results of InO in R/R ALL were also shown in an additional phase 1/2 study reported by DeAngelo DJ and colleagues (DeAngelo et al. 2017). In a subgroup of 16 patients with R/R Ph+ ALL, CR/CRi was achieved in 56% of patients, all of them were MRD negative. The median PFS and OS were 4.4 months and 7.4 months, respectively (Stock et al. 2021). As for other TKIs, in a phase 1/2 study, InO in combination with a second-generation TKI, bosutinib, approved for the treatment of patients with chronic myeloid leukemia (CML) was evaluated in 18 patients with R/R Ph+ leukemias (ALL Ph+, $n = 16$; lymphoid blast phase of CML, $n = 2$) (Jain et al. 2021). This combination seemed to be very effective, resulting in 83% of CR/CRi, 61% of MRD negativity, and 56% of CMR. The median duration of response and OS were 7.7 and 13.5 months, respectively, and 6 patients proceeded to alloHSCT. In a recent different phase 2 study, Jabbour E and colleagues showed once more high effectiveness of InO in eradicating MRD in Ph+ ALL patients being in CR (MRD negativity was achieved in 18 (69%) patients) (Jabbour et al. 2024b). The 2-year RFS and OS rates were 54% and 60%, respectively. Although the combination of InO with TKI seems to be effective for R/R Ph+ ALL, the data are very limited and based mainly on case reports (Micheva et al. 2021; Liu et al. 2024).

Finally, CAR-T has emerged as a new treatment option for patients with R/R ALL. Tisagenlecleucel, a novel CAR-T product directed against the CD19 antigen, was approved by the Food and Drug Administration (FDA) for children and young adults, up to the age of 25 years, with R/R ALL based on the ELIANA study (Maude et al. 2018). Among 18 patients diagnosed with R/R Ph+ ALL, all patients achieved CR with a 2-year RFS and OS at 88% for both. Subsequently, a second CAR-T CD19 product, brexucabtagen autoleucel, was approved for adults with ALL based on the ZUMA-3 in which 80% of the 15 patients with R/R Ph+ ALL achieved CR/CRi (Shah et al. 2021). The 6-month RFS and 12-month OS were 60% and 93%, respectively. Based on ELIANA and ZUMA-3 trials, CD19 CAR-T seems to be as favorable or even better in patients with R/R Ph+ in comparison to Ph- ALL regarding CR and OS outcome parameters. Lastly, an innovative treatment using CAR-T in combination with CD19+ feeding autologous T cells (FTC) and TKI (imatinib or dasatinib) as a consolidation therapy was assessed in a phase 1/2 study in 15 patients with de novo Ph+ ALL (Chen et al. 2023). After induction chemotherapy combined with TKI, the patients received a single cycle of CAR-T infusion and another three cycles of CAR-T and CD19+ FTC infusions followed by TKI. All patients achieved CR while 92% of patients had CMR. With a median follow-up of 27 months, the RFS and OS were 84% and 83%, respectively. To sum up, the management algorithm and paradigm of Ph+ ALL have been drastically altered in recent years and are currently based on TKIs and humoral and cellular immunotherapy offering a chemotherapy-free therapeutic avenue questioning the role of chemotherapy and transplantation which used to be for years the backbone of anti-Ph+ ALL therapy.

8.4 Allogeneic Hematopoietic Stem Cell Transplantation in the Era of New Therapies

Allogeneic hematopoietic stem cell transplantation played a crucial role in the management of Ph+ ALL. In the pre-TKI era, it was the only curative therapeutic option concurring this devastating form of leukemia doubling the 5-year OS from about 20 to 40% (Fielding et al. 2009). Therefore, alloHSCT was considered an integral standard therapy for all eligible Ph+ ALL patients achieving CR1. However, with the advent of TKIs and new immunotherapies, the role of alloHSCT has become debatable.

In the UKALLXII/ECOG2993 prospective trial evaluating imatinib in Ph+ ALL patients, the authors confirmed the beneficial role of alloHSCT in the era of first-generation TKI (Fielding et al. 2014). The 4-year OS for patients who received myeloablative alloHSCT from a sibling donor or matched unrelated donor (MUD) was 50% compared to 19% for those treated with chemotherapy alone. The benefits of alloHSCT were also shown regarding additional outcome parameters including EFS (46% vs. 14%) and RFS (69% vs. 18%, respectively). The difference in favor of alloHSCT holds also in the patients who received mismatched unrelated, cord blood, and haploidentical transplants as well as those treated with reduced intensity conditioning before alloHSCT. In another phase 2 multicenter study from Japan, Fujisawa S. and colleagues demonstrated the advantage of alloHSCT in patients with Ph+ ALL receiving imatinib-based chemotherapy, especially in patients under 55 years of age (Fujisawa et al. 2017). The 3-year OS and EFS in patients who received alloHSCT at CR1 were 75% and 72%, while in those who did not receive an alloHSCT, they were 40% and 21%, respectively (Fujisawa et al. 2017).

The role of alloHSCT in patients treated with chemotherapy-free protocol including dasatinib plus prednisone was evaluated in a recently reported Chinese phase 2 multicenter study (Xie et al. 2023a). The authors reported excellent results in the transplanted group ($n = 17$) with 2-year DFS and OS at 100%, which were significantly better than those in the non-transplanted patients (33% and 45%, respectively). The combination of dasatinib with hyper-CVAD followed by alloHSCT and maintenance therapy with dasatinib was studied by the MD Anderson group, resulting in a 12-month RFS of 71% and OS of 87% following transplant (Ravandi et al. 2016). In a landmark analysis, colleagues from MD Anderson were able to show the superior outcomes of Ph+ ALL patients who received alloHSCT consolidation in comparison to those who did not. Interestingly, in contrast to these results, in another somewhat older phase 2 study also by Ravandi F and colleagues from MD Anderson, the investigators failed to show any survival benefit with alloHSCT in patients with Ph+ ALL in CR1 being treated with dasatinib and chemotherapy (Ravandi et al. 2015) with the caveat that the number of transplanted patients enrolled in the study was relatively small ($n = 12$) and the transplant-related mortality was extremely high of 58%, which is unusual (Ravandi et al. 2015). Similarly, in another retrospective analysis from the State, analyzing results in Ph+ ALL patients aged 44–45 years receiving chemotherapy plus dasatinib the authors showed no differences in the 3-year RFS and OS between the 30 patients who underwent alloHSCT (70.5% and 76%, respectively) and the 40 non-transplanted patients (80.1% and 71.3%, respectively) (Chang et al. 2019). No differences in EFS and OS between transplanted ($n = 19$) and non-transplanted ($n = 41$) patients after dasatinib plus chemotherapy treatment were also shown in the group of children and young adults up to 30 years (the 5-year OS were 88% vs. 83%, $p = 0.71$ and the 5-year EFS were 60% vs. 61%, $p = 0.84$, respectively) (Slayton et al. 2018).

In contrast, an advantage and an important therapeutic role for alloHSCT was shown in several studies including patients treated with first or second-generation TKI. A very recent big study by Hu L and Chinese investigators confirmed the benefits of alloHSCT in Ph+ ALL patients achieving CR1 with imatinib or dasatinib with a very convincing highly significant statistical difference (Hu et al. 2024). In the transplanted patients ($n = 216$), the 4-year DFS and OS rates were 68% and 72%, respectively, while in the non-transplanted patients ($n = 76$), the figures were 24% and 47%, respectively. Of note, in a rather new meta-analysis of 12 studies including a total of 404 patients who underwent alloHSCT followed by TKI maintenance therapy and 635 patients treated with TKI (mainly first and second-generation ones) without transplantation, Shahzad and colleagues demonstrated a trend towards poor OS, DFS, and higher relapse rate in the patients that did not undergo alloHSCT (Shahzad et al. 2023). These findings are in line with the results of another recently published meta-analysis from China that included 39 studies involving 5054 patients, in which patients treated with chemotherapy plus TKI (mainly first and second-generation) and achieving CR1 were compared to patients treated with TKI and alloHSCT (Xie et al. 2023b). Once more, analyzing the whole cohort the authors demonstrated an advantage for transplantation as alloHSCT significantly prolonged DFS and OS (the estimated 5-year DFS of the transplanted and non-transplanted patients were 59% and 42%, respectively, and the estimated 5-year OS were 60% and 36%, respectively). However, of note, the survival rates in patients achieving CMR within 3 months of induction therapy initiation were comparable in the transplanted and non-transplanted groups most probably indicating that in patients that achieve deep response transplant may be of less importance and may be deferred. However, this assumption although tempting is still controversial and needs confirmation preferably in a well-designed randomized 2-arm study. As an example, in contrast to the above-mentioned study, in a recent published study on behalf of the Japan Adult Leukemia Study Group, the advantage of alloHSCT in the era of first and second-generation TKIs was shown also in Ph+ ALL patients who achieved CMR within 3 months of diagnosis (the 5-year adjusted OS and RFS were 73% and 70% in the transplanted patients, whereas they were 50% and 20% in the non-transplanted cohort) (Nishiwaki et al. 2024). We believe most colleagues would tend to agree that the advantage of alloHSCT in patients treated with imatinib is rather evident, while its role in the era of second and third-generation TKIs, especially in combination with new immunotherapies is becoming questionable. Supporting this notion, in a single-arm phase 2 study with ponatinib in combination with hyper-CVAD, in young Ph+ ALL cohort (the median age of transplanted patients was 36 years) there were no differences in OS between patients who underwent or did not undergo alloHSCT (Jabbour et al. 2018a; Kantarjian et al. 2023). It is conceivable that this was mainly due to the greater potency of ponatinib in inducing MRD negativity, which is the strongest factor dictating treatment outcomes in ALL. Indeed, in an additional meta-analysis by investigators from China of 15 studies including 473 patients treated with TKI-based regimens followed by alloHSCT and 486 who received TKI (first, second, or third-generation TKIs) plus chemotherapy without transplantation, the authors demonstrated no difference in OS between both groups; however, patients who underwent alloHSCT had better RFS and non-relapsed survival (NRS) with a cost of higher NRM (Zeng et al. 2021). These results are in agreement with a retrospective analysis of 230 patients who attained CMR within 3 months after TKI (first, second, or third generation)-based regimens in 5 transplant centers in the United States (Ghobadi et al. 2022). In a multivariable analysis, alloHSCT was not associated with improved OS and RFS when compared to transplant-free therapy. However, alloHSCT was associated with a lower cumulative incidence of relapse (CIR) and higher NRM. The achievement of CMR at 3-month therapy, defined as the absence of a detectable BCR-ABL1 transcript with a sensitivity of 0.01%, was shown to be associated with longer median OS and RFS compared with those who achieved a lesser response (127 vs. 38 months, $p = 0.01$; and 126 vs. 18 months, $p = 0.01$, respectively) (Short et al. 2016). In a study by Sasaki K and colleagues from MD Anderson, CMR within 3 months from diagnosis was achieved in 52%, 35%, and 13% of patients treated with ponatinib, dasatinib, and imatinib, respectively (Sasaki et al. 2021). The study did not reveal any additional survival benefits for alloHSCT in patients who achieved CMR at 3 months from diagnosis. Moreover, since ponatinib was the only significant favorable independent factor predicting progression and death, the authors suggest that ponatinib-based regimens may safely negate the necessity of alloHSCT in patients who achieve a 3-month CMR. Similarly, there was no difference in favor of alloHSCT in 53 patients treated with the combination of dasatinib and blinatumomab (Foà et al. 2024). In the group of 24 patients who underwent alloHSCT following dasatinib-blinatumomab therapy, 83.3% of patients were alive in CR at a median follow-up of 49 months, while in the non-transplanted group ($n = 29$), 96.5% of patients remained in persistent CR at a median follow-up of 48 months. It is worth noting, that 93.1% of the patients in the transplanted group achieved CMR compared to only 45.8% in patients who did not undergo alloHSCT which may indirectly suggest, that ALL Ph+ patients who do not achieve CMR after combined therapy of TKI and blinatumomab are the ones who probably benefit from alloHSCT and therefore transplantation should be still recommended in this high-risk group. Most recently, Kantarjian H et al. showed outstanding results with the simultaneous combination of ponatinib and blinatumomab in 60 patients with newly diagnosed Ph+ ALL patients (Kantarjian et al. 2024). This chemotherapy-free treatment resulted in a very high rate of CMR (83%) with 98% of patients receiving MRD-negativity by next-generation sequencing (NGS). Although almost all (97%) patients did not proceed to alloHSCT, the estimated 3-year OS and EFS rates were 91% and 77%, respectively. This indicates, that the combination of ponatinib and blinatumomab as frontline therapy will most probably reduce the need for alloHSCT in the majority of patients with the caveat that accessibility and affordability may turn out to be an issue.

Regardless of the type of frontline therapy, alloHSCT may still play a role in Ph+ ALL patients having ACAs at diagnosis. The presence of ACAs such as complex karyotype, +der (22), +8, and−7 in Ph+ ALL is common and ranges from 40% to 80% (Dombret et al. 2002; Wetzler et al. 2004; Short et al. 2017). In patients treated with TKI-based therapy followed by alloHSCT, the presence of ACAs at diagnosis, was associated with higher CIR and lower PFS (Zheng et al. 2024). Additionally, monosomy 7 correlated with inferior OS (HR 3.02, 95% CI 1.28–7.13, $p = 0.012$) (Zheng et al. 2024). Although some ACAs are associated with poor prognosis regardless of the use of TKI and alloHSCT, there is no targeted therapy to eliminate ACAs (Wetzler et al. 2004; Zheng et al. 2024). Thus, next-generation TKI combined with immunotherapy followed by alloHSCT and TKI-based maintenance therapy seems to be the most attractive and promising option in this setting. Likewise, the presence of some co-mutations was also shown to impact treatment results in Ph+ ALL. The IKZF1 deletion, being the primary co-mutation with BCR-ABL1 oncogene, is present in around 50–70% cases (Slayton et al. 2018; Roberts et al. 2012). Even in the era of TKI-based therapy, the IKZF deletion was found to be significantly associated with unfavorable OS and high CIR (Sasaki et al. 2022). In a phase 2 study assessing dasatinib plus prednisone as induction and consolidation therapy in patients with Ph+ ALL, the presence of IKZF1 mutation was associated with a significantly higher risk of relapse (HR 3.97, 95% CI 1.25–12.59, $p = 0.019$) (Xie et al. 2023a). This may suggest that early bridging to alloHSCT may be the best option for this group of high-risk leukemic patients. The negative impact of IKZF mutation was also shown in patients treated with dasatinib combined with blinatumomab (Foà et al. 2020). With a median follow-up of 18 months, DFS in patients without IKZF mutation was 100%, while in those with IKZF mutation or with IKZF mutation with additional genetic aberrations (i.e., $IKZF^{plus}$) it was 92%, and 64%, respectively ($p = 0.03$). Although the difference was also shown in OS (100%, 93%, and 82%, respectively), it did not reach statistical significance ($p = 0.15$), which suggests that treatment intensification with blinatumomab may overcome at least in part the negative prognostic impact of the IKZF mutation. Another molecular abnormality, CDKN2A/2B gene deletion, accounting for approximately 40% of concurrent mutations in Ph+ ALL is also associated with poor prognosis for long-term outcomes despite TKI-based therapy and alloHSCT (Xu et al. 2016; Pfeifer et al. 2018). However, the efficacy of the combination of TKI with blinatumomab in those patients is unknown. Therefore, further studies on the role of new therapies and alloHSCT in patients with co-mutations are needed to tailor treatments to individual patients.

In conclusion, we believe that alloHSCT still plays a role in the management of Ph+ ALL. In geographical areas in which only first-generation TKI are available or affordable, alloHSCT should be most probably offered for all patients achieving CR1. The incorporation of second and third-generation TKIs in frontline treatments may limit the indications for alloHSCT to patients not achieving CR after the second or subsequent line of therapy but also to patients not achieving CMR within 3 months from diagnosis. Given the high rate of patients receiving CMR after frontline therapy with next-generation TKI and blinatumomab combination and the introduction of CAR-T cells, the percentage of patients that will be referred to transplantation is expected to further decrease. An additional group of patients that still benefit from alloHSCT is patients with high-risk mutations and genetic profiles including those with IKZF, although results should be improved. The patient categories that still benefit from alloHSCT are listed in Table 8.1. Finally, the maintenance therapy with TKI post-alloHSCT plays an important role in reducing the incidence of relapse. However, the type and dose of TKI to be used as a maintenance therapy as well as its duration are still undetermined and should be dictated by their side effects, patient comorbidities, and the genetic and molecular characteristics of the disease. The question of whether or not a patient achieving MRD negativity post CAR-T cell therapy will benefit from consolidation with alloHSCT is highly controversial and may

Table 8.1 The groups of Ph+ ALL patients who still benefit from alloHSCT

(a) Patients in ≥ CR2
(b) Patients who do not achieve MRD negativity after induction and consolidation therapy
(c) Patients treated with first generation TKI in frontline therapy
(d) Patients with adverse additional cytogenetic abnormalities or co-mutations (i.e., IKZF mutation)

Ph+ ALL Philadelphia positive acute lymphoblastic leukemia, *alloHSCT* allogeneic hematopoietic stem cell transplantation, *CR* complete remission, *MRD* measurable residual disease, *TKI* tyrosine kinase inhibitor

Table 8.2 The unanswered and questionable issues in the context of alloHSCT in Ph+ ALL patients

(a) The choice of TKI and its dose for maintenance therapy after alloHSCT
(b) Timing of TKI initiation after alloHSCT
(c) Duration of TKI maintenance therapy after alloHSCT
(d) Possibility of chemotherapy-free treatment before alloHSCT
(e) AlloHSCT as a consolidation therapy after CAR-T cell therapy
(f) Possibility of replacement of alloHSCT with CAR-T cell therapy in the high-risk group of Ph+ ALL patients

Ph+ ALL Philadelphia positive acute lymphoblastic leukemia, *alloHSCT* allogeneic hematopoietic stem cell transplantation, *TKI* tyrosine kinase inhibitor, *CAR-T cell* chimeric antigen receptor T cell

depend on the patient's age being adult or pediatric, the CAR-T construct dictating the persistence of the CAR-T cells, previous alloHSCT, as well as the loss versus continuation of B cell aplasia and probably most important, the NGS based MRD assessment results at day 28 (Marinos and Heslop 2022; Pulsipher et al. 2022). Even more controversial is the speculation whether CAR-T cells will be able to eventually replace alloHSCT in the extremely high-risk group of ALL patients that still benefit from a transplant. These questions are beyond the scope of the current chapter. The controversies about management of Ph+ ALL patients in the context of alloHSCT and new therapies are shown in Table 8.2. Further studies assessing the role of alloHSCT in Ph+ ALL are needed to determine the group of patients who will continue to benefit from transplantation in the era of next generation TKIs and new immunotherapies.

References

Advani AS, Moseley A, O'Dwyer KM, Wood BL, Park J, Wieduwilt M et al (2023) Dasatinib/prednisone induction followed by blinatumomab/dasatinib in Ph+ acute lymphoblastic leukemia. Blood Adv 7(7):1279–1285. https://doi.org/10.1182/bloodadvances.2022008216

Aldoss I, Stiller T, Cao TM, Palmer JM, Thomas SH, Forman SJ et al (2015) Impact of additional cytogenetic abnormalities in adults with Philadelphia chromosome-positive acute lymphoblastic leukemia undergoing allogeneic hematopoietic cell transplantation. Biol Blood Marrow Transplant 21:1326–1329. https://doi.org/10.1016/j.bbmt.2015.03.021

Assi R, Kantarjian H, Short NJ, Daver N, Takahashi K, Garcia-Manero G et al (2017) Safety and efficacy of Blinatumomab in combination with a tyrosine kinase inhibitor for the treatment of relapsed Philadelphia chromosome-positive leukemia. Clin Lymphoma Myeloma Leuk 17(12):897–901. https://doi.org/10.1016/j.clml.2017.08.101

Badar T, Szabo A, Advani A, Wadleigh M, Arslan S, Khan MA et al (2020) Real-world outcomes of adult B-cell acute lymphocytic leukemia patients treated with blinatumomab. Blood Adv 4(10):2308–2316. https://doi.org/10.1182/bloodadvances.2019001381

Bergfelt Lennmyr E, Engvall M, Barbany G, Fogelstrand L, Rhodin H, Hallböök H (2021) Cytogenetic aberrations in adult acute lymphoblastic leukemia-A population-based study. EJHaem 2(4):813–817. https://doi.org/10.1002/jha2.300

Burmeister T, Schwartz S, Bartram CR, Gökbuget N, Hoelzer D, Thiel E; GMALL study group (2008) Patients' age and BCR-ABL frequency in adult B-precursor ALL: a retrospective analysis from the GMALL study group. Blood 112:918–919. https://doi.org/10.1182/blood-2008-04-149286

Candoni A, Chiusolo P, Lazzarotto D, Sartor C, Dargenio M, Chiaretti S et al (2024) Ponatinib as a prophylactic or pre-emptive strategy to prevent cytological relapse after allogeneic stem cell transplantation in patients with Philadelphia chromosome-positive Acute Lymphoblastic Leukemia transplanted in complete cytological remission. Cancers (Basel) 16(11):2108. https://doi.org/10.3390/cancers16112108

Chalandon Y, Thomas X, Hayette S, Cayuela JM, Abbal C, Huguet F et al (2015) Randomized study of reduced-intensity chemotherapy combined with imatinib in adults with Ph-positive acute lymphoblastic leukemia. Blood 125(24):3711–3719. https://doi.org/10.1182/blood-2015-02-627935

Chang J, Douer D, Aldoss I, Vahdani G, Jeong AR, Ghaznavi Z et al (2019) Combination chemotherapy plus dasatinib leads to comparable overall survival and relapse-free survival rates as allogeneic hematopoietic stem cell transplantation in Philadelphia positive acute lymphoblastic leukemia. Cancer Med 8(6):2832–2839. https://doi.org/10.1002/cam4.2153. Epub 2019 Apr 23

Chen H, Liu KY, Xu LP, Liu DH, Chen YH, Zhao XY et al (2012) Administration of imatinib after allogeneic hematopoietic stem cell transplantation may improve disease-free survival for patients with Philadelphia chromosome-positive acute lymphobla stic leukemia. J Hematol Oncol 5:29. Published 2012 Jun 8. https://doi.org/10.1186/1756-8722-5-29

Chen H, Xu LP, Zhang XH, Wang Y, Chen YH, Yan CH et al (2022) Safety and outcomes of maintenance therapy with third-generation tyrosine kinase inhibitor after allogeneic hematopoietic cell transplantation in Philadelphia chromosome-positive acute lymphoblastic leukemia patients with T315I mutation. Leuk Res 121:106930. https://doi.org/10.1016/j.leukres.2022.106930. Epub 2022 Aug 17

Chen LY, Gong WJ, Li MH, Zhou HX, Xu MZ, Qian CS et al (2023) Anti-CD19 CAR T-cell consolidation therapy combined with CD19+ feeding T cells and TKI for Ph+ acute lymphoblastic leukemia. Blood Adv 7(17):4913–4925. https://doi.org/10.1182/bloodadvances.2022009072

Chiaretti S, Ansuinelli M, Vitale A, Elia L, Matarazzo M, Piciocchi A et al (2021) A multicenter total therapy strategy for de novo adult Philadelphia chromosome-positive acute lymphoblastic leukemia patients: final results of the GIMEMA LAL1509 protocol. Haematologica 106(7):1828–1838. https://doi.org/10.3324/haematol.2020.260935

Couturier MA, Thomas X, Raffoux E, Huguet F, Berthon C, Simand C et al (2021) Blinatumomab + ponatinib for relapsed/refractory Philadelphia chromosome-positive acute lymphoblastic leukemia in adults. Leuk Lymphoma 62(3):620–629. https://doi.org/10.1080/10428194.2020.1844198. Epub 2020 Nov 6

Daver N, Thomas D, Ravandi F, Cortes J, Garris R, Jabbour E et al (2015) Final report of a phase II study of imatinib mesylate with hyper-CVAD for the front-line treatment of adult patients with Philadelphia chromosome-positive acute lymphoblastic leukemia. Haematologica 100(5):653–661. https://doi.org/10.3324/haematol.2014.118588

de Labarthe A, Rousselot P, Huguet-Rigal F, Delabesse E, Witz F, Maury S et al; Group for Research on Adult Acute Lymphoblastic Leukemia (GRAALL) (2007) Imatinib combined with induction or consolidation chemotherapy in patients with de novo Philadelphia chromosome-positive acute lymphoblastic leukemia: results of the GRAAPH-2003 study. Blood 109(4):1408–1413

DeAngelo DJ, Stock W, Stein AS, Shustov A, Liedtke M, Schiffer CA et al (2017) Inotuzumab ozogamicin in adults with relapsed or refractory CD22-positive acute lymphoblastic leukemia: a phase 1/2 study. Blood Adv 1(15):1167–1180. https://doi.org/10.1182/bloodadvances.2016001925

Dombret H, Gabert J, Boiron JM, Rigal-Huguet F, Blaise D, Thomas X et al; de Traitement de la Leucémie Aiguë Lymphoblastique de l'Adulte (GET-LALA Group) (2002) Outcome of treatment in adults with Philadelphia chromosome-positive acute lymphoblastic leukemia–results of the prospective multicenter LALA-94 trial. Blood 100(7):2357–2366. https://doi.org/10.1182/blood-2002-03-0704

Druker BJ, Tamura S, Buchdunger E, Ohno S, Segal GM, Fanning S et al (1996) Effects of a selective inhibitor of the Abl tyrosine kinase on the growth of Bcr-Abl positive cells. Nat Med 2(5):561–566

Faderl S, Kantarjian HM, Talpaz M, Estrov Z (1998) Clinical significance of cytogenetic abnormalities in adult acute lymphoblastic leukemia. Blood 91(11):3995–4019

Fielding AK, Rowe JM, Richards SM, Buck G, Moorman AV, Durrant IJ et al (2009) Prospective outcome data on 267 unselected adult patients with Philadelphia chromosome-positive acute lymphoblastic leukemia confirms the superiority of allogeneic transplantation over chemotherapy in the pre-imatinib era: results from the International ALL Trial MRC UKALLXII/ECOG2993. Blood 113:4489–4496. https://doi.org/10.1182/blood-2009-01-199380

Fielding AK, Rowe JM, Buck G, Foroni L, Gerrard G, Litzow MR et al (2014) UKALLXII/ECOG2993: the addition of imatinib to a standard treatment regimen enhances long-term outcomes in Philadelphia-positive acute lymphoblastic leukemia. Blood 123(6):843–850. https://doi.org/10.1182/blood-2013-09-529008. Epub 2013 Nov 25

Foà R, Vitale A, Vignetti M, Meloni G, Guarini A, De Propris MS et al; GIMEMA Acute Leukemia Working Party (2011) Dasatinib as first-line treatment for adult patients with Philadelphia chromosome-positive acute lymphoblastic leukemia. Blood 118(25):6521–6528. https://doi.org/10.1182/blood-2011-05-351403. Epub 2011 Sep 19

Foà R, Bassan R, Vitale A, Elia L, Piciocchi A, Puzzolo MC et al; GIMEMA investigators (2020) Dasatinib-Blinatumomab for Ph-positive Acute Lymphoblastic Leukemia in adults. N Engl J Med 383(17):1613–1623. https://doi.org/10.1056/NEJMoa2016272

Foà R, Bassan R, Elia L, Piciocchi A, Soddu S, Messina M et al (2024) Long-term results of the Dasatinib-Blinatumomab Protocol for adult Philadelphia-positive ALL. J Clin Oncol 42(8):881–885. https://doi.org/10.1200/JCO.23.01075. Epub 2023 Dec 21

Fujisawa S, Mizuta S, Akiyama H, Ueda Y, Aoyama Y, Hatta Y et al (2017) Phase II study of imatinib-based chemotherapy for newly diagnosed BCR-ABL-positive acute lymphoblastic leukemia. Am J Hematol 92(4):367–374. https://doi.org/10.1002/ajh.24653. Epub 2017 Feb 21

Ghobadi A, Slade M, Kantarjian H, Alvarenga J, Aldoss I, Mohammed KA et al (2022) The role of allogeneic transplant for adult Ph+ ALL in CR1 with complete molecular remission: a retrospective analysis. Blood 140(20):2101–2112. https://doi.org/10.1182/blood.2022016194

Giebel S, Czyz A, Ottmann O, Baron F, Brissot E, Ciceri F et al (2016) Use of tyrosine kinase inhibitors to prevent relapse after allogeneic hematopoietic stem cell transplantation for patients with Philadelphia chromosome-positive acute lymphoblastic leukemia: a position statement of the Acute Leukemia Working Party of the European Society for Blood and marrow transplantation. Cancer 122(19):2941–2951. https://doi.org/10.1002/cncr.30130. Epub 2016 Jun 16

Giebel S, Labopin M, Peric Z, Passweg J, Blaise D, Salmenniemi U et al (2024) Impact of the type of tyrosine kinase inhibitor (imatinib or dasatinib) used before allo-HCT on outcome of patients with Philadelphia-positive acute lymphoblastic leukemia. A study on behalf of the Acute Leukemia Working Party of the EBMT. Transplant Cell Ther S2666–6367(24):00547–00545. https://doi.org/10.1016/j.jtct.2024.07.016

Guan F, Yang L, Chen Y, Shi J, Song X, Lai X et al (2024) Comparison of long-term outcomes between imatinib and dasatinib prophylaxis after allogeneic stem cell transplantation in patients with Philadelphia-positive acute lymphoblastic leukemia: a multicenter retrospective study. Cancer 130(12):2139–2149. https://doi.org/10.1002/cncr.35232

Hanif A, Wang ES, Thompson JE, Baron JI, Walsh MD, Griffiths EA (2018) Combining blinatumomab with targeted therapy for BCR-ABL mutant relapsed/refractory acute lymphoblastic leukemia. Leuk Lymphoma 59(8):2011–2013. https://doi.org/10.1080/10428194.2017.1411595

Hu L, Li Z, Yang S, Zhao T, Duan W, Qin Y et al (2024) Allogeneic stem cell transplantation is still a highly curative therapy in adults with Philadelphia chromosome-positive acute lymphoblastic leukaemia. Ann Hematol 103:3745. https://doi.org/10.1007/s00277-024-05682-3

Jabbour E, Short NJ, Ravandi F, Huang X, Daver N, DiNardo CD et al (2018a) Combination of hyper-CVAD with ponatinib as first-line therapy for patients with Philadelphia chromosome-positive acute lymphoblastic leukaemia: long-term follow-up of a single-center, phase 2 study. Lancet Haematol 5(12):e618–e627. https://doi.org/10.1016/S2352-3026(18)30176-5

Jabbour E, DerSarkissian M, Duh MS, McCormick N, Cheng WY, McGarry LJ et al (2018b) Efficacy of Ponatinib versus earlier generation tyrosine kinase inhibitors for front-line treatment of newly diagnosed Philadelphia-positive Acute Lymphoblastic Leukemia. Clin Lymphoma Myeloma Leuk 18(4):257–265. https://doi.org/10.1016/j.clml.2018.02.010. Epub 2018 Feb 17

Jabbour E, Short NJ, Jain N, Huang X, Montalban-Bravo G, Banerjee P et al (2023) Ponatinib and blinatumomab for Philadelphia chromosome-positive acute lymphoblastic leukaemia: a US, single-centre, single-arm, phase 2 trial. Lancet Haematol 10(1):e24–e34. https://doi.org/10.1016/S2352-3026(22)00319-2. Epub 2022 Nov 16

Jabbour E, Kantarjian HM, Aldoss I, Montesinos P, Leonard JT, Gómez-Almaguer D et al (2024a) Ponatinib vs Imatinib in frontline Philadelphia chromosome-positive Acute Lymphoblastic Leukemia: a randomized clinical trial. JAMA 331(21):1814–1823. https://doi.org/10.1001/jama.2024.4783

Jabbour E, Haddad FG, Short NJ, Senapati J, Jain N, Sasaki K et al (2024b) Phase 2 study of inotuzumab ozogamicin for measurable residual disease in acute lymphoblastic leukemia in remission. Blood 143(5):417–421. https://doi.org/10.1182/blood.2023022330

Jain N, Maiti A, Ravandi F, Konopleva M, Daver N, Kadia T et al (2021) Inotuzumab ozogamicin with bosutinib for relapsed or refractory Philadelphia chromosome positive acute lymphoblastic leukemia or lymphoid blast phase of chronic myeloid leukemia. Am J Hematol 96(8):1000–1007. https://doi.org/10.1002/ajh.26238. Epub 2021 May 28

Kantarjian HM, DeAngelo DJ, Stelljes M, Martinelli G, Liedtke M, Stock W et al (2016) Inotuzumab Ozogamicin versus standard therapy for acute lymphoblastic Leukemia. N Engl J Med 375(8):740–753. https://doi.org/10.1056/NEJMoa1509277. Epub 2016 Jun 12

Kantarjian HM, DeAngelo DJ, Stelljes M, Liedtke M, Stock W, Gökbuget N et al (2019) Inotuzumab ozogamicin versus standard of care in relapsed or refractory acute lymphoblastic leukemia: final report and long-term survival follow-up from the randomized, phase 3 INO-VATE study. Cancer 125(14):2474–2487. https://doi.org/10.1002/cncr.32116. Epub 2019 Mar 28

Kantarjian H, Short NJ, Jain N, Sasaki K, Huang X, Haddad FG et al (2023) Frontline combination of ponatinib and hyper-CVAD in Philadelphia chromosome-positive acute lymphoblastic leukemia: 80-months follow-up results. Am J Hematol 98(3):493–501. https://doi.org/10.1002/ajh.26816

Kantarjian H, Short NJ, Haddad FG, Jain N, Huang X, Montalban-Bravo G et al (2024) Results of the simultaneous combination of Ponatinib and Blinatumomab in Philadelphia chromosome-positive ALL. J Clin Oncol 42:4246-4251 https://doi.org/10.1200/JCO.24.00272

Kim DY, Joo YD, Lim SN, Kim SD, Lee JH, Lee JH et al (2015) Nilotinib combined with multiagent chemotherapy for newly diagnosed Philadelphia-positive acute lymphoblastic leukemia. Blood 126(6):746–756. https://doi.org/10.1182/blood-2015-03-636548

Lee KH, Lee JH, Choi SJ, Lee JH, Seol M, Lee YS et al (2005) Clinical effect of imatinib added to intensive combination chemotherapy for newly diagnosed Philadelphia chromosome-positive acute lymphoblastic leukemia. Leukemia 19(9):1509–1516. https://doi.org/10.1038/sj.leu.2403886

Leotta S, Markovic U, Pirosa MC, Stella S, Tringali S, Martino M et al (2021) The role of ponatinib in adult BCR-ABL1 positive acute lymphoblastic leukemia after allogeneic transplantation: a real-life retrospective multicenter study. Ann Hematol 100(7):1743–1753. https://doi.org/10.1007/s00277--021-04504-0. Epub 2021 Mar 28

Liu H, Xuan L, Lin R, Deng L, Fan Z, Nie D et al (2021) A new pre-emptive TKIs strategy for preventing relapse based on BCR/ABL monitoring for Ph+ALL undergoing allo-HCT: a prospective clinical cohort study. Leukemia 35(7):2054–2063. https://doi.org/10.1038/s41375-020-01090-4

Liu T, Wang C, Fu Y, Yang YP, Tan YH (2024) Olverembatinib combined with inotuzumab ozogamicin in relapsed refractory Philadelphia chromosome-positive acute lymphoblastic leukemia: a case report. Medicine (Baltimore) 103(29):e38985. https://doi.org/10.1097/MD.0000000000038985

Marinos A, Heslop HE (2022) Should all CAR-T therapy for acute lymphoblastic leukemia be consolidated with an allogeneic stem cell transplant? Best Pract Res Clin Hematol 35:101414

Martinelli G, Boissel N, Chevallier P, Ottmann O, Gökbuget N, Topp MS et al (2017) Complete hematologic and molecular response in adult patients with relapsed/refractory Philadelphia chromosome-positive B-precursor acute lymphoblastic leukemia following treatment with Blinatumomab: results from a Phase II, single-arm, multicenter study. J Clin Oncol. 2017 Jun 1;35(16):1795-1802. doi: 10.1200/JCO.2016.69.3531. Epub 2017 Mar 29. Erratum in: J Clin Oncol. 2017 Aug 10;35(23):2722. doi: 10.1200/JCO.2017.74.7550. Erratum in: J Clin Oncol. 2017 Aug 20;35(24):2856. doi: 10.1200/JCO.2017.75.0489

Martinelli G, Boissel N, Chevallier P, Ottmann O, Gökbuget N, Rambaldi A et al (2021) Long-term follow-up of blinatumomab in patients with relapsed/refractory Philadelphia chromosome-positive B-cell precursor acute lymphoblastic leukaemia: final analysis of ALCANTARA study. Eur J Cancer 146:107–114. https://doi.org/10.1016/j.ejca.2020.12.022. Epub 2021 Feb 13

Martinelli G, Papayannidis C, Piciocchi A, Robustelli V, Soverini S, Terragna C et al (2022) INCB84344-201: Ponatinib and steroids in frontline therapy for unfit patients with Ph+ acute lymphoblastic leukemia. Blood Adv 6(6):1742–1753. https://doi.org/10.1182/bloodadvances.2021004821

Maude SL, Laetsch TW, Buechner J, Rives S, Boyer M, Bittencourt H et al (2018) Tisagenlecleucel in children and young adults with B-cell lymphoblastic leukemia. N Engl J Med 378(5):439–448. https://doi.org/10.1056/NEJMoa1709866

Micheva I, Gerov V, Dimitrova S, Efraim M (2021) Efficacy of Inotuzumab Ozogamicin plus Ponatinib followed by allogeneic stem cell transplantation in a patient with relapsed Philadelphia chromosome-positive Acute Lymphoblastic Leukemia. Case Rep Hematol 2021:1717506. Published 2021 May 26. https://doi.org/10.1155/2021/1717506

Nagel I, Bartels M, Duell J, Oberg HH, Ussat S, Bruckmueller H et al (2017) Hematopoietic stem cell involvement in BCR-ABL1-positive ALL as a potential mechanism of resistance to blinatumomab therapy. Blood 130(18):2027–2031. https://doi.org/10.1182/blood-2017-05-782888

Nanno S, Matsumoto K, Nakamae M, Okamura H, Nishimoto M, Hirose A et al (2020) Effect of prophylactic post-transplant ponatinib administration on outcomes in patients with Philadelphia chromosome-positive acute lymphoblastic leukemia. Clin Lymphoma Myeloma Leuk 20(12):813–819.e1. https://doi.org/10.1016/j.clml.2020.07.005. Epub 2020 Jul 16

Nishiwaki S, Sugiura I, Fujisawa S, Hatta Y, Atsuta Y, Doki N et al; Japan Adult Leukemia Study Group (2024) Utility of allogeneic stem cell transplantation for adult Ph+ALL with complete molecular remission. Am J Hematol 99(5):806–815. https://doi.org/10.1002/ajh.27237. Epub 2024 Feb 5

O'Hare T, Shakespeare WC, Zhu X, Eide CA, Rivera VM, Wang F et al (2009) AP24534, a pan-BCR-ABL inhibitor for chronic myeloid leukemia, potently inhibits the T315I mutant and overcomes mutation-based resistance. Cancer Cell 16(5):401–412. https://doi.org/10.1016/j.ccr.2009.09.028

Oliver G, Ottmann HP, Cayuela J-M, Spiekermann K, Beck J, Jung WE et al (2014) Nilotinib (Tasigna®) and chemotherapy for first-line treatment in elderly patients with De Novo Philadelphia chromosome/BCR-ABL1 positive Acute Lymphoblastic Leukemia (ALL): a trial of the European Working Group for Adult ALL (EWALL-PH-02). Blood 124(21):798. https://doi.org/10.1182/blood.V124.21.798.798

Ottmann O, Dombret H, Martinelli G, Simonsson B, Guilhot F, Larson RA et al (2007) Dasatinib induces rapid hematologic and cytogenetic responses in adult patients with Philadelphia chromosome-positive acute lymphoblastic leukemia with resistance or intolerance to imatinib: interim results of a phase 2 study. Blood 110(7):2309–2315. https://doi.org/10.1182/blood--2007-02-073528. Epub 2007 May 11

Pfeifer H, Wassmann B, Bethge W, Dengler J, Bornhäuser M, Stadler M et al (2013) Randomized comparison of prophylactic and minimal residual disease-triggered imatinib after allogeneic stem cell transplantation for BCR-ABL1-positive acute lymphoblastic leukemia. Leukemia 27(6):1254–1262. https://doi.org/10.1038/leu.2012.352

Pfeifer H, Raum K, Markovic S, Nowak V, Fey S, Obländer J et al (2018) Genomic CDKN2A/2B deletions in adult Ph+ ALL are adverse despite allogeneic stem cell transplantation. Blood 131(13):1464–1475. https://doi.org/10.1182/blood-2017-07-796862. Epub 2018 Jan 18

Platanias LC (2011) Mechanisms of BCR-ABL leukemogenesis and novel targets for the treatment of chronic myeloid leukemia and Philadelphia chromosome-positive acute lymphoblastic leukemia. Leuk Lymphoma 52(Suppl 1):2–3. https://doi.org/10.3109/10428194.2010.546922

Pulsipher MA, Han X, Maude SL, Laetsch TW, Qayed M, Rives S et al (2022) Next-generation sequencing of minimal residual disease for predicting relapse after Tisagenlecleucel in children and young adults with Acute Lymphoblastic Leukemia. Blood Cancer Discov 3(1):66–81. https://doi.org/10.1158/2643-3230.BCD-21-0095

Ravandi F, O'Brien SM, Cortes JE, Thomas DM, Garris R, Faderl S et al (2015) Long-term follow-up of a phase 2 study of chemotherapy plus dasatinib for the initial treatment of patients with Philadelphia chromosome-positive acute lymphoblastic leukemia. Cancer 121(23):4158–4164. https://doi.org/10.1002/cncr.29646. Epub 2015 Aug 26

Ravandi F, Othus M, O'Brien SM, Forman SJ, Ha CS, Wong JYC et al (2016) US Intergroup Study of Chemotherapy Plus Dasatinib and allogeneic stem cell transplant in Philadelphia chromosome positive ALL. Blood Adv 1(3):250–259. https://doi.org/10.1182/bloodadvances.2016001495

Ribera JM, García O, Montesinos P, Brunet S, Abella E, Barrios M et al (2012) Treatment of young patients with Philadelphia chromosome-positive acute lymphoblastic leukaemia using increased dose of imatinib and deintensified chemotherapy before allogeneic stem cell transplantation. Br J Haematol 159(1):78–81. https://doi.org/10.1111/j.1365-2141.2012.09240.X. Epub 2012 Jul 24

Ribera JM, García-Calduch O, Ribera J, Montesinos P, Cano-Ferri I, Martínez P et al (2022) Ponatinib, chemotherapy, and transplant in adults with Philadelphia chromosome-positive acute lymphoblastic leukemia. Blood Adv 6(18):5395–5402. https://doi.org/10.1182/bloodadvances.2022007764

Ribera JM, Prawitz T, Freitag A, Sharma A, Dobi B, Rizzo F et al (2023) Ponatinib vs. Imatinib as frontline treatment for Philadelphia chromosome-positive Acute Lymphoblastic Leukemia: a matching adjusted indirect comparison. Adv Ther 40(7):3087–3103. https://doi.org/10.1007/s12325-023-02497-y. Epub 2023 May 19

Ribera JM, Morgades M, Ribera J, Montesinos P, Cano-Ferri I, Martínez P et al (2024) Ponalfil trial for adults with Philadelphia chromosome-positive acute lymphoblastic leukemia: long-term results. Hema 8(4):e67. Published 2024 Apr 2. https://doi.org/10.1002/hem3.67

Roberts KG, Morin RD, Zhang J, Hirst M, Zhao Y, Su X et al (2012) Genetic alterations activating kinase and cytokine receptor signaling in high-risk acute lymphoblastic leukemia. Cancer Cell 22(2):153–166. https://doi.org/10.1016/j.ccr.2012.06.005

Rousselot P, Coudé MM, Gokbuget N, Gambacorti Passerini C, Hayette S, Cayuela JM et al; European Working Group on Adult ALL (EWALL) group (2016) Dasatinib and low-intensity chemotherapy in elderly patients with Philadelphia chromosome-positive ALL. Blood 128(6):774–782. https://doi.org/10.1182/blood-2016-02-700153. Epub 2016 Apr 27

Sasaki K, Jabbour EJ, Ravandi F, Short NJ, Thomas DA, Garcia-Manero G et al (2016) Hyper-CVAD plus ponatinib versus hyper-CVAD plus dasatinib as frontline therapy for patients with Philadelphia chromosome-positive acute lymphoblastic leukemia: a propensity score analysis. Cancer 122(23):3650–3656. https://doi.org/10.1002/cncr.30231. Epub 2016 Aug 1

Sasaki K, Kantarjian HM, Short NJ, Samra B, Khoury JD, Kanagal Shamanna R et al (2021) Prognostic factors for progression in patients with Philadelphia chromosome-positive acute lymphoblastic leukemia in complete molecular response within 3 months of therapy with tyrosine kinase inhibitors. Cancer 127(15):2648–2656. https://doi.org/10.1002/cncr.33529. Epub 2021 Apr 1

Sasaki Y, Kantarjian HM, Short NJ, Wang F, Furudate K, Uryu H et al (2022) Genetic correlates in patients with Philadelphia chromosome-positive acute lymphoblastic leukemia treated with Hyper-CVAD plus dasatinib or ponatinib. Leukemia 36(5):1253–1260. https://doi.org/10.1038/s41375-021-01496-8. Epub 2022 Feb 7. Erratum in: Leukemia. 2022 May; 36(5):1448. https://doi.org/10.1038/s41375-022-01568-3

Shah NP, Tran C, Lee FY, Chen P, Norris D, Sawyers CL (2004) Overriding imatinib resistance with a novel ABL kinase inhibitor. Science (New York, NY) 305(5682):399–401. https://doi.org/10.1126/science.1099480

Shah BD, Ghobadi A, Oluwole OO, Logan AC, Boissel N, Cassaday RD et al (2021) KTE-X19 for relapsed or refractory adult B-cell acute lymphoblastic leukaemia: phase 2 results of the single-arm, open-label, multicentre ZUMA-3 study. Lancet 398(10299):491–502. https://doi.org/10.1016/S0140-6736(21)01222-8

Shahzad M, Hussain A, Tariq E, Anwar I, Faisal MS, Syed L et al (2023) Outcomes of tyrosine kinase inhibitors maintenance therapy with or without allogeneic hematopoietic stem cell transplantation in Philadelphia chromosome positive Acute Lymphoblastic Leukemia in first complete remission: a systematic review and meta-analysis. Clin Lymphoma Myeloma Leuk 23(3):178–187. https://doi.org/10.1016/j.clml.2023.01.002. Epub 2023 Jan 12

Shi T, Xie M, Chen L, Yuan W, Wang Y, Huang X et al (2022) Distinct outcomes, ABL1 mutation profile, and transcriptome features between p190 and p210 transcripts in adult Philadelphia-positive acute lymphoblastic leukemia in the TKI era. Exp Hematol Oncol 11(1):13. https://doi.org/10.1186/s40164-022-00265-2

Short NJ, Jabbour E, Sasaki K, Patel K, O'Brien SM, Cortes JE et al (2016) Impact of complete molecular response on survival in patients with Philadelphia chromosome-positive acute lymphoblastic leukemia. Blood 128(4):504–507. https://doi.org/10.1182/blood--2016-03-707562. Epub 2016 May 27

Short NJ, Kantarjian HM, Sasaki K, Ravandi F, Ko H, Cameron Yin C et al (2017) Poor outcomes associated with +der (22)t(9;22) and −9/9p in patients with Philadelphia chromosome-positive acute lymphoblastic leukemia receiving chemotherapy plus a tyrosine kinase inhibitor. Am J Hematol 92(3):238–243. https://doi.org/10.1002/ajh.24625. Epub 2017 Feb 3

Slayton WB, Schultz KR, Kairalla JA, Devidas M, Mi X, Pulsipher MA et al (2018) Dasatinib plus intensive chemotherapy in children, adolescents, and young adults with Philadelphia chromosome-positive acute lymphoblastic leukemia: results of children's oncology group trial AALL0622. J Clin Oncol 36(22):2306–2314. https://doi.org/10.1200/JCO.2017.76.7228. Epub 2018 May 29

Stock W, Martinelli G, Stelljes M, DeAngelo DJ, Gökbuget N, Advani AS et al (2021) Efficacy of inotuzumab ozogamicin in patients with Philadelphia chromosome-positive relapsed/refractory acute lymphoblastic leukemia. Cancer 127(6):905–913. https://doi.org/10.1002/cncr.33321. Epub 2020 Nov 24

Sugiura I, Doki N, Hata T, Cho R, Ito T, Suehiro Y et al (2022) Dasatinib-based 2-step induction for adults with Philadelphia chromosome-positive acute lymphoblastic leukemia. Blood Adv 6(2):624–636. https://doi.org/10.1182/bloodadvances.2021004607

Thomas DA, Faderl S, Cortes J, O'Brien S, Giles FJ, Kornblau SM et al (2004) Treatment of Philadelphia chromosome-positive acute lymphocytic leukemia with hyper-CVAD and imatinib mesylate. Blood 103(12):4396–4407

Tokarski JS, Newitt JA, Chang CY, Cheng JD, Wittekind M, Kiefer SE et al (2006) The structure of Dasatinib (BMS-354825) bound to activated ABL kinase domain elucidates its inhibitory activity against imatinib-resistant ABL mutants. Cancer Res 66(11):5790–5797. https://doi.org/10.1158/0008-5472.Can-05-4187

Towatari M, Yanada M, Usui N, Takeuchi J, Sugiura I, Takeuchi M et al; Japan Adult Leukemia Study Group (2004) Combination of intensive chemotherapy and imatinib can rapidly induce high-quality complete remission for a majority of patients with newly diagnosed BCR-ABL-positive acute lymphoblastic leukemia. Blood 104(12):3507–3512

Wassmann B, Pfeifer H, Goekbuget N, Beelen DW, Beck J, Stelljes M et al (2006) Alternating versus concurrent schedules of imatinib and chemotherapy as front-line therapy for Philadelphia-positive acute lymphoblastic leukemia (Ph+ ALL). Blood 108(5):1469–1477

Wetzler M, Dodge RK, Mrózek K, Stewart CC, Carroll AJ, Tantravahi R et al (2004) Additional cytogenetic abnormalities in adults with Philadelphia chromosome-positive acute lymphoblastic leukaemia: a study of the cancer and Leukaemia group B. Br J Haematol 124(3):275–288. https://doi.org/10.1046/j.1365-2141.2003.04736.x

Xie M, Lu Y, Ouyang G, Li X, Shi T, Yang M et al (2023a) Dasatinib plus prednisone as induction and consolidation for adults with Ph-positive acute lymphoblastic leukaemia: a single-arm, multicentre, phase 2 trial. Br J Haematol 202(6):1119–1126. https://doi.org/10.1111/bjh.18975

Xie M, Shi T, Jiang Q, Jia Y, Zhou D, Tong H et al (2023b) Chemotherapy with the use of next-generation tyrosine kinase inhibitors based on measurable residual disease has the potential to avoid hematopoietic stem cell transplantation in treatment for adults with Philadelphia chromosome-positive acute lymphoblastic leukemia. Cancer 129(10):1523–1536. https://doi.org/10.1002/cncr.34710. Epub 2023 Mar 7

Xu N, Li YL, Li X, Zhou X, Cao R, Li H et al (2016) Correlation between deletion of the CDKN2 gene and tyrosine kinase inhibitor resistance in adult Philadelphia chromosome-positive acute lymphoblastic leukemia. J Hematol Oncol 9:40. https://doi.org/10.1186/s13045-016-0270-5

Yurkiewicz IR, Muffly L, Liedtke M (2018) Inotuzumab ozogamicin: a CD22 mAb-drug conjugate for adult relapsed or refractory B-cell precursor acute lymphoblastic leukemia. Drug Des Devel Ther 12:2293–2300. https://doi.org/10.2147/DDDT.S150317

Zeng Q, Xiang B, Liu Z (2021) Comparison of allogeneic hematopoietic stem cell transplantation and TKI combined with chemotherapy for adult Philadelphia chromosome-positive acute lymphoblastic leukemia: a systematic review and meta-analysis. Cancer Med 10(24):8741–8753. https://doi.org/10.1002/cam4.4413. Epub 2021 Nov 11

Zheng J, Zhao Y, Luo Y, Yu J, Lai X, Wang J et al (2024) Impact of additional cytogenetic aberrations at diagnosis on prognosis of adult patients with Philadelphia chromosome-positive acute lymphoblastic leukemia undergoing allogeneic hematopoietic cell transplantation: a retrospective analysis. Ann Hematol 103(8):2983–2991. https://doi.org/10.1007/s00277--024-05871-0. Epub 2024 Jul 4

Zimmerman Z, Maniar T, Nagorsen D (2015) Unleashing the clinical power of T cells: CD19/CD3 bi-specific T cell engager (BiTE®) antibody construct blinatumomab as a potential therapy. Int Immunol 27(1):31–37. https://doi.org/10.1093/intimm/dxu089

Myelofibrosis: Timing of Transplantation and Management of Splenomegaly

9

Nicolaus Kröger, Kristin Rathje, and Nico Gagelmann

Abstract

Allogeneic hematopoietic stem cell transplantation (HSCT) is a potentially curative treatment option for patients with myelofibrosis. Because of inherent risk of the procedure, optimal timing in the course of the disease is crucial in order to balance risk and benefit. Besides optimal timing, reducing spleen size—a hallmark of the disease—is needed to ensure timely and sufficient engraftment after HSCT.

Disease- and transplant-specific risk scores have been developed to predict the outcome of myelofibrosis with (MTSS) and without transplantation (DIPSS, MIPSS, and MYSEC), and current recommendation is to offer HSCT in primary or secondary myelofibrosis patients up to 70 years of age with intermediate-2 or high-risk DIPSS or MYSEC score or high risk according to MIPSS70 and a low or intermediate risk according to MTSS. Regarding the spleen size, the current European recommendation is for patients with splenomegaly greater than 5 cm below the lower costal margin or splenomegaly-related symptoms to receive a spleen-directed treatment, ideally with JAK-inhibitor and in case of resistance to use second-generation JAK inhibitor or spleen irradiation close to the transplant or splenectomy.

Keywords

Myelofibrosis · JAK inhibitor · Spleen irradiation · Timing of allogeneic stem cell transplantation

9.1 Introduction

Primary myelofibrosis (PMF) and post-polycythemia vera or -essential thrombocythemia myelofibrosis (post-PV/ET MF) are Philadelphia chromosome-negative hematological diseases classified as myeloproliferative neoplasms with heterogeneous clinical courses and survival between less than 2 years and more than 10 years. Hallmarks of the disease are splenomegaly, constitutional symptoms, and progressive bone marrow failure. Basic research in the last decade has provided more insight into pathogenesis and molecular genetics of the disease (Arber et al. 2022). The development and approval of JAK inhibitors has broadened the treatment options for myelofibrosis and improved quality of life by reducing spleen size and improving constitutional symptoms, although this treatment is not targeting the underlying

N. Kröger (✉) · K. Rathje · N. Gagelmann
Department of Stem Cell Transplantation, University Medical Center Hamburg, Hamburg, Germany
e-mail: n.kroeger@uke.de

© The Author(s), under exclusive license to Springer Nature Switzerland AG 2025

K. Ballen (ed.), *Clinical Management of Challenges in Allogeneic Stem Cell Transplant*, Advances in Experimental Medicine and Biology 1475, https://doi.org/10.1007/978-3-031-84988-6_9

malignant stem cells and thus is not curative (Verstovsek et al. 2012; Harrison et al. 2012).

Allogeneic hematopoietic stem cell transplantation (HSCT) is a potentially curative treatment option for patients with myelofibrosis but due to its complexity as well as therapy-associated morbidity and mortality not all myelofibrosis patients are suitable candidates for this procedure. However, in recent years due to improvement in the prevention and treatment of infectious complications and graft-versus-host disease as well as the introduction of less toxic, reduced-intensity conditioning regimen, numbers of HSCT, which has become standard of care treatment for higher risk myelofibrosis patients, are steadily increasing (McLornan et al. 2021; Kroger et al. 2015, 2024).

However, due to the aforementioned risk of the procedure and the heterogeneous courses of the disease, a careful balancing between risk and benefit is necessary in order to determine optimal timing in the course of the disease with the lowest risk of therapy-related mortality and the highest chance of cure. Besides optimal timing a proper preparation for the transplantation is needed to lower transplant related complications by reducing spleen size, iron overload, and constitutional symptoms among others.

9.2 Risk Stratification

To allow prognostication in this heterogeneous disease several scoring systems have been established which allow a more individualized counseling of patient and selection of therapies (Passamonti and Mora 2023). While the International Prognostic Scoring System (IPSS) was developed to determine the prognosis at time of diagnosis, the Dynamic International Prognostic Scoring System (DIPSS) allows to determine survival at any time period during the course of the disease (Passamonti et al. 2010). This scoring system was further developed by adding cytogenetics and transfusion dependency (DIPSS plus) (Gangat et al. 2011). The Mutation-Enhanced International Prognostic Scoring System (MIPSS70) included unfavorable mutations such as ASXL1, EZH2, IDH1, IDH2, and SRSF2 among other factors for patients up to 70 years who were potentially eligible for allogeneic stem cell transplantation (Guglielmelli et al. 2018). A further revision termed MIPSS70-plus incorporated the mutation U2AF1 as additional unfavorable factor (MIPSS70-plus version 2.0) (Tefferi et al. 2018a). A genetically inspired scoring system which was exclusively based on molecular and cytogenetic variables was also introduced (Tefferi et al. 2018b). For post-PV/ET myelofibrosis the Myelofibrosis Secondary to PV and ET Prognostic Model (MYSEC-PM) was developed (Passamonti et al. 2017). In clinical practice DIPSS, DIPSS-plus, MIPSS70, MIPSS70-plus 2.0, and MYSEC-PM are constantly used and are listed in Table 9.1 including the variables, the scoring as well as the median survival according to the different risk scores.

Other unfavorable mutations such as TP53 have been shown to impact survival in myelofibrosis patients (Grinfeld et al. 2018; Santos et al. 2020). In transplant setting only TP53 mutated configuration was associated with worse outcome and leukemic transformation after transplantation (Gagelmann et al. 2023). Due to lack of prospective, randomized trials these disease-specific scores have been used to compare patients who did or did not receive HSCT from transplant or non-transplant registries with statistical methods such as left truncation. In both studies from Europe and the United States for low-risk patients, according to DIPSS, there was no advantage for an early HSCT while a clear benefit for immediate transplantation was seen for DIPSS intermediate-2 and high-risk patients. Uncertainty remained for intermediate-1 patients because the comparison was inconclusive with no statistical significance (Kroger et al. 2015; Gowin et al. 2020).

Because DIPSS, MIPSS, and MYSEC are disease-specific risk scores which do not take transplant-specific risk factors into account the Myelofibrosis Transplant Scoring System (MTSS) was developed to properly counsel patients on transplant outcome. Beside disease-specific risk factors also patient-specific (Karnofsky score and age) and transplant-specific

Table 9.1 Disease-specific scoring systems

	Disease-specific				
	PMF				Post ET/PV MF
Factor	DIPSS (scoring points)	DIPSS+ (scoring points)	MIPSS70 (scoring points)	MIPSS70+ v.2 (scoring points)	MYSEC-PM (scoring points)
Hemoglobin (g/dl)	<10 (2)	<10 (2)	<10 (1)	Moderate anemia[a] (1)	<11 (2)
				Severe (2)	
Leukocytes ($\times 10^9$/l)	>25 (1)	>25 (1)	>25 (2)		
Platelets ($\times 10^9$/l)		<100 (1)	<100 (2)		<150 (1)
Circulating blasts (%)	≥1 (1)	≥1 (1)	≥2 (1)	≥2 (2)	≥3 (1)
Constitutional symptoms	Yes (1)	Yes (1)	Yes (1)	Yes (2)	Yes (1)
Fibrosis grade			≥2 (1)		
Age (years)	>65 (1)	>65 (1)			Years x 0.15
RBC transfusion-dependent		Yes (1)			
Cytogenetics		Unfavorable[b] (1)		Unfavorable[c] (3)	
				VHR (4)	
Driver mutation			Non-*CALR* type 1 (1)	Non-*CALR* type 1 (2)	Non-*CALR* (2)
Other mutations			HMR present[d] (1)	HMR present[e] (2)	
			≥2 HMR mutations (2)	≥2 HMR mutations (3)	
Score	Low (0)	Low (0)	Low (0–1)	Very low (0)	Low (<11)
	Intermediate-1 (1–2)	Intermediate-1 (1)	Intermediate (2–4)	Low (1, 2)	Intermediate-1 (11–13)
	Intermediate-2 (3–4)	Intermediate-2 (2–3)	High (≥5)	Intermediate (3, 4)	Intermediate-2 (14–15)
	High (≥5)	High (4–6)		High (5–8)	High (≥16)
				Very high (≥9)	
Survival	(median, 5-year %)	(median, 5-year %)	(median, 5-year %)	(median, 5-year %)	(median, 5-year %)
	Low (NR, 95)	Low (15 y, 90)	Low (NR, 96)	Very low (NR, 95)	Low (NR, 90)
	Intermediate-1 (14 y, 80)	Intermediate-1 (7 y, 70)	Intermediate (6 y, 67)	Low (1 y, 85)	Intermediate-1 (9 y, 80)
	Intermediate-2 (4 y, 40)	Intermediate-2 (3 y, 40)	High (3 y, 34)	Intermediate (8 y, 70)	Intermediate-2 (4 y, 45)
	High (2 y, 5)	High (1 y, 10)		High (y, 40)	High (2 y, 20)
				Very high (2 y, 10)	

Abbreviations: y, years; RBC, red blood cell; PMF, primary myelofibrosis; SMF, secondary myelofibrosis post polycythemia vera and essential thrombocythemia; HMR, high molecular risk; VHR, very high risk

[a]Moderate (<8 g/dl in women and < 9 g/dL in men), severe (8–9.9 g/dl in women and 9–10.9 g/dl in men)

[b]Complex karyotype or sole or 2 abnormalities including +8, −7/7q-, i(17q), −5/5q-, 12p-, inv(3) or 11q23 rearrangement

[c]Unfavorable (chromosomal abnormalities except VHR or sole 13q-, +9, 20q-, chromosome 1 translocation/duplication or sex chromosome alterations including -Y, VHR (Single/multiple abnormalities of −7, i(17q), inv(3)/3q21, 12p-/12p11.2, 11q-/11q23, +21, or other autosomal trisomies except +8/9)

[d]HMR (presence and number of *ASXL1, EZH2, SRSF2,* or *IDH1/2*)

[e]Original HMR and *U2AF1Q157*

Table 9.2 Myelofibrosis Transplant Scoring System (MTSS)

	Transplant
Factor	MTSS (scoring points)
Leukocytes ($\times 10^9$/l)	>25 (1)
Platelets ($\times 10^9$/l)	<150 (1)
Karnofsky status (%)	<90 (1)
Age (years)	≥57 (1)
Driver mutation	Non-*CALR/MPL* (2)
Other mutations	*ASXL1* (1)
Donor relation	Mismatched unrelated (2)
Score	Low (0–2)
	Intermediate (3, 4)
	High (5)
	Very high (6–9)
Survival	(median, 5-year %)
	Low (NR, 90)
	Intermediate (14 y, 77)
	High (5 y, 50)
	Very high (1 y, 34)

To predict outcome after allogeneic stem cell transplantation: in brackets: scoring points

factor (HLA match of the donor) were included in this model (see Table 9.2). The MTSS was better in predicting survival after transplantation than DIPSS, MIPSS, or MYSEC and can be used for both PMF and post-PV/ET MF (Gagelmann et al. 2019). Besides predicting probability of survival the MTSS was also useful regarding non-relapse mortality. According to MTSS, the 5-year survival and non-relapse mortality was 90% and 10% for low, 78% and 22% for intermediate, 63% and 36% for high, and 43% and 57% for very high, respectively (Gagelmann et al. 2019).

The current recommendations for HSCT from the European Society for Blood and Marrow Transplantation (EBMT) and the European LeukemiaNet (ELN) is to combine disease-specific and transplant-specific scores (Kroger et al. 2024): *Patients with primary or secondary myelofibrosis and intermediate-2 or high risk DIPSS or MYSEC score or high risk according to MIPSS70 and a low or intermediate risk according to MTSS are candidates for allogeneic stem cell transplantation.*

For patients with DIPSS intermediate-1 or intermediate risk according to MIPSS70 or MIPSS70-plus and low risk transplant score according to MTSS the indication for HSCT should be balanced between patients' preferences, actual treatment options including clinical trials, and other risk factors which are not included in the current risk scores such as TP53 mutation. While age is a critical factor in disease- and transplant-specific scores patients aged >70 years should not be excluded from the curative option of HSCT per se. Single center experiences as well as large registry data support transplantation in elderly patients after careful selection especially regarding comorbidities and performance status (Daghia et al. 2019; Ballen et al. 2010; Hernandez-Boluda et al. 2021; Polverelli et al. 2024; Gagelmann et al. 2024a), thus an HSCT should be offered on an individual basis (Kroger et al. 2024). More data investigating geriatric and social assessments are needed to consider individual eligibility for an HSCT.

9.3 Timing of Stem Cell Transplantation

Optimal timing to perform HSCT for candidates is in the absence of well-designed prospective trials difficult to assess. In a multistate model comparing transplant to non-transplant approaches in the pre-JAK inhibitor era months of survival peaked at 9.7 months from diagnosis in DIPSS high-risk patients and at 16.6 months in intermediate-2 patients, while in intermediate-1 and low risk DIPSS patients, no gain in life expectancy was at 20.5 and 29–45 months, respectively (Cipkar et al. 2022), leading to the recommendation for upfront transplantation for patients diagnosed with DIPSS intermediate-2 and high risk, whereas the procedure can be delayed for patients with DIPSS intermediate-1 and low risk (Kroger et al. 2024).

However, most of the risk models as well as multi-state models were developed in patients without JAK inhibition therapy, which has now become the standard care of treatment in symptomatic myelofibrosis patients. This should be acknowledged in new risk models. One of these new models is the response to ruxolitinib at 6 months model (RR6) which is based on evaluation of drug doses, reduction of spleen size, and number of red blood cell transfusion during the first 6 months after initiation of ruxolitinib treatment. RR6 allows to identify ruxolitinib-pretreated patients with a poor prognosis who are potential candidates for HSCT independent from other disease-specific risk models (Maffioli et al. 2022). Taking this model into account the EBMT/ELN recommendation for optimal timing is: (Kroger et al. 2024)

- *Immediate for patients with intermediate-2-risk and high-risk DIPSS score, whereas allogeneic HSCT can be delayed for low-risk or intermediate-1-risk disease.*
- *Transplantation-eligible patients on JAK inhibitors should be assessed for response and after 6 months of therapy patients falling into the high-risk category of the response to ruxolitinib after 6 months (RR6) model should receive timely evaluation for transplantation.*

9.4 Pre-transplantation Management

Optimal pre-transplant management is crucial for the success of an HSCT. It is worth noting that anemia which is a risk factor in all disease-specific risk scores did not have significant impact on outcome after HSCT. Even if not specifically shown for myelofibrosis iron overload at transplantation is in general associated with a high risk of toxicity and non-relapse mortality (Alessandrino et al. 2010; Isidori et al. 2021). Reducing iron overload prior to transplantation by chelation is recommended but since effective chelation might take several weeks, HSCT should not be postponed if a suitable donor is in place to avoid disease progression, which may lower the curative potential.

Because deferasirox suppresses carbide plasma iron to decrease free radicals which are considered to be the main cause of organ toxicity post HSCT, it can be used during conditioning regimen but because of its interference with busulfan therapeutic drug monitoring for busulfan is needed (Essmann et al. 2023; Essmann et al. 2022).

Similar to anemia, the number of blasts, which is also part of all disease-specific risk scores, is not included in MTSS, and clinical studies have shown that between 0 and 10% of blasts no increase of relapse is seen after HSCT. Risk of relapse increased beyond 10% of blasts which explained the relatively good outcome after HSCT for patients in accelerated phase myelofibrosis (Gagelmann et al. 2022). A specific treatment to reduce blast count prior to HSCT is—outside of clinical studies—not justified unless progressing, accelerated, or blastic phase is present (Kroger et al. 2024).

9.5 Role of JAK Inhibitors and Management of Splenomegaly

Ruxolitinib, the first JAK1/2 inhibitor approved for patients with myelofibrosis, was shown to not only generate control of constitutional symptoms, but also significantly reduce spleen size, which is associated with posttransplant complications such as delayed engraftment, potential graft failure, and/or poor graft function (Jain et al. 2022; Kunte et al. 2022; Alchalby et al. 2016) (Table 9.3).

A large EBMT study reported on higher non-relapse mortality in patients with large spleen size (>15 cm) prior to transplantation (Polverelli et al. 2021), while a larger single center study

Table 9.3 Management of splenomegaly prior to stem cell transplantation in myelofibrosis

Advantages	JAK-inhibition	Splenectomy	Splenic irradiation
	Approved drugs	Leading to faster engraftment	Easy to perform prior to conditioning therapy
	Improving also constitutional symptoms		Rapid response
	Easy to administer		Spleen reduction in JAK-inhibitor refractory patients
	2nd generation of JAK inhibition is effective in case of ruxolitinib resistance		May reduce risk of relapse
	Can be combined with other drugs (e.g., BET inhibitors)		
Disadvantages	Expensive and not available in all countries	May delay transplantation due to post-surgery complications	Severe cytopenia if done weeks before HSCT
		Mortality risk	Only few data available
		Concern of high relapse incidence	Potential higher liver toxicity

investigating spleen size by computer tomography showed an association between spleen size and risk of relapse (Luther et al. 2023). Taking the available data of splenomegaly and outcome after HSCT into account there is a clear rationale to reduce spleen size prior to HSCT in myelofibrosis patients. There are three options to reduce spleen size before HSCT:

1. JAK inhibitor or other drugs.
2. Splenectomy.
3. Splenic irradiation.

9.5.1 JAK Inhibitors or Other Drugs

A large EBMT retrospective study investigated the impact of pretransplant ruxolitinib on outcome of HSCT in myelofibrosis patients. In this study, relapse incidence and event-free survival were significantly improved in patients who received HSCT while responding to ruxolitinib in comparison to those who failed or lost response to ruxolitinib or were never exposed to ruxolitinib (Kroger et al. 2021). Other JAK inhibitors such as fedratinib, pacritinib, or momelotinib are also able to generate spleen size reduction but have not been investigated systematically prior to HSCT.

In order to avoid a withdrawal syndrome after cessation of JAK inhibition prior to HSCT and to take advantage of the immunosuppressive activity of JAK inhibitors, some transplant centers do not discontinue JAK inhibition prior to transplantation, but continue JAK inhibition throughout the transplantation process until stable engraftment to facilitate ensured engraftment and reduce incidence of graft-versus-host disease (GVHD) (Kroger et al. 2018, 2023). Ruxolitinib in combination with BET inhibitors showed increased spleen size reduction in comparison to ruxolitinib alone and might be a further option to reduce spleen size prior to HSCT but approval for the drug is pending (Saenz et al. 2017; Mascarenhas et al. 2023).

Thus the recommendation from EBMT/ELN according to spleen size reduction is: (Kroger et al. 2024)

- *All patients with splenomegaly greater than 5 cm below the lower costal margin or splenomegaly-related symptoms who are candidates for allogeneic HSCT should receive a spleen-directed treatment, ideally with JAK-inhibitor.*

- *JAK-inhibitor weaning with drug stop at the time of, or after, the stat of the conditioning regimen seems to be safe to reduce cytokine rebound syndrome and cardiac complications.*

9.5.2 Splenectomy

The benefit of splenectomy prior to transplantation is uncertain. In a large EBMT study, splenectomy was associated with faster engraftment and lower non-relapse mortality but with a higher risk of relapse (Polverelli et al. 2021). While a survival benefit was reported only for a subgroup of patients with spleen size >15 cm (Polverelli et al. 2021), splenectomy previously showed a higher risk of relapse compared to myelofibrosis without splenectomy (Kroger et al. 2009). Furthermore, splenectomy is also associated with significant morbidity and also mortality up to 8% due to post surgery complications such as bleeding or thrombosis (Tefferi et al. 2000; Mesa et al. 2006).

A small study comparing 39 splenectomized with 46 non-splenectomized myelofibrosis patients showed no difference in NRM and in relapse incidence, but multivariate analysis showed improved event-free and overall survival for splenectomized patients. In this study leukocytes and platelets engraftment was similar, but half of the splenectomized patients developed post-surgical complications, mostly bleeding and thrombosis (Robin et al. 2017).

9.5.3 Splenic Irradiation

Splenic irradiation prior to HSCT has been reported in small studies, reporting a transient reduction in spleen size and a reduction of disease-related symptoms (Helbig et al. 2019; Bales et al. 2023) but major concerns are severe hematotoxicity and increased liver toxicity. However, since responses to splenic irradiation are rapid, the high myelotoxicity with long-lasting neutropenia, anemia, and thrombocytopenia can be reduced if splenic irradiation is performed immediately prior to or as part of the conditioning regimen. In an international collaborative study, results from 59 patients with myelofibrosis who received splenic irradiation within 2 weeks prior to conditioning therapy were analyzed. A reduction in spleen size with a median of 5 cm (range 4.1–6.3) was seen in 95% of the patients. The median dose was 7 Gy (range 3–12) with no correlation between total dose and spleen response or hematotoxicity (Gagelmann et al. 2024b). In a propensity score comparing splenic irradiation with splenectomy and immediate transplantation, relapse rate was significantly lower after splenic irradiation in comparison to splenectomy (12% vs. 38%) (Gagelmann et al. 2024b).

The current recommendation from EBMT/ELN in patients who failed to respond properly to ruxolitinib is: (Kroger et al. 2024)

- *In case of splenomegaly response (less than 5 cm below lower costal margin), proceed with allogeneic HSCT; in other circumstances, second-line options are recommended (alternative JAK inhibitors or novel agents, splenectomy, or splenic irradiation), particularly when spleen is palpable more than 15 cm below the lower costal margin.*

Persistence of splenomegaly post allograft can cause poor graft function and splenomegaly post allograft was a major risk factor for poor graft function (Alchalby et al. 2016) and single case reported rapid resolution of pancytopenia after post-transplant splenectomy (Robin et al. 2010).

References

Alchalby H, Yunus DR, Zabelina T, Ayuk F, Kroger N (2016) Incidence and risk factors of poor graft function after allogeneic stem cell transplantation for myelofibrosis. Bone Marrow Transplant 51(9):1223–1227. https://doi.org/10.1038/bmt.2016.98

Alessandrino EP, Della Porta MG, Bacigalupo A et al (2010) Prognostic impact of pre-transplantation transfusion history and secondary iron overload in patients with myelodysplastic syndrome undergoing allogeneic stem cell transplantation: a GITMO study. Haematologica 95(3):476–484. https://doi.org/10.3324/haematol.2009.011429

Arber DA, Orazi A, Hasserjian RP et al (2022) International consensus classification of myeloid neoplasms and acute Leukemias: integrating morphologic, clinical, and genomic data. Blood 140(11):1200–1228. https://doi.org/10.1182/blood.2022015850

Bales JR, Kim HT, Portillo R et al (2023) Splenic irradiation prior to allogeneic hematopoietic cell transplantation for patients with myelofibrosis. Bone Marrow Transplant 58(4):459–461. https://doi.org/10.1038/s41409-023-01913-9

Ballen KK, Shrestha S, Sobocinski KA et al (2010) Outcome of transplantation for myelofibrosis. Biol Blood Marrow Transplant 16(3):358–367. https://doi.org/10.1016/j.bbmt.2009.10.025

Cipkar C, Kumar S, Thavorn K, Kekre N (2022) Optimal timing of allogeneic stem cell transplantation for primary myelofibrosis. Transplant Cell Ther 28(4):189–194. https://doi.org/10.1016/j.jtct.2022.01.018

Daghia G, Zabelina T, Zeck G et al (2019) Allogeneic stem cell transplantation for myelofibrosis patients aged >/=65 years. Eur J Haematol 103(4):370–378. https://doi.org/10.1111/ejh.13294

Essmann S, Dadkhah A, Janson D, Wolschke C, Ayuk F, Kroger NM, Langebrake C (2022) Iron chelation with deferasirox increases busulfan AUC during conditioning chemotherapy prior to allogeneic stem cell transplantation. Transplant Cell Ther 28(2):115 e1–115 e5. https://doi.org/10.1016/j.jtct.2021.11.003

Essmann S, Heestermans M, Dadkhah A et al (2023) Iron chelation with Deferasirox suppresses the appearance of labile plasma iron during conditioning chemotherapy prior to allogeneic stem cell transplantation. Transplant Cell Ther 29(1):42 e1–42 e6. https://doi.org/10.1016/j.jtct.2022.10.002

Gagelmann N, Ditschkowski M, Bogdanov R et al (2019) Comprehensive clinical-molecular transplant scoring system for myelofibrosis undergoing stem cell transplantation. Blood 133(20):2233–2242. https://doi.org/10.1182/blood-2018-12-890889

Gagelmann N, Wolschke C, Salit RB et al (2022) Reduced intensity hematopoietic stem cell transplantation for accelerated-phase myelofibrosis. Blood Adv 6(4):1222–1231. https://doi.org/10.1182/bloodadvances.2021006827

Gagelmann N, Badbaran A, Salit RB et al (2023) Impact of TP53 on outcome of patients with myelofibrosis undergoing hematopoietic stem cell transplantation. Blood 141(23):2901–2911. https://doi.org/10.1182/blood.2023019630

Gagelmann N, Schuh C, Zeiser R et al (2024a) Allogeneic hematopoietic cell transplantation for Myelofibrosis aged 70 years or older: A study from the German registry for stem cell transplantation. Transplant Cell Ther 30(10):1011 e1–1011 e13. https://doi.org/10.1016/j.jtct.2024.07.026

Gagelmann N, Hobbs GS, Campodonico E et al (2024b) Splenic irradiation for myelofibrosis prior to hematopoietic cell transplantation: A global collaborative analysis. Am J Hematol 99(5):844–853. https://doi.org/10.1002/ajh.27252

Gangat N, Caramazza D, Vaidya R et al (2011) DIPSS plus: a refined dynamic international prognostic scoring system for primary myelofibrosis that incorporates prognostic information from karyotype, platelet count, and transfusion status. J Clin Oncol 29(4):392–397. https://doi.org/10.1200/JCO.2010.32.2446

Gowin K, Ballen K, Ahn KW et al (2020) Survival following allogeneic transplant in patients with myelofibrosis. Blood Adv 4(9):1965–1973. https://doi.org/10.1182/bloodadvances.2019001084

Grinfeld J, Nangalia J, Baxter EJ et al (2018) Classification and personalized prognosis in myeloproliferative neoplasms. N Engl J Med 379(15):1416–1430. https://doi.org/10.1056/NEJMoa1716614

Guglielmelli P, Lasho TL, Rotunno G et al (2018) MIPSS70: mutation-enhanced international prognostic score system for transplantation-age patients with primary myelofibrosis. J Clin Oncol 36(4):310–318. https://doi.org/10.1200/JCO.2017.76.4886

Harrison C, Kiladjian JJ, Al-Ali HK et al (2012) JAK inhibition with ruxolitinib versus best available therapy for myelofibrosis. N Engl J Med 366(9):787–798. https://doi.org/10.1056/NEJMoa1110556

Helbig G, Wieczorkiewicz-Kabut A, Markiewicz M et al (2019) Splenic irradiation before allogeneic stem cell transplantation for myelofibrosis. Med Oncol 36(2):16. https://doi.org/10.1007/s12032-019-1245-5

Hernandez-Boluda JC, Pereira A, Kroger N et al (2021) Determinants of survival in myelofibrosis patients undergoing allogeneic hematopoietic cell transplantation. Leukemia 35(1):215–224. https://doi.org/10.1038/s41375-020-0815-z

Isidori A, Loscocco F, Visani G et al (2021) Iron toxicity and chelation therapy in hematopoietic stem cell transplant. Transplant Cell Ther 27(5):371–379. https://doi.org/10.1016/j.jtct.2020.11.007

Jain T, Tsai HL, DeZern AE et al (2022) Post-transplantation cyclophosphamide-based graft- versus-host disease prophylaxis with nonmyeloablative conditioning for blood or marrow transplantation for myelofibrosis. Transplant Cell Ther 28(5):259 e1–259 e11. https://doi.org/10.1016/j.jtct.2022.02.004

Kroger N, Holler E, Kobbe G et al (2009) Allogeneic stem cell transplantation after reduced-intensity conditioning in patients with myelofibrosis: a prospective, multicenter study of the Chronic Leukemia Working Party of the European Group for Blood and Marrow Transplantation. Blood 114(26):5264–5270. https://doi.org/10.1182/blood-2009-07-234880

Kroger N, Giorgino T, Scott BL et al (2015) Impact of allogeneic stem cell transplantation on survival of patients less than 65 years of age with primary myelofibrosis. Blood 125(21):3347–3350.; quiz 3364. https://doi.org/10.1182/blood-2014-10-608315

Kroger N, Shahnaz Syed Abd Kadir S, Zabelina T, Badbaran A, Christopeit M, Ayuk F, Wolschke C (2018) Peritransplantation ruxolitinib prevents acute graft-versus-host disease in patients with myelofibrosis undergoing allogenic stem cell transplantation. Biol Blood Marrow Transplant 24(10):2152–2156. https://doi.org/10.1016/j.bbmt.2018.05.023

Kroger N, Sbianchi G, Sirait T et al (2021) Impact of prior JAK-inhibitor therapy with ruxolitinib on outcome after allogeneic hematopoietic stem cell transplantation for myelofibrosis: a study of the CMWP of EBMT. Leukemia 35(12):3551–3560. https://doi.org/10.1038/s41375-021-01276-4

Kroger N, Wolschke C, Gagelmann N (2023) How I treat transplant-eligible patients with myelofibrosis. Blood 142(20):1683–1696. https://doi.org/10.1182/blood.2023021218

Kroger N, Bacigalupo A, Barbui T et al (2024) Indication and management of allogeneic haematopoietic stem-cell transplantation in myelofibrosis: updated recommendations by the EBMT/ELN International Working Group. Lancet Haematol 11(1):e62–e74. https://doi.org/10.1016/S2352-3026(23)00305-8

Kunte S, Rybicki L, Viswabandya A et al (2022) Allogeneic blood or marrow transplantation with haploidentical donor and post-transplantation cyclophosphamide in patients with myelofibrosis: a multicenter study. Leukemia 36(3):856–864. https://doi.org/10.1038/s41375-021-01449-1

Luther M, Henes FO, Zabelina T et al (2023) Spleen volume and length determined by computed tomography impact outcome after allogeneic stem cell transplantation for myelofibrosis. Bone Marrow Transplant 58(7):755–761. https://doi.org/10.1038/s41409-023-01968-8

Maffioli M, Mora B, Ball S et al (2022) A prognostic model to predict survival after 6 months of ruxolitinib in patients with myelofibrosis. Blood Adv 6(6):1855–1864. https://doi.org/10.1182/bloodadvances.2021006889

Mascarenhas J, Kremyanskaya M, Patriarca A et al (2023) MANIFEST: Pelabresib in combination with ruxolitinib for Janus kinase inhibitor treatment-naive myelofibrosis. J Clin Oncol 41(32):4993–5004. https://doi.org/10.1200/JCO.22.01972

McLornan D, Eikema DJ, Czerw T et al (2021) Trends in allogeneic haematopoietic cell transplantation for myelofibrosis in Europe between 1995 and 2018: a CMWP of EBMT retrospective analysis. Bone Marrow Transplant 56(9):2160–2172. https://doi.org/10.1038/s41409-021-01305-x

Mesa RA, Nagorney DS, Schwager S, Allred J, Tefferi A (2006) Palliative goals, patient selection, and perioperative platelet management: outcomes and lessons from 3 decades of splenectomy for myelofibrosis with myeloid metaplasia at the Mayo Clinic. Cancer 107(2):361–370. https://doi.org/10.1002/cncr.22021

Passamonti F, Mora B (2023) Myelofibrosis. Blood 141(16):1954–1970. https://doi.org/10.1182/blood.2022017423

Passamonti F, Cervantes F, Vannucchi AM et al (2010) A dynamic prognostic model to predict survival in primary myelofibrosis: a study by the IWG-MRT (International Working Group for Myeloproliferative Neoplasms Research and Treatment). Blood 115(9):1703–1708. https://doi.org/10.1182/blood-2009-09-245837

Passamonti F, Giorgino T, Mora B et al (2017) A clinical-molecular prognostic model to predict survival in patients with post polycythemia vera and post essential thrombocythemia myelofibrosis. Leukemia 31(12):2726–2731. https://doi.org/10.1038/leu.2017.169

Polverelli N, Mauff K, Kroger N et al (2021) Impact of spleen size and splenectomy on outcomes of allogeneic hematopoietic cell transplantation for myelofibrosis: A retrospective analysis by the chronic malignancies working party on behalf of European society for blood and marrow transplantation (EBMT). Am J Hematol 96(1):69–79. https://doi.org/10.1002/ajh.26020

Polverelli N, Bonneville EF, de Wreede LC et al (2024) Impact of comorbidities and body mass index on the outcomes of allogeneic hematopoietic cell transplantation in myelofibrosis: a study on behalf of the chronic malignancies working party of EBMT. Am J Hematol 99(5):993–996. https://doi.org/10.1002/ajh.27262

Robin M, Esperou H, de Latour RP, Petropoulou AD, Xhaard A, Ribaud P, Socie G (2010) Splenectomy after allogeneic haematopoietic stem cell transplantation in patients with primary myelofibrosis. Br J Haematol 150(6):721–724. https://doi.org/10.1111/j.1365-2141.2010.08276.x

Robin M, Zine M, Chevret S et al (2017) The impact of splenectomy in myelofibrosis patients before allogeneic hematopoietic stem cell transplantation. Biol Blood Marrow Transplant 23(6):958–964. https://doi.org/10.1016/j.bbmt.2017.03.002

Saenz DT, Fiskus W, Manshouri T et al (2017) BET protein bromodomain inhibitor-based combinations are highly active against post-myeloproliferative neoplasm secondary AML cells. Leukemia 31(3):678–687. https://doi.org/10.1038/leu.2016.260

Santos FPS, Getta B, Masarova L et al (2020) Prognostic impact of RAS-pathway mutations in patients with myelofibrosis. Leukemia 34(3):799–810. https://doi.org/10.1038/s41375-019-0603-9

Tefferi A, Mesa RA, Nagorney DM, Schroeder G, Silverstein MN (2000) Splenectomy in myelofibrosis with myeloid metaplasia: a single-institution experience with 223 patients. Blood 95(7):2226–2233. https://www.ncbi.nlm.nih.gov/pubmed/10733489

Tefferi A, Finke CM, Lasho TL, Hanson CA, Ketterling RP, Gangat N, Pardanani A (2018a) U2AF1 mutation types in primary myelofibrosis: phenotypic and prognostic distinctions. Leukemia 32(10):2274–2278. https://doi.org/10.1038/s41375-018-0078-0

Tefferi A, Guglielmelli P, Nicolosi M et al (2018b) GIPSS: genetically inspired prognostic scoring system for primary myelofibrosis. Leukemia 32(7):1631–1642. https://doi.org/10.1038/s41375-018-0107-z

Verstovsek S, Mesa RA, Gotlib J et al (2012) A double-blind, placebo-controlled trial of ruxolitinib for myelofibrosis. N Engl J Med 366(9):799–807. https://doi.org/10.1056/NEJMoa1110557

Hematopoietic Stem Cell Transplantation in Sickle Cell Disease

10

Mohammed Essa, Mohsen Alzahrani, Ali D. Alahmari, and Mahmoud Aljurf

Abstract

Sickle cell disease (SCD) is the most common inherited hemoglobinopathy worldwide with more than 300,000 babies in the world born every year with this disease. The clinical presentation and severity can be variable depending on many factors such as disease genotype, geographical location, environmental factors, and inheritance of other genetic abnormalities. The survival of patients with SCD have improved over last 3 decades where the expected median survival is more than 50 years for patients living in the developed countries. This improved survival is secondary to simple prophylactic and therapeutic interventions such as blood transfusions, prophylactic antibiotics, and vaccination. Disease modifying agents such as hydroxyurea contributed to the improved all disease outcome, survival, and quality of life. Over last 2 decades, curative options such as allogeneic stem cell transplant have gained popularity and increased evidence of safety and efficacy as the only curative option for SCD. However, there are many challenges to consider in patients who undergo hematopoietic stem cell transplantation (HSCT) that may affect the outcome. In this chapter, multiple challenges will be discussed including the indications of HSCT, choosing the appropriate donor and how to prevent and manage the unique or common post-transplant complications. Elaborative sections will focus on conditioning regimens choices in matched related donor HSCT. In addition, challenges in regards to various approaches of alternative donor transplant will be thoroughly discussed. Finally, long term effects and recommended follow up will be described.

Keywords

Sickle cell disease · Hemoglobinopathy · Stem cell transplant · Haploidentical · Gene therapy

M. Essa (✉)
King Abdullah Specialist Children's Hospital, King Saud bin Abdulaziz University for Health Sciences, Ministry of National Guard Health Affairs, Riyadh, Saudi Arabia
e-mail: essamo@mngha.med.sa

M. Alzahrani
King Abdulaziz Medical City, King Saud bin Abdulaziz University for Health Sciences, Ministry of National Guard Health Affairs, Riyadh, Saudi Arabia
e-mail: Alzahranimo5@mngha.med.sa

A. D. Alahmari · M. Aljurf
Department of Oncology, King Faisal Specialist Hospital and Research Centre, Riyadh, Saudi Arabia
e-mail: aalahmari7@kfshrc.edu.sa; maljurf@kfshrc.edu.sa

© The Author(s), under exclusive license to Springer Nature Switzerland AG 2025
K. Ballen (ed.), *Clinical Management of Challenges in Allogeneic Stem Cell Transplant*, Advances in Experimental Medicine and Biology 1475, https://doi.org/10.1007/978-3-031-84988-6_10

10.1 Introduction

Sickle cell disease (SCD) is the most common inherited hemoglobinopathy worldwide. SCD predominantly affects people of African, Mediterranean, Indian, and Middle Eastern descent. However, due to migration, it is now found worldwide. The epidemiology of SCD varies globally, largely reflecting the prevalence of the sickle cell trait, which provides some resistance to malaria. It is estimated that over 200,000 infants are born with the disease each year in Sub-Saharan Africa (Piel et al. 2013a, b).

SCD affects approximately 100,000 people in the United States, predominantly among African Americans. The prevalence of the sickle cell trait in African Americans is about 8%. SCD poses significant public health challenges, particularly in regions with high prevalence but limited resources. The life expectancy of patients with SCD is much lower compared to normal population and many patients do not reach adulthood in some areas around the world (Piel et al. 2013a, b; GBD 2021 Sickle Cell Disease Collaborators 2023).

Efforts to implement newborn screening programs, improve access to healthcare, and increase awareness are crucial in managing the disease's impact and prevent complications.

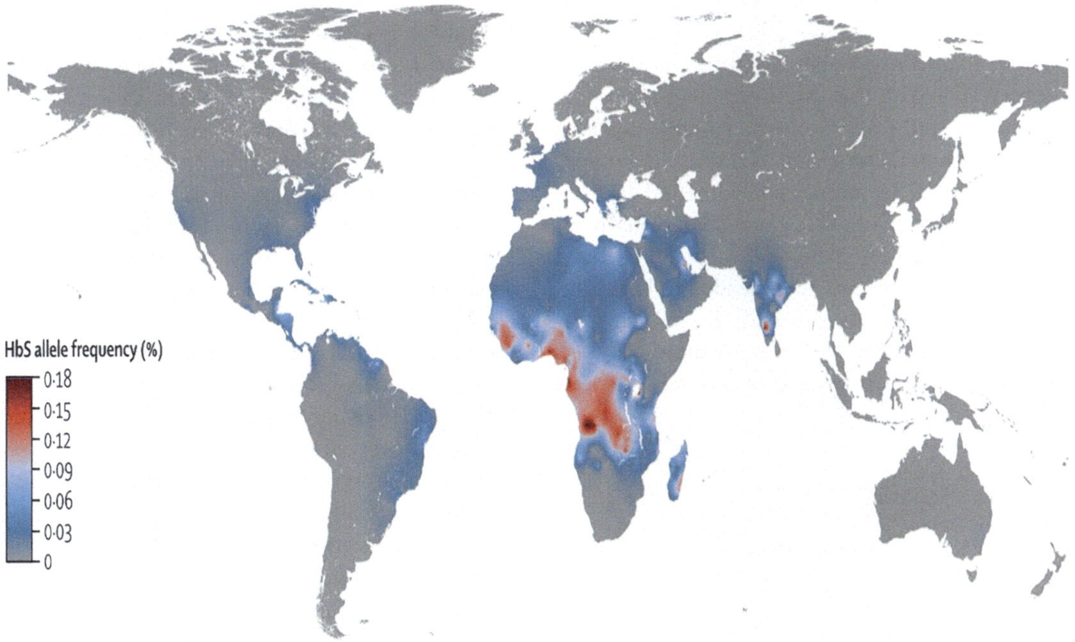

SCD is caused by a single-nucleotide substitution that leads to a propensity toward hemoglobin polymerization that leads to sickling. SCD is characterized by anemia, hemolysis, painful crises, recurrent infections, chronic microinfarcts, and inflammatory process affecting multiple organs resulting in organs dysfunction such, cerebral vasculopathy, cerebrovascular disease such as cerebral infarcts, Moyamoya disease, chronic lung disease, pulmonary hypertension, hepatic, renal, and bone complications. The severity of the disease depends on the sickle cell variants in addition to other disease modifiers, such as the presence of hereditary persistent hemoglobin F (HPHF) or alfa thalassemia coinheritance. Disease modifying therapy especially hydroxyurea and recently approved therapies such as L-glutamine have influenced the care of patients with SCD and improved outcome but yet not considered curative options. Voxelotor has been recently withdrawn due to real world data safety concerns. (Kapoor et al. 2018; Vichinsky et al. 2019; Niihara et al. 2018; Telen et al. 2020).

10.2 Indications and Challenges in Choosing Patients

Unfortunately, in areas with high prevalence, access to comprehensive care, including pain management, hydroxyurea therapy, and curative treatments like stem cell transplantation, is often limited. Allogeneic HCT remains the only curative therapy for SCD aiming to switch patients' marrow to produce normal donor HbA before reaching irreversible end organ damage. The recently approved Gene therapy, where a patient's own stem cells are genetically modified or edited to produce an efficient modified hemoglobin or shift the hemoglobin production towards the fetal hemoglobin is very promising and transforming the care the patients with SCD (Johnson et al. 1984; Gluckman et al. 2017; Walters et al. 1996; Hsieh et al. 2009, 2011; Cappelli et al. 2024).

Several barriers to HSCT include the high number of patients, lack of suitable donors, lack of access to care, disease complications, comorbidities and limited understanding of HSCT by patients or non-specialized health care providers. In addition, the well-known risks of developing HSCT related complications may influence patients' decisions such as preparative regimens toxicity, GVHD, graft failure, and treatment-related mortality (Cappelli et al. 2024).

The number of transplants has increased since the first one in 1984 to above 1500 (Johnson et al. 1984) with increment in EFS and OS exceeding 90% in different types of HSCT (Gluckman et al. 2017). The survival is better in pediatric population exceeding 94%, while in adult patients was 81% (Gluckman et al. 2017). Many factors contributed to this decreased survival outcome in adult patients including poor tolerance to myeloablative conditioning, chronic organ damage such as nephropathy, hepatopathy; higher risk of neurological complications such as seizure, PRES, and ICH; finally, higher risk of graft failure in adult patients secondary to alloimmunization (Gluckman et al. 2017; Kanter et al. 2021).

Overall, The presence of complications or severe manifestations of the disease despite hydroxyurea are the main indications to proceed with HSCT; earlier identification of those patients is a key factor for the best outcome (Cappelli et al. 2024).

There is an increasing evidence that reversibility affecting almost all organs, including cerebral perfusion and osteonecrosis is possible or at least halting progression. Patients with CNS events including stroke, silent stroke, CNS vasculopathy, and Moyamoya disease are the strongest indications for undergoing curative allogeneic HSCT including haploidentical HSCT (Kanter et al. 2021; Hernigou et al. (1997); Powars et al. 1978; Shenoy 2013).

Individuals who develop recurrent acute pain crises have very poor quality of life and have high rate of acute care visits. In addition, patients with frequent pain events requiring acute care and opioids experience more morbidity and early mortality than those with less frequent acute pain crises. Patients with recurrent VOCs defined by 2 or more admission per year for last 2 years and refractory for allogeneic HSCT (Ballas et al. 2012; Yawn et al. 2014; Platt et al. 1991).

Patients with SCD may experience recurrent episodes of Acute chest syndrome (ACS), which can be life threatening or fatal. ACS or progressive chronic lung damage with pulmonary hypertension is a leading cause of mortality in adults with SCD and poses a significant risk. Recurrent acute chest syndrome (ACS) despite hydroxyurea with good drug adherence is also considered to be a standard indication for allogeneic HSCT (Rostami et al. 2024).

Other indications for HSCT include complications such as multiple joints (≥2) AVN, recurrent priapism, sickle hepatopathy, and sickle nephropathy. In regards to pulmonary hypertension, there is no agreement as to whether or not patients with this condition should proceed to transplantation. Some degree of renal dysfunction should not be seen as a reason to avoid transplantation despite the use of nephrotoxic drugs such as calcineurin inhibitors or fludarabine; nephropathy will be discussed in a separate section in this chapter (Kanter et al. 2021; Rostami et al. 2024; Alzahrani et al. 2021; Hernigou et al. 1997) (Tables 10.1 and 10.2).

Table 10.1 Benefits and risks of HSCT

Category	Benefits	Risks
Cure potential	Potential to cure sickle cell disease Pediatric outcome >95% Adult outcome >90%	Not guaranteed; some patients may still experience symptoms or graft failure
Reduction in complications	Reduction in pain crises, decreased need for blood transfusions	Risk of graft failure
Improved life expectancy	Increased survival rates	Immediate risks from transplant procedure Transplant related toxicity (infection, bleeding, organ dysfunction, death) TRM up to 5%
Quality of life	Improved quality of life, less chronic pain	Long-term complications (e.g., graft-versus-host disease)
Immunological effects	Restoration of normal hemoglobin production	Immune system suppression required, increasing infection risk
Donor selection	Matched sibling donor transplants have higher success rates	Difficulty in finding a matched donor, especially for minorities
Gene therapy and editing	Ongoing research may offer a safer alternative to traditional transplants	High cost Affect fertility Long-term effects unknown

Table 10.2 Risk of HSCT by age groups and donor type

Age group (years)	Type of HSCT (Conditioning regimens)	Risks and special considerations
<3	HSCT is not usually offered	The severity of SCD is not usually clear. Patient is at risks of severe complications
3–10	Matched related donor transplants have higher success rates (Myeloablative regimens)	Viral infections are common Challenging pharmacokinetics Higher risk of veno-occlusive disease with younger age
10–18	Matched related donor transplant Alternative donor transplant (Myeloablative regimens or non-myeloablative regimens) Gene-therapy as per licensing authorities age limit (Myeloablative Busulfan regimen)	Few irreversible organ damage prior to HSCT Iron overload in older age Assents and detailed understanding of the process is essential Options of fertility preservation Less veno-occlusive disease compared to younger age group unless hepatopathy is present

Adapted from Kanter et al. (2021); Rostami et al. (2024); Alzahrani et al. (2021); Hernigou et al. (1997); Frangoul et al. (2021) and de la Fuente et al. (2020)

10.3 Matched Family Donor HSCT

10.3.1 Age and Organ Function

Age at transplantation is an important predictor of survival. Patients younger than 5 years have an excellent OS and EFS therefore matched sibling HSCT should be proposed early in life, before the development of complications. It must be mentioned that at a cutoff of 15 years, the incidence of cGVHD in matched sibling donors can rise beyond 20%. This has likely changed after including serotherapy in the preparative regimen.

In a large retrospective analysis of HSCT in patients with SCD receiving myeloablative regimen, mortality occurred in 5–20% depending on age. The probabilities of OS and EFS are summarized below:

For patients <16 y of age: OS, 95% EFS, 93%
For patients ≥16 y of age: OS, 81% EFS, 77%

In a more recent analysis of the data by CIBMTR, a higher incidence of mortality was observed in patients >13 y of age following myeloablative conditioning and MSD HSCT (Eapen M et al. 2019; Kanter et al. 2021;Cappelli et al. 2019).

Adult patients may face additional challenges such as more severe disease complications, organs dysfunction, or failure or comorbidities that may affect tolerance to myeloablative regimens. Pre-transplant conditioning regimens are often adjusted to account for these factors. It is expected that once end organ damage is significant, between the age of 30 and 40, MAC regimens is expected to be toxic. RIC regimens are more popular to be used in older adult and has been offered up to the age of 60.

10.3.2 Family Donor Clearance

It is very crucial to make sure that the family donor is free from the disease whether HbSS or other variants by doing hemoglobin electrophoresis. Sickle cell anemia traits (carriers) are allowed to donate based on many publications and it is safe to use GCSF as a mobilizing agent in these donors (Rostami et al. 2024; Alzahrani et al. 2021). The patient should be aware about the carrier status of the donor. Converting the patient into carrier status is ethically acceptable. G6PD deficiency is another prevalent enzymopathy that may manifest with hemolysis in patients and their donors. Many patients agree to trade of SCD with G6PD deficiency if it is the only available donor.

Screening for donor-directed antibodies against RBC antigens is usually recommended because of the risk of pure red cell aplasia after HSCT; this could be attained by also doing cross match (Cappelli et al. 2019).

10.3.3 General Challenges Post HSCT

10.3.3.1 Posterior Reversible Encephalopathy Syndrome (PRES)

PRES is a neurological syndrome usually manifested with hypertension, seizures, behavioral changes and altered visual disturbances in addition to MRI abnormalities especially in the occipital lobe. It is reported to be ranging from 9% to 18% of patients with haemoglobinopathies compared to other HSCT indications (1–5%).

It is expected that the risk is higher in patients with Moyamoya disease with additional common risk factors such as the use of calcineurin inhibitors and labile blood pressure. Developing seizures might be very dangerous to put the patient at risk of intracranial bleeding. The current strategies to mitigate and decrease the risk of neurological complications include: judicious blood pressure control (avoid hypertension), optimize serum electrolytes levels i.e., magnesium levels and Platelets count >50,000/ul. The routine use of prophylactic antiseizure medications for 6 months such as levetiracetam is very common. Many centers use sirolimus instead of calcineurin inhibitors to reduce the risk further. But still antiseizure prophylaxis should continue as long as the patient is taking sirolimus. In an event of PRES, it's crucial to optimize supportive therapy as the outcome is usually favorable. Stopping CNI and replacing it with alternatives is a common practice in such events although difficult to study (Cappelli et al. 2024; Kanter et al. 2021; Thompson et al. 2013).

10.3.3.2 CNS Vasculopathy

CNS Vasculopathy is usually a main indication to proceed with transplant as HSCT may halt the progression or even improve the degree of vasculopathy in long term follow-up. The degree of pathology varies from mild narrowing to extensive mechanical abnormalities in conjunction with inflammation. However, those patients are also at high risk of developing PRES, bleeding, and acute cerebrovascular accidents. Keeping the platelets at higher range usually more than 50×10^9/uL prior to engraftment is preferred in addition to all PRES prevention strategies that was mentioned earlier. Some patients may need evaluation for neurosurgical intervention prior to transplant such as Encephaloduroarteriosynangiosis (EDAS), which transposes scalp arteries onto the surface of the brain; Aspirin is usually prescribed for long time post such procedure. The benefit of this procedure required more vigorous research. Some centers prefer to delay EDAS procedure for at least 6 months post HSCT or delay HSCT for 6–12 months if the procedure was done earlier (Cappelli et al. 2024; Kanter et al. 2021; Ngo et al. 2014; Thompson et al. 2013; DeBaun et al. 2020).

10.3.3.3 Veno-Occlusive Disease of the Liver (VOD) or Sinusoidal Obstruction Syndrome (SOS)

VOD is a common complication after myeloablative preparative regimens especially in the presence of previous liver impairment. Many patients with SCD have a degree of hepatopathy related directly to SCD end organ damage or secondary to Iron overload.

If Iron overload is suspected, Liver T2* MRI and cardiac MRI assessment is usually attained similar to beta thalassemia patients.

Although defibrotide prophylaxis is not usually recommended per latest evidence, it might be considered in high risk patients after weighing the benefits and risks or as early intervention if VOD develops(Grupp et al. 2023).

Severe liver impairment with high comorbidity index or cirrhosis might be a relative contraindication unless the decision was use a RIC regimen. Recent reports have shown promising results. Non-myeloablative regimen using alemtuzumab and TBI 300 cGy in the setting of MRD allogeneic transplant was reported to be safe and successfully reverse the features of chronic liver disease and severe hyperbilirubinemia in some patients with sickle hepatopathy (Saib et al. 2020).

10.3.3.4 Nephropathy

Adult patients are at risk of severe nephropathy leading to chronic renal impairment requiring dialysis. Non-Myeloablative regimens using alemtuzumab/TBI regimen has been reported to be safe approach for successful allogeneic MRD transplant in these patients with acceptable results (Hsieh et al. 2011). It is still considered very risky to transplant dialysis patients with many post HSCT medications. However, small studies showed excellent tolerance to chemotherapy free conditioning regimen (Saib et al. 2020).

10.3.3.5 GVHD

The development of GVHD after HSCT is a major significant concern for patients and physicians. The risk of chronic GVHD might be perceived as worse than SCD for some patients. After MSD HSCT and myeloablative conditioning, the incidence of acute GVHD (grade 2–4) for patients $\leq$16 and >16 years is 12% and 16%, respectively, and chronic GVHD incidence is 15 and 23%, respectively. The use of routine serotherapy has decreased the incidence of GVHD in comparison to transplant for malignant diseases.

Lower GVHD rate has been seen confirmed in patients who received RI non-myeloablative regimen such as TBI/Alemtuzumab regimen in the expense of higher incidence of mixed donor chimerism and higher risk of graft failure compared to patients who receive myeloablative conditioning.

Haploidentical allogeneic transplant is of increasing popularity in the sitting of SCD over last decade with many supporting prospective studies of this excellent outcome reaching more than 90% for EFS and 95% OS (Kassim et al. 2024). Post-transplant cyclophosphamide GVHD prophylaxis regimens has resulted in significant reduction of GVHD rate of less than 15% in these studies. MUD allogeneic SCT approach has been tested in a prospective clinical trial and was associated with high rate of chronic and extensive chronic GVHD with high TRM reaching up to 28% (Shenoy et al. 2016). Based on this observation, the MUD option should be carefully optimized with GVHD prophylaxis.

10.3.3.6 Preparation for HSCT

Hydroxyurea usually intensified for last 2 months pre transplant and must be discontinued before starting the conditioning regimen, and exchange transfusion pre transplant is often initiated to lower HbS level <30% hence to minimize the risk of unexpected sickle cell crises. Patients and their families should be counseled about the risk of infertility; options for fertility preservation should be offered before initiating the transplantation process. Other pre transplant work up is according to center standards in addition disease specific work up including brain imaging (preferable MRI) with referral to neurology if any features of CNS vasculopathy, assessment of iron overload by measuring ferritin, MRI liver and cardiac if iron overload present. Liver biopsy maybe undertaken if patient is going to undergo myeloablative regimen to role features that will exclude patients from transplant such as presence of liver cirrhosis, fibrosis or inflammation. Iron chelation can be intensified pre transplant to reduce risk of liver VOD secondary to conditioning regimen Alzahrani et al. 2021.

10.3.4 Conditioning Regimen

10.3.4.1 Myeloablative

See Table 10.3.

To date, the use of myeloablative conditioning (MAC) regimens is the gold standard for HLA matched sibling HSCT (EFS: 73–96%, OS: 91–100%) despite the risk of transplant-related toxicity, especially when patients have end organ damage due to their SCD status. Historically, Busulfan-based conditioning regimen including cyclophosphamide or Fludarabine has been used but with high toxicities and GvHD risk; therefore, the addition of serotherapy (ATG) has been recommended to lower the GvHD risk. The use of targeted myeloablative busulfan levels is suggested to ensure appropriate levels and as a strategy to reduce VOD risk. In recent years, fludarabine, treosulfan, and thiotepa (FTT) emerged as a frequently used reduced toxicity MAC conditioning in SCD in children and adults with excellent results. Measuring treosulfan levels is not routinely required. The use of myeloablative regimen is usually restricted to pediatric patients and young fit adult patients. Infertility and long term transplant complications are the main disadvantages of using this approach. Table 10.3 summarizes key studies (Cappelli et al. 2024; Dedeken L et al. 2014 ;Kanter et al. 2021; Bernaudin et al. 2020, Rostami et al. 2024).

10.3.4.2 Reduced Intensity and Non-myeloablative Regimen

Reduced intensity conditioning (RIC) regimens have been explored to decrease toxicity allowing a stable mixed chimerism. The aim of this approach is to reduce organ toxicity in already impaired organs and preserve fertility.

Recently, encouraging outcomes have been observed with FLU-based RIC regimens or after a chemotherapy-free regimen (ALEM-TBI 300 cGy; DFS = 92% and OS = 100%). Mixed donor chimerism as low 20% in transplanted patients is considered sufficient to produce donor-type hemoglobin and reverse the red blood cell SCD phenotype, and minimize the risk of GvHD, confirming the therapeutic efficacy of mixed chimerism for hemoglobinopathies (Hsieh et al. 2011; Cappelli et al. 2024).

The use of pre-conditioning immunosuppression (maximum tolerated dose of HU for at least 2 months prior to transplant) may reduce the risk of rejection. In cases of a mixed donor chimerism post HSCT, split chimerism analyses are important for a proper understanding of the hematopoietic compartments, where a full myeloid chimerism encourages a watch-and-wait strategy. This approach avoids unnecessary and dangerous interventions to achieve a full donor chimerism,

Table 10.3 Myeloablative conditioning regimen in patients with SCD

	No.	Age range (years)	Graft source (*N*)	Donor (*N*)	Conditioning regimen	GvHD prophylaxis	Acute GvHD (grade II–IV)	Chronic GvHD	OS %	EFS %
Bernaudin et al.	87	2–22	BM: 74 PB: 1 UCB: 12	MRD	MAC: BU + CY ± ATG	CSA ± MTX	20%	Limited: 11 % Extensive: 2.4%	6 years: 93.1%	6 years: 86.1%
Panepinto et al.	67	2–27	BM: 54 PB: 9 UCB: 4	MRD	MAC: BU + CY	CSA + MTX Tacrolimus T cell depletion Non	10%	22% Limited: 9 patients Extensive: 3 patients	5 years: 97%	5 years: 85%
Strocchio et al.	30	1.7–18.8	BM: 22 PB: 1 UCB: 4 BM + UCB: 3	MRD: 24 MUD: 6	MAC: BU + Thio + Flu Treo + Thio + Flu ± ATG	CSA ± MTX	7% in the Bu- and and 0% in the Treo-group	7% in the Bu- and 0% in the Treo-group	100%	5-years: 93% MRD: 96% MUD: 83%
Krishnamurti et al.	22	17–36	BM	MRD: 17 MUD: 5	MAC: BU + Flu + ATG	CNI + MTX	18%	27%	1 year: 91% MRD: 94% MUD: 80%	1 year: 86% MRD: 94% MUD: 60%
Kogel et al.	25	1–21	BM: 21 PB: 4	MRD: 17 MUD: 8	MAC: BU + CY Flu + Thio + Treo Flu + Thio + Alk	CSA + MTX CSA + MMF CSA Tacrolimus + MTX CSA + MMF + MTX	Grade I–II: 8 patients	No	100%	88%
Dedeken et al.	50	1.7–15.3	BM: 39 PB: 1 UCB: 3 BM + UCB: 7	MRD	MAC: BU + CY ± ATG	CSA + MTX CSA CSA + MMF	Grade I –II: 12% Grade III–IV: 10%	Mild in 10 patients	8 years: 94.1%	8 years: 85.6%
Bernaudin et al.	234	2.2–28.9	BM: 195 PB: 1 UCB: 30 BM + UCB: 8	MRD	MAC: BU + CY ± ATG	CSA ± MTX	20.1%	10.5%	97%	93.9%

Rostami et al. (2024)

Alk alkeran (melphalan), *ATG* anti-thymocyte globulin, *BM* bone marrow, *BU* busulfan, *CNI* calcineurin inhibitor, *CSA* cyclosporine, *CY* cyclophosphamide, *EFS* event-free survival, *Flu* fludarabine, *GvHD* graft versus host disease, *MAC* myeloablative conditioning, *MMF* mycophenolate mofetil, *MRD* matched related donor, *MUD* matched unrelated donor, *MTX* methotrexate, *OS* overall survival, *PB* peripheral blood, *Thio* thiotepa, *Treo* treosulfan, *UCB* umblical cord blood which is not mandatory in case of a non-malignant myeloid disease (Alzahrani et al. 2021; Cappelli et al. 2024).

10.4 Alternative Donor Transplant

10.4.1 Matched Unrelated Donor (MUD)

Finding a matched unrelated donor (MUD) through donor registries can be challenging, particularly for patients of African descent, due to lower representation in registries. However, when a good match is found, GvHD risk remains very high. Whereas in many diseases, outcome does not differ much between MSD and MUD, a recent retrospective multicenter study in SCD found significant differences in pivotal outcome parameters such as OS and GVHD in SCD. EFS, OS, and also the incidence of cGVHD is significantly higher compared to an age matched sibling donor with a similar cutoff for worse outcome at age ≥13 years. Mortality in one study of almost 23% with MUD compared to <10% in MSD suggests the need to improve the outcome especially GVHD risk. Abatacept might be a promising agent in this population. Further studies are warranted to address this significant complication (Shenoy et al. 2016; Eapen M 2019).

Shenoy et al. reported a poor outcome on 29 children aged 4–19 years, transplanted with a MUD. The 2-year OS and EFS was 79% and 69%, respectively. The incidence of aGvHD grade II–IV was 28% and cGvHD was 62%, with 38% classified as extensive, most likely caused by insufficient immunosuppressive therapy. More recently, Cappelli et al. reported an improved outcome from MUD-transplanted SCD patients.

10.4.2 Cord Blood

Family-directed cord blood banking could represent a useful stem cell resource for families with a child affected by SCD (Rafii et al.). However, the benefit and risk ratio of using cord blood vs BM from a MSD for a child with SCD and an indication of HSCT is not well understood.

There were three studies that reported retrospective comparisons of outcomes after UCB vs BM transplantation in patients with SCD. One study evaluated individuals with all types of hemoglobinopathies and included those with SCD (UCB, $n = 30$) (Eapen et al. 2019). Bernaudin et al. reported the comparison of 30 UCB vs 195 BM transplantations in patients with SCD, and a recent report by CIBMTR compared outcomes of 34 UCB vs 311 BM transplantations (Kanter et al. 2021; Locatelli et al. (2013); Bernaudin et al. 2020).

These studies reported a significantly longer median time to neutrophil and platelet recovery after HSCT with a UCB donor when compared with BM donor, but without a significantly increased risk of infections or primary graft failure. There were no deaths observed after UCB transplantation, and there was a trend toward lower acute and chronic GVHD after UCB transplantation. OS and EFS were similar for all cell sources. The current recommendation to use UCB is limited to matched family UCB transplantation because of the limited studies(Kanter et al. 2021).

10.4.3 Haploidentical Donor

The development of both in vivo and ex vivo T-cell depletion strategies has facilitated haploidentical HSCT as a solution with easier availability of donors. This approach was more appealing especially that the outcome in MUD-HSCT is discouraging. Initial small studies showed high rates of GVHD up to 25–29% of the patients (Eapen et al. 2019).

Post transplant cyclophosphamide-haplo-HSCT approach is becoming more popular due to accessibility. Bolaños-Meade et al. (2012) reported 0% GvHD outcome of non myeloablative conditioning regimen containing ATG, cyclophosphamide, fludarabine, 2 Gy total body irradiation, in addition to mycophenolate mofetil, tacrolimus was switched to sirolimus for GvHD prevention to decrease PRES. This approach reported high graft failure rate at 43% in 14 young adult SCD patients. Bolaños-Meade et al. (2019) modified the regimen by increasing the dose of TBI to 400 cGy. Of the 17 participants enrolled, including both children and adults 71% had SCD. A total of 76% developed full donor myeloid engraftment, with 18% having mixed chimerism and 6% having primary graft failure. Another group modified the original regimen by adding pre-conditioning with hydroxyurea and azathioprine over 2 months in addition to thiotepa but macrophage activation syndrome (MAS) developed in 12% resulting in death in 8% of the study population that included 87 patients (Cappelli et al. 2024; de la Fuente et al. 2019). Due to these unexpected results, BMT CTN1507 pursued a trial with DFS and OS of 93% and 100% in 15 children, respectively. Viral reactivations were reported commonly in both studies but mostly without developing end organ disease. Vanderbilt Global Haploidentical BMT Learning Collaborative group has recently confirmed the improvement of engraftment in participants to at least 80%. In a recent international multicenter learning collaborative consortium, phase 2 study of related nonmyeloablative haploidentical BMT added thiotepa to the conditioning regimen for children and adults with SCD. A total of 70 participants were included. Graft failure occurred in (11.4%) and only in participants aged <18 years who had autologous recovery. After a median follow-up of 2.4 years, EFS was 82.6%. The 1-year grade 3 to 4 acute graft-versus-host disease (GVHD) rate was 10%, and the 2-year moderate–severe chronic GVHD rate was 10%. Five participants (7.1%) died from infectious complications. This study demonstrated that nonmyeloablative haploidentical BMT with thiotepa and PTCy is a readily available curative therapy for most adults, even those with organ damage, compared to the more expensive myeloablative gene therapy and gene editing. Additional strategies are required for children to decrease graft failure rates (Kassim et al. 2024) (Table 10.4).

A single-center experience with TCRαβ-depleted and CD19-depleted grafts conditioned with fludarabine treosulfan thiotepa and ATG showed rapid engraftment and a low incidence of GvHD with OS and DFS comparable to MSD HCT but delayed immune reconstitution led to more frequent infections (Foell et al. 2020). This approach is currently evaluated in a prospective stratified trial (NCT04201210).

HSCT with an in vitro T-cell- (CD3+ or TCRαβ+ cells) and B-cell (CD19)-depleted peripheral blood stem cells graft is a well-established haploidentical transplant modality. The OS in the T-haplo-HSCT was 90%. None developed grade III–IV acute GvHD; 20% developed mild or moderate chronic GvHD. These results demonstrate the feasibility, safety, and efficacy for adult advanced stage SCD patients. However, the delay of immune reconstitution after T cell depleted-haplo-HSCT can lead to a higher incidence of viral reactivations with infectious complications being the major causes of morbidity and mortality after T-haplo-HSCT. Sixty-five percent in the T-haplo-HSCT group presented with viral reactivations (EBV, CMV, and BKV). The median follow-up was 3.9 years with an OS and DFS of 84% and 69%, respectively. Twenty percent developed mild/moderate cGvHD in T cell depleted-haplo-HSCT and the majority were beyond 20 years of age (Foell et al. 2019).

The risk of graft rejection increases as the conditioning regimen intensity decreases and if an alternative donor (HLA mismatched or unrelated) is used. Although less frequent than GVHD, the risk of graft failure may carry a larger emotional burden than other HSCT complications and should be thoroughly discussed with potential patients and families.

Based on these studies, the haploidentical allogeneic HCT can be considered as an acceptable 2nd option for those patients who don't have a MRD. Most of the centers use the approach of

Table 10.4 Nonmyeloablative/RIC conditioning regimen in patients with SCD

	No.	Age range (years)	Graft source	Donor	Conditioning regimen	GvHD prophylaxis	Acute GvHD (grade II–IV)	Chronic GvHD	OS %	EFS %
Hsieh et al.	30	17–65	PB	MRD	NMA: Alemtuzumab + TBI	Sirolimus	No	No	97%	87%
Saraf et al.	13	17–40	PB	MRD	NMA: Alemtuzumab + TBI	Sirolimus	No	No	100%	92%
Al-Zahrani et al.	18	14–39	PB	MRD	NMA: Alemtuzumab + TBI	Sirolimus	No	No	100%	88.9%
Ozdogu et al.	20	20–45	PB	MRD	NMA: Flu + CY + BU + TBI + ATG	Sirolimus	5%	No	100%	100%
Alzahrani et al.	110	14–43	PB	MRD	NMA: Alemtuzumab + TBI	Sirolimus	Skin grade I–II in 3 patients	No	97%	87.3%
Krishnamurti et al.	7	6–18	BM	MSD	RIC: BU(2) + Flu + ATG + TLI	CSA + MMF	Skin grade II in I patient	Limited skin in I patient	100%	86%

Rostami et al. (2024)

Alk alkeran (melphalan), *ATG* anti-thymocyte globulin, *BM* bone marrow, *BU* busulfan, *CSA* cyclosporins, *CY* cyclophosphamide, *EFS* event-free survival, *Flu* fludarabine, *GvHD* graft versus host disease, *NMA* nonmyeloablative conditioning, *MMF* mycophenolate mofetil, *MRD* matched related donor, *MSD* matched sibling donor, *OS* overall survival, *PB* peripheral blood, *RIC* reduced intensity conditioning, *TBI* total body irradiation, *TLI* total lymphoid irradiation

PTCy as GVHD prophylaxis and pre transplant intensification of the preparative regimen by adding thiotepa and ATG to FluCYTBI or increasing the TBI dose from 200 to 400 cGy.

10.5 Gene Therapy as an Alternative Choice to Manage Hemoglobinopathy

Lentiviral vector based approaches carrying a modified beta globin or gamma globin gene are under investigation for safety and efficacy. In 2021, lentiviral vectors trials were temporarily suspended due to the observation of three cases of leukemia after gene therapy with Lentiglobin. The clustered regularly interspaced short palindromic repeats (CRISPR)-Cas9 nuclease system is another strategy used targeting the erythroid-specific enhancer region of BCL11A editing to increase the production of fetal hemoglobin HbF. Using this strategy, the CLIMB 121 SCD study was published recently leading to FDA, EMA, and other international agencies approval of these gene therapy product. CD34+ cells collection after G-CSF stimulation is currently avoided in SCD patient for severe complications, whereas plerixafor (CXCR4 chemokine receptor antagonist) alone or in combination with GRO-β (CXCR2 agonist) has been used successfully. Based on these studies, gene therapy can be acceptable options for young patients with severe disease complications and who can tolerate myeloablative regimen. The disadvantage of this approach is acute toxicities from Busulfan, risk of VOD, cost, and lack of long term follow up. The longest follow up in gene therapy studies didn't exceed 2 years and accordingly long term safety of this approach is still to be confirmed (Frangoul et al. 2021), Ashorobi et al. 2024).

Organs dysfunction in many adult patients is a barrier and make these patients ineligible for the recently approved myeloablative gene therapy options, due to concerns for increased treatment-related mortality.

10.6 Long-Term Effects

While stem cell transplantation offers a potential cure for sickle cell disease, it poses significant risks to fertility, primarily due to the conditioning regimens. Fertility preservation should be a key consideration in the pre-transplant planning process. Advances in transplantation techniques and fertility preservation strategies continue to evolve, providing hope for better outcomes for individuals with SCD seeking to maintain their fertility post-treatment(Balduzzi et al. 2017).

1. Malignancy

   The incidence of a therapy-related malignancy, particularly after myeloablative allogeneic HSCT, seems to be low overall. Risks after less intense conditioning regimens are not known at this time. This risk should be discussed with the patients and families. Risk of developing myeloid malignancy post gene therapy has been highlighted and long term follow up of this approach will guide this approach.
2. Chronic Pain

   Many patients will continue to suffer from chronic pain symptoms even if they have good graft function and full chimerism. This may last for few years post-transplant. It is important to rule out objective causes for such as AVN of joints, infection, and osteomyelitis. Drug dependency can be a difficult to assess and many patients will need referral to psychology and mental health.

10.7 Future Directions:

1. Improving Transplant Protocols

   Reduced-Intensity or non-myeloablative Conditioning: Efforts are being made to reduce the intensity of the chemotherapy required before a transplant, which can make the procedure safer and more tolerable, especially for older patients or those with other health issues and end organs damage.

2. Haploidentical Transplantation

   It is already expanding faster in both techniques including selective ex-vivo T cell depletion and the use of PTCY, efforts are being focused or decreasing infections, GVHD and graft rejection rates, in addition to accessibility to newer antiviral agents and viral specific T cells.
3. MUD and Cord Blood Transplants

   Efforts are ongoing in MUD to decrease GVHD rates like the use of PTCY or abatacept as GvHD prophylaxis. While efforts in cord blood transplant are focusing in improving the methods to expand the number of stem cells from a single cord blood unit to enhance engraftment.
4. Gene Therapy and Gene Editing

   Researchers and pharmaceutical companies are working on techniques to expand on Gene addition and Gene Editing technologies like CRISPR-Cas9 and hopefully make it available for patients without matched donor.
5. Expanding Access

   Efforts are underway to study the barriers to transplant access, particularly among minority populations who are disproportionately affected by SCD and improve health care systems.
6. Long-Term Outcomes and Quality of Life

   Studies are ongoing to improve the long-term follow-up of transplant recipients to better understand and manage late complications, such as chronic GVHD, organ damage, or secondary cancer and improve the overall quality of life for transplant survivors, addressing issues like mental health, pain management, and social reintegration.

These research areas are critical in advancing the treatment of sickle cell disease, moving closer to more effective and widely accessible cures.

References

Alzahrani M, Damlaj M, Jeffries N, Alahmari B, Singh A, Rondelli D, Tisdale JF, Saraf SL, Hsieh MM (2021) Non-myeloablative human leukocyte antigen-matched related donor transplantation in sickle cell disease: outcomes from three independent centres. Br J Haematol 192:761–768. https://doi.org/10.1111/bjh.17311

Ashorobi D, Naha K, Bhatt R (2024) Hematopoietic stem cell transplantation in sickle cell disease. [Updated 2023 Jul 19]. In: StatPearls [Internet]. StatPearls Publishing, Treasure Island

Balduzzi A, Dalle JH, Jahnukainen K et al (2017) Fertility preservation issues in pediatric hematopoietic stem cell transplantation: practical approaches from the consensus of the Pediatric Diseases Working Party of the EBMT and the International BFM Study Group. Bone Marrow Transplant 52(10):1406–1415. https://doi.org/10.1038/bmt.2017.147

Ballas SK, Gupta K, Adams-Graves P (2012) Sickle cell pain: a critical reappraisal. Blood 120(18):3647–3656

Bernaudin F, Dalle JH, Bories D et al (2020) Société Française de Greffe de Moelle et de Thérapie Cellulaire. Long-term event-free survival, chimerism and fertility outcomes in 234 patients with sickle-cell anemia younger than 30 years after myeloablative conditioning and matched-sibling transplantation in France. Haematologica 105(1):91–101

Bolaños-Meade J, Fuchs EJ, Luznik L et al (2012) HLA-haploidentical bone marrow transplantation with posttransplant cyclophosphamide expands the donor pool for patients with sickle cell disease. Blood 120:4285–4291

Bolaños-Meade J, Cooke KR, Gamper CJ et al (2019) Effect of increased dose of total body irradiation on graft failure associated with HLA-haploidentical transplantation in patients with severe haemoglobinopathies: a prospective clinical trial. Lancet Haematol 6:e183–e193. https://pubmed.nc

Cappelli B, Volt F, Tozatto-Maio K, Scigliuolo GM, Ferster A, Dupont S, Simões BP, Al-Seraihy A, Aljurf MD, Almohareb F, Belendez C, Matthes S, Dhedin N, Pondarre C, Dalle JH, Bertrand Y, Vannier JP, Kuentz M, Lutz P, Michel G, Rafii H, Neven B, Zecca M, Bader P, Cavazzana M, Labopin M, Locatelli F, Magnani A, Ruggeri A, Rocha V, Bernaudin F, de La Fuente J, Corbacioglu S, Gluckman E, Eurocord, the Cellular Therapy and Immunobiology Working Party (CTIWP), and the Paediatric Diseases Working Party (PDWP) of the EBMT (2019) Risk factors and outcomes according to age at transplantation with an HLA-identical sibling for sickle cell disease. Haematologica 104(12):e543–e546

Cappelli B, Gluckman E, Corbacioglu S, de la Fuente J, Abboud MR (2024) Hemoglobinopathies (sickle cell disease and thalassemia). In: Sureda A, Corbacioglu S, Greco R, Kröger N, Carreras E (eds) The EBMT handbook. Springer, Cham. https://doi.org/10.1007/978-3-031-44080-9_80

de la Fuente J, Dhedin N, Koyama T et al (2019) Haploidentical bone marrow transplantation with post-transplantation cyclophosphamide plus thiotepa improves donor engraftment in patients with sickle cell anemia: results of an international learning collaborative. Biol Blood Marrow Transplant 25(6):1197–1209

de la Fuente J, Amrolia P et al (2020) The role of haematopoietic stem cell transplantation for sickle cell disease in the era of targeted disease-modify-

ing therapies and gene editing. Lancet Haematol 7(12):e902–e911
DeBaun MR, Jordan LC, King AA et al (2020) American Society of Hematology 2020 guidelines for sickle cell disease: prevention, diagnosis, and treatment of cerebrovascular disease in children and adults. Blood Adv 4(8):1554–1588. https://doi.org/10.1182/bloodadvances.2019001142
Dedeken L, Lê PQ et al (2014) Haematopoietic stem cell transplantation for severe sickle cell disease in childhood: a single centre experience of 50 patients. Br J Haematol 165:402–408. https://doi.org/10.1111/bjh.12737
Eapen M, Brazauskas R, Walters MC et al (2019) Effect of donor type and conditioning regimen intensity on allogeneic transplantation outcomes in patients with sickle cell disease: a retrospective multicentre, cohort study. Lancet Haematol 6(11):e585–e596
Foell J, Kleinschmidt K, Jakob M, Troeger A, Corbacioglu S (2020) Alternative donor: αß/CD19 T-cell-depleted haploidentical hematopoietic stem cell transplantation for sickle cell disease. Hematol Oncol Stem Cell Ther 13(2):98–105. https://doi.org/10.1016/j.hemonc.2019.12.006
Frangoul H, Altshuler D, Cappellini MD et al (2021) CRISPR-Cas9 gene editing for sickle cell disease and β-thalassemia. N Engl J Med 384(3):252–260. https://doi.org/10.1056/NEJMoa2031054
GBD 2021 Sickle Cell Disease Collaborators (2023) Global, regional, and national prevalence and mortality burden of sickle cell disease, 2000–2021: a systematic analysis from the Global Burden of Disease Study 2021 [published correction appears in Lancet Haematol. 2023 Aug;10(8):e574. https://doi.org/10.1016/S2352-3026(23)00215-6]. Lancet Haematol 10(8):e585–e599. https://doi.org/10.1016/S2352-3026(23)00118-7
Gluckman E, Cappelli B, Bernaudin F et al (2017) Eurocord, the Pediatric Working Party of the European Society for Blood and Marrow Transplantation, and the Center for International Blood and Marrow Transplant Research. Sickle cell disease: an international survey of results of HLA-identical sibling hematopoietic stem cell transplantation. Blood 129(11):1548–1556
Grupp SA et al (2023) Defibrotide plus best standard of care compared with best standard of care alone for the prevention of sinusoidal obstruction syndrome (HARMONY): a randomised, multicentre, phase 3 trial. Lancet Haematol 10(5):e333–e345
Hernigou P, Bernaudin F, Reinert P, Kuentz M, Vernant JP (1997) Bone-marrow transplantation in sickle-cell disease. Effect on osteonecrosis: a case report with a four-year follow-up. J Bone Joint Surg Am 79(11):1726–1730. https://doi.org/10.2106/00004623-199711000-00015
Hsieh MM, Kang EM, Fitzhugh CD et al (2009) Allogeneic hematopoietic stem-cell transplantation for sickle cell disease. N Engl J Med 361:2309–2317
Hsieh MM, Fitzhugh CD, Tisdale JF (2011) Allogeneic hematopoietic stem cell transplantation for sickle cell disease: the time is now. Blood 118:1197–1207. https://pubmed.ncbi.nlm.nih.gov/21628400
Johnson FL, Look AT, Gockerman J et al (1984) Bone-marrow transplantation in a patient with sickle-cell anemia. N Engl J Med 311:780–783
Kanter J, Liem RI, Bernaudin F et al (2021) American Society of Hematology 2021 guidelines for sickle cell disease: stem cell transplantation. Blood Adv 5(18):3668–3689
Kapoor S, Little JA, Pecker LH (2018) Advances in the treatment of sickle cell disease. Mayo Clin Proc 93(12):1810–1824. https://doi.org/10.1016/j.mayocp.2018.08.001
Kassim AA, de la Fuente J, Nur E, Wilkerson KL, Alahmari AD, Seber A, Bonfim C, Simões BP, Alzahrani M, Eckrich MJ, Horn B, Hanna R, Dhedin N, Rangarajan HG, Gouveia RV, Almohareb F, Aljurf M, Essa M, Alahmari B, Gatwood K, Connelly JA, Dovern E, Rodeghier M, DeBaun MR (2024) An international learning collaborative phase 2 trial for haploidentical bone marrow transplant in sickle cell disease. Blood 143(25):2654–2665. https://doi.org/10.1182/blood.2023023301
Krishnamurti L, Kharbanda S, Biernacki MA, Zhang W, Baker KS, Wagner JE, Wu CJ (2008) Stable long-term donor engraftment following reduced-intensity hematopoietic cell transplantation for sickle cell disease. Biol Blood Marrow Transplant 14(11):1270–1278
Locatelli F, Kabbara N, Ruggeri A et al (2013) Eurocord and European Blood and Marrow Transplantation (EBMT) group. Outcome of patients with hemoglobinopathies given either cord blood or bone marrow transplantation from an HLA-identical sibling. Blood 122(6):1072–1078
Martin AS et al (2022) Long-term survival after hematopoietic cell transplant for sickle cell disease compared to the United States population. Transplant Cell Ther 28(6):325.e1–325.e7
Ngo S, Bartolucci P, Lobo D et al (2014) Causes of death in sickle cell disease adult patients: old and new trends [abstract]. Blood 124(21):2715
Niihara Y, Miller ST, Kanter J et al (2018) A phase 3 trial of l-glutamine in sickle cell disease. N Engl J Med 379(3):226–235. https://doi.org/10.1056/NEJMoa1715971
Piel FB, Hay SI, Gupta S et al (2013a) Global burden of sickle cell anaemia in children under five, 2010–2050: modelling based on demographics, excess mortality, and interventions. PLoS Med 10:e1001484
Piel FB et al (2013b) Global epidemiology of sickle haemoglobin in neonates: a contemporary geostatistical model-based map and population estimates. Lancet 381(9861):142–151
Platt OS, Thorington BD, Brambilla DJ et al (1991) Pain in sickle cell disease. Rates and risk factors. N Engl J Med 325(1):11–16

Powars D, Wilson B, Imbus C, Pegelow C, Allen J (1978) The natural history of stroke in sickle cell disease. Am J Med 65(3):461–471

Rostami T, Rad S, Rostami MR et al (2024) Hematopoietic stem cell transplantation in sickle cell disease: a multidimentional review. Cell Transplant 33. https://doi.org/10.1177/09636897241246351

Saib I et al (2020) EBMT abstract

Shenoy S (2013) Hematopoietic stem-cell transplantation for sickle cell disease: current evidence and opinions. Ther Adv Hematol 4(5):335–344

Shenoy S, Eapen M, Panepinto JA, Logan BR, Juan W, Abraham A, Brochstein J, Chaudhury S, Godder K, Haight AE, Kasow KA, Leung K, Andreansky M, Bhatia M, Dalal J, Haines H, Jaroscak J, Lazarus HM, Levine JE, Krishnamurti L, Margolis D, Megason GC, Yu LC, Pulsipher MA, Gersten I, DiFronzo N, Horowitz MM, Walters MC, Kamani N (2016) A trial of unrelated donor marrow transplantation for children with severe sickle cell disease. Blood 128(21):2561–2567. https://doi.org/10.1182/blood-2016-05-715870

Telen MJ (2020) Curative vs targeted therapy for SCD: does it make more sense to address the root cause than target downstream events? Blood Adv 4(14):3457–3465

Thompson D, Harrington Y, de la Fuente J (2013) The incidence of posterior reversible encephalopathy syndrome is increased in BMT for Haemoglobinopathies. Blood 122(21):4579. https://doi.org/10.1182/blood.V122.21.4579.4579

Vichinsky E, Hoppe CC, Ataga KI et al (2019) A phase 3 randomized trial of voxelotor in sickle cell disease. N Engl J Med 381(6):509–519. https://doi.org/10.1056/NEJMoa1903212

Walters MC, Patience M, Leisenring W et al (1996) Bone marrow transplantation for sickle cell disease. N Engl J Med 335:369–376. https://pubmed.ncbi.nlm.nih.gov/8663884

Yawn BP, Buchanan GR, Afenyi-Annan AN et al (2014) Management of sickle cell disease: summary of the 2014 evidence-based report by expert panel members. JAMA 312(10):1033–1048

Pharmacological Maintenance Therapies After Allogeneic Hematopoietic Cell Transplantation

11

Ritika Walia and Yi-Bin Chen

Abstract

Disease relapse is the leading cause of failure for patients with hematological malignancies who have received allogeneic hematopoietic cell transplantation (HCT). Maintenance therapy administered after HCT is a promising strategy to reduce the incidence of relapse and enhance the curative potential of HCT. Historically, maintenance therapy was not possible due to off-target toxicities from conventional chemotherapy. With the advent of specifically targeted agents—such as those targeting BCR-ABL, FLT3, IDH1, IDH2 and Menin—therapy given in the immediate period after HCT while patients remain in remission has become a reality. Research investigations and clinical applications of this approach have greatly increased in recent years, with a rapidly expanding number of available therapeutic agents waiting to be tested in the post-transplant setting. However, many questions and challenges remain regarding the feasibility and clinical impact of maintenance therapy after HCT. In this chapter, we review the published data as well as ongoing studies regarding the rapidly evolving field of maintenance therapy after HCT.

Keywords

Maintenance · Allogeneic hematopoietic cell transplantation · Relapse

11.1 Introduction

The risk of disease relapse after allogeneic hematopoietic cell transplantation (HCT) remains relatively unchanged despite substantial progress in mitigating transplant related mortality (Bejanyan et al. 2015). Treatment of disease relapse after HCT has traditionally yielded dismal rates of durable remission, and attention has increasingly focused on the prevention of disease relapse after HCT. Several methods of reducing relapse have been proposed including: better pre-HCT treatment, better patient selection, intensification of conditioning, improved donor selection, and the use of maintenance therapy after HCT. Maintenance therapy is defined as therapy initiated and continued while a patient remains in complete remission. Maintenance therapy is a rather new concept, only able to be considered due to an emerging number of less toxic or targeted therapies which possess significantly less

R. Walia
University of Massachusetts, Worcester, MA, USA

Y.-B. Chen (✉)
Massachusetts General Hospital, Boston, MA, USA
e-mail: ychen6@mgb.org

© The Author(s), under exclusive license to Springer Nature Switzerland AG 2025

K. Ballen (ed.), *Clinical Management of Challenges in Allogeneic Stem Cell Transplant*, Advances in Experimental Medicine and Biology 1475, https://doi.org/10.1007/978-3-031-84988-6_11

off-target toxicity compared to traditional chemotherapeutics. The ideal agent for maintenance therapy after HCT is one which is convenient to take, has relatively minimal toxicity, does not cause significant cytopenias, does not have significant drug-drug interactions, and does not increase the risk of graft-vs.-host disease (GVHD).

Pre-emptive therapy is a type of maintenance therapy—but with therapy given only upon detection of minimal or measurable residual disease (MRD). The choice to pursue maintenance or pre-emptive therapy after HCT revolves around the underlying disease, sensitivity of an available MRD assay, kinetics of disease relapse after MRD detection, available interventions, and, certainly, toxicity (see Table 11.1). Due to its indiscriminate nature, maintenance therapy inherently overtreats those patients who otherwise would not have relapsed; whereas pre-emptive therapy would restrict such therapy to those clearly at a higher risk for relapse.

While becoming more common, research investigating maintenance therapy after HCT has been challenging. Interpretation of results from retrospective or single-arm maintenance trials has been hampered by a relatively high drop-out rate after HCT due to competing risks such as infection, GVHD, relapse, and patient choice, making comparisons to published historical controls difficult due to this selection bias. Formal landmark analyses are one way to correct for this, but difficult to carry out on historically published controls. In terms of prospective research, industry has frequently been reluctant to support post-HCT trials before regulatory approval of their respective agent in another treatment setting due to the multiple adverse events encountered after HCT. After approval, while early phase single arm trials have been relatively easy to enroll, prospective randomized controlled trials have been quite difficult due to fear of disease relapse and reluctance of both patient and provider to agree to randomization to a potential placebo or no maintenance arm. In addition, while HCT practices have improved over the last few years, the multiple competing risks and events which can occur after HCT can make conduct of such trials challenging (see Table 11.2). Indeed, only four prospective randomized trials of maintenance therapy after HCT have been conducted to date (see Table 11.3).

11.2 Ph+ ALL, or Advanced Stage CML—Maintenance with BCR-ABL TKI

The advent of tyrosine kinase inhibitors (TKI) targeting the BCR-ABL1 fusion protein has dramatically improved treatment outcomes in chronic myeloid leukemia (CML), such that most

Table 11.1 Different treatment settings after HCT

Setting after HCT	Drawback
Frank relapse	Clonal evolution/other mechanisms of resistance may have already occurred Therapies may be less effective
Pre-emptive	Requires sensitive MRD assay which allows time to intervene
Maintenance	Overtreats patients who otherwise would not have relapsed and subjects them to potential toxicity

Table 11.2 Challenges in conducting clinical trials for maintenance therapy after HCT

Competing risks including opportunistic infections, GVHD, and other HCT-related complications
Access to agents as industry is often reluctant to conduct trials on patients after HCT
Difficulty in interpreting retrospective analyses given inherent selection bias of those who were well enough to initiate maintenance therapy
Reluctance of patients/providers to enroll patients on randomized studies which include a control arm without active drug

Table 11.3 Prospective randomized trials of maintenance therapy after HCT

Trial	Agent	Eligible population	Arms	Notes
Burchert et al. *JCO* 2020;38(26):2993–3002 (SORMAIN)	Sorafenib	FLT3-ITD mutated AML in CR undergoing HCT	Sorafenib ($n = 43$) Placebo ($n = 40$) Therapy for 24 months	Few received FLT3 TKI during induction Trial closed early due to poor accrual 2y RFS benefit 85% vs. 53% ($p = 0.002$) with Sorafenib 2y OS benefit 90.5% vs. 66.2% ($p = 0.007$) with Sorafenib
Xuan et al. *Lancet Oncol* 2020;21(9):1201–1212	Sorafenib	Young FLT3-ITD mutated AML in CR1 undergoing MAC HCT	Sorafenib ($n = 100$) Control ($n = 102$) Therapy for 6 months	Few received FLT3 TKI during induction No placebo All younger, MAC patients 1 yr relapse rate of 7% vs. 24.5% ($p = 0.001$) in favor of Sorafenib 2 yr LFS benefit 78.9% vs. 56.6% ($p < 0.001$) in favor of Sorafenib 2 yr OS benefit 82.1% vs. 68% ($p = 0.012$) in favor of Sorafenib
Levis M et al. *JCO* 2024;42: 1766–1775 (MORPHO, BMT CTN 1506)	Gilteritinib	FLT3-ITD mutated AML in CR1 undergoing HCT	Gilteritinib ($n = 178$) Placebo ($n = 178$) Therapy for 24 months	International study with results varying significantly by region 2 yr RFS of 77.2% vs. 69.9% ($p = 0.0518$) 2 yr OS of 80.7% vs. 77.5% ($p = 0.439$) Benefit restricted to those with detectable peri-HCT FLT3-ITD based MRD Placebo arm performed better than expected
Oran B et al. *Blood Adv* 2020;4(21):5580–5588	Azacitidine	High-risk MDS/CMML/CML undergoing HCT at a single center	Azacitidine ($n = 93$) for 12 4-week cycles SOC ($n = 94$)	Single center, no placebo Median # of Aza cycles = 4 Median RFS of 2.07 vs. 1.28 years ($p = 0.43$) Median OS of 2.52 vs. 2.56 years ($p = 0.85$) Control arm performed better than expected Challenging to give Aza therapy as designed

patients with chronic phase (CP) CML enjoy a duration and quality of life comparable to the unaffected population without HCT. Unfortunately, prognosis remains poor for the 5% of patients who present in advanced phase CML, defined as accelerated phase (AP) or blast phase (BP) CML, and for the ~7% of patients who progress despite TKI therapy (Senapati et al. 2023). Acute lymphoblastic leukemia comprises approximately 20% of all adult leukemias, and among these, the presence of the Philadelphia chromosome (Ph+) is a traditional adverse prognostic indicator for relapse after HCT and inferior overall survival (Abou Dalle et al. 2019). Currently, HCT remains the mainstay of therapy in Ph+ ALL in first complete remission following initial therapy with a chemotherapy-TKI combination and for BP-CML, and is used in cases of TKI failure in CP and AP-CML (DeFilipp et al. 2016). Given the well-established efficacy and familiar toxicity profiles of BCR-ABL targeting TKIs, these have been widely used as maintenance therapy after HCT and described in multiple case series and several prospective single-arm studies. The overall mechanism of action of BCR-ABL TKIs is thought to be direct antileukemic cytotoxicity.

A retrospective, single-center study of 26 patients receiving posttransplant maintenance with imatinib, dasatinib, nilotinib, or ponatinib observed 5-year OS and RFS estimates of 78% and 61%, respectively, with a 35% rate of GVHD which accounted for 3 deaths (DeFilipp et al. 2016). A subsequent retrospective CIBMTR registry analysis was conducted by the same group on 390 patients with CML, of which 89 patients (23%) received maintenance TKI therapy starting at a median of 100 days post-transplant, including dasatinib ($n = 50$), imatinib ($n = 27$), and nilotinib ($n = 27$). No significant difference was noted between the two groups in OS and leukemia free survival (LFS). No significant between-group difference was observed in the incidence of chronic GVHD nor in outcome based on disease status at transplantation (DeFilipp et al. 2020). The authors concluded that the broad application of maintenance TKI following HCT does not seem to be beneficial for patients with CML, and the decision to initiate maintenance TKI therapy should be individualized based on patient and disease specific factors. In another retrospective analysis of 197 patients from the CIBMTR registry, post-transplant maintenance TKIs did not significantly impact relapse; again, only 43/197 study patients (21%) received maintenance TKI in this cohort (Bachanova et al. 2014). Finally, in a retrospective analysis of 102 adults and 11 children with Ph+ ALL over 10 years in the MD Anderson registry, 32 patients (39%) received maintenance TKI (imatinib in all but one patient who received dasatinib) and no significant impact on PFS or OS was observed with TKI use (Kebriaei et al. 2012).

In contrast, a retrospective registry study from the Acute Leukemia Working Party of the European Group for Blood and Marrow Transplantation (EBMT) assessed the outcome of 473 adult patients with Ph+ ALL who underwent HCT over a 10-year period. Sixty patients received maintenance TKIs post-transplant, in whom a statistically significant improvement in OS (HR 0.42), LFS (HR 0.44), and relapse incidence (RI) (HR 0.40) was observed. The large registry nature of the study precluded a full analysis of other outcomes, specific TKI, duration and toxicity (Brissot et al. 2015).

A prospective study of 22 patients with CML in first chronic phase (CP1) undergoing HCT evaluated the impact of a time limited course of maintenance of imatinib on the kinetics of relapse posttransplant. Twenty-one to twenty-two (95%) patients started imatinib on day +35 and continued until 1 year after HCT. No relapses were observed during the study period. At the time of discontinuation of imatinib, seven patients were in molecular remission and the remaining had a BCR-ABL ratio less than 0.5%. After discontinuation of imatinib, 15/22 patients were noted to have relapsed at a median of 17 months after transplantation and required rescue with donor lymphocyte infusion (DLI). Three-year OS in this study was calculated at 87%, with 15/22 (68%) of patients having achieved molecular remission (Olavarria et al. 2007).

A phase II trial of 30 patients with newly diagnosed Ph+ ALL examined the safety and efficacy of imatinib administration before and after transplantation. Twenty-seven of thirty patients (90%) achieved a complete remission, of which 16 underwent allogeneic HCT. Seven of sixteen (44%) patients were unable to start maintenance imatinib due to transplant related mortality ($n = 4$), severe GVHD ($n = 1$), and patient choice ($n = 2$). Nine of sixteen (56%) patients were able to receive post-transplant maintenance on imatinib, for a median of 9 months duration. Imatinib therapy was interrupted or dose reduced due to relapse ($n = 3$), severe cGVHD ($n = 2$), grade 3–4 toxicity ($n = 2$), non-relapse death ($n = 1$), and patient choice ($n = 1$) (Ribera et al. 2010).

A phase I/II, open-label, prospective study examined the safety and efficacy of the second generation TKI nilotinib as maintenance after HCT in advanced CML and Ph+ ALL. A total of 22 patients (15 with CML and 7 with Ph+ ALL) underwent allogeneic HCT, 16 of whom were given nilotinib maintenance at median day +38. Three patients who died shortly after HCT, one with early disease progression, one with severe acute GVHD, and one who refused treatment did not go on to receive nilotinib maintenance after transplantation. The maximal tolerated dose was determined to be 200 mg every 12 hours. Early toxicities included hematologic, gastrointestinal, and cutaneous, whereas hepatic, pancreatic, and cardiac toxicities occurred later during treatment. The 2-year OS and PFS rates were 55% and 45%, respectively, for all patients undergoing transplant, and 69% and 56% for the 16 patients on nilotinib maintenance. The 2-year non-relapse mortality (NRM) and relapse rates for all patients undergoing transplant were 32% and 23%, respectively. Moreover, 9 of 12 patients with MMR early after HCT converted to CMR with nilotinib maintenance, and only 1 of 11 patients with CMR ultimately relapsed while on maintenance (Shimoni et al. 2015). A subsequent study of nilotinib maintenance assessed the feasibility of administering nilotinib at a dose of ≥400 mg/day for 85% of the interval between days +81 and +365. It was observed that only 23% of the pre-transplant intention-to-treat population met the feasibility endpoint. At the end of the analysis, all 13 patients who completed nilotinib therapy at the target dose were alive without relapse, suggesting efficacy of nilotinib maintenance (Carpenter et al. 2017).

The likelihood of acquired tyrosine kinase domain mutations conferring resistance in patients who have received pre-transplantation TKI therapy should also be considered as having a substantial impact on outcomes. The T315I mutation confers resistance to first- and second-generation TKIs and is identified predominantly at the time of relapse. In a prospective cohort study spanning 10 years, 26 Ph+ ALL patients with the T315I mutation who received HCT were enrolled. Nineteen patients received ponatinib therapy after transplant, of which 12 received prophylactic maintenance for a median of 12 months and 7 received pre-emptive treatment based on detection of MRD for a median of 3 months. Adverse events (AE) were observed in 13 of 19 patients (69.4%), with grade 3/4 toxicity occurring only in 3 patients; therapy was dose reduced in 2 of these three patients and discontinued in one. The median cumulative incidence of relapse (CIR) at 5 years was 18.5%, 57.1%, and 71.4% in the prophylactic, preemptive, and non-maintenance groups, respectively. The 5-year DFS was observed to be 81.5%, 0%, and 14.3% in the prophylactic, preemptive, and non-maintenance groups, respectively, and the median 5-year OS was observed to be 91.7%, 0%, and 28.6%, respectively. Due to small sample size and poor response to ponatinib, outcomes were notably poor in the preemptive group; however, DFS and OS were impressively high and relapse rates low in the maintenance group compared to the non-ponatinib group (Chen et al. 2022).

Several studies have sought to determine whether a maintenance vs. a pre-emptive approach to posttransplant BCR-ABL TKI therapy is preferable. It is known that detection of MRD via PCR post-transplant predicts imminent relapse. A prospective randomized controlled trial of 55 patients with Ph+ ALL or BP-CML assigned patients equally to receive either prophylactic imatinib or MRD-triggered imatinib. Overall, early, prophylactic imatinib (started at a median of 48 days after HCT), significantly reduced the incidence of molecular recurrence, with a 45.6% probability of remaining PCR negative at 46 months vs 27.4% in the MRD-triggered arm. At a median follow-up after HCT of 30.3 months and 32.4 months in the prophylactic and MRD-triggered arms respectively, 44 patients (82%) remained in CR, with no difference between treatment arms (81% vs 78%). At 5 years after HCT, the probabilities of ongoing remission, DFS, OS, and EFS were 82.6%, 69.4%, 77.3%, and 62.5%, respectively, with no statistically significant difference between the study arms. As such, the study concluded that post-transplant imatinib, either prophylactic or MRD-triggered, reduced the likelihood of molecular relapse and improved post-transplant outcomes (Pfeifer et al. 2012).

Overall, no prospective controlled randomized trials have ever shown a definitive benefit of maintenance TKI after HCT in either CML or Ph+ ALL. Moreover, large registry analyses are unable to distinguish between specific TKI, date of initiation, and duration of therapy, and smaller single arm studies do not provide firm conclusions. However, the implications of disease relapse and the familiarity with TKI administration make maintenance TKI use after HCT fairly routine across most centers in the United States. Employing a pre-emptive approach, rather than maintenance for all, mandates close monitoring with BCR-ABL based MRD testing given TKI can easily be added if MRD becomes detectable. If used, maintenance for two years after HCT appears to be a reasonable duration given that is the peak period of risk for disease relapse.

11.3 Ph$^{-negative}$ B-ALL

Unlike its Ph+ counterparts, B-ALL patients without the Philadelphia chromosome do not possess a single unique targetable mutation. Maintenance therapeutic approaches for these patients have focused on novel targeted agents against B-cell antigens. Blinatumomab is an approved bispecific engager targeting CD19 and was studied in a phase 2 single center trial of 23 patients as maintenance therapy after HCT. Therapy was safe with low rates of cytokine release syndrome or neurotoxicity. Correlative analysis suggested that benefit from blinatumomab depended on the T-cell profile of the recipient after HCT (Gaballa et al. 2022). Inotuzumab is an approved antibody drug conjugate targeting CD22. It was studied in a phase 1 maintenance trial at low doses given its prior association with hepatic sinusoidal obstruction syndrome (SOS). The maximum tolerated dose was identified at 0.6 mg/m^2 given in 28-day cycles with no cases of SOS observed at any dose tested. Compelling rates of 1-year PFS (89%) and OS (94%) were observed (Metheny et al. 2024). Larger trials using these agents are certainly merited. With the advent of highly sensitive MRD assays for B-ALL, such as those next generation sequencing techniques to detect immunoglobulin receptor gene rearrangements, future maintenance trials should use MRD status to define eligibility, duration of therapy as well as efficacy endpoints.

11.4 MDS/AML

Maintenance therapy after HCT is being most actively studied in myeloid diseases such as MDS and AML given these conditions are the leading indications for HCT, their significant risk for disease relapse after HCT and the recent approval of several new targeted agents in other therapeutic settings.

11.4.1 Hypomethylating Agents (HMAs)

Hypomethylating agents (HMA), including azacitidine (Aza) and decitabine, have well established efficacy in AML/MDS (DiNardo et al. 2020). HMA are thought to silence tumor-suppressing genes through epigenetic modification and may also enhance the graft-versus-leukemia effect (GVL) through increased expression of tumor antigens. Initially, a small retrospective study established the safety of low dose subcutaneous azacitidine for maintenance ($n = 8$) and treatment of relapsed disease; eight patients who received azacitidine for maintenance remained in CR for a median of 17 months posttransplant (Jabbour et al. 2009). A similar dose finding study of oral decitabine identified a dose of 10 mg/m^2 as the optimal dose due to high rates of grade 3/4 hematologic toxicity with higher doses (Pusic et al. 2015). Subsequently, 30 patients undergoing HCT for MDS/AML were enrolled in a prospective phase I/II study of a novel oral formulation of azacitidine (CC-486) as maintenance therapy post HCT. The dosing comprised of four cohorts varying dose and duration. Based on observed efficacy and tolerability, the 14-day 400 mg cohort was expanded to a total of 19 patients. The 1-year rate of relapse was 21%, and lower (13%) in the 14-day dosing group. The one-year relapse and progression free survival (RPFS) in the latter group was 72%. Estimated 1-year survival rates in the whole group was >80% and median OS was not reached. The rate of severe chronic GVHD was lower than expected at 10%, supporting the hypothesis that Aza may modulate alloreactive T-cells potentially reducing the incidence of GVHD. The 14-day 400 mg dose was well tolerated, with no new safety signals and low rates of discontinuation due to treatment related adverse effects (TRAE). Overall, the authors concluded that once daily oral azacitidine was safe and potentially effective as post-HCT maintenance therapy with obvious benefits in convenience given the oral formulation (de Lima et al. 2018). Recently, a phase 1B trial using a reduced dose of oral decitabine/cedazuridine as maintenance therapy after HCT for patients with MDS/CMML finished accrual and results are awaited (NCT04980404).

The benefit of maintenance HMA after HCT has been observed in several retrospective analyses. A single-center retrospective analysis of 136 patients, of whom 76 (56%) received post-transplant maintenance with azacitidine ($n = 36$) or decitabine ($n = 40$), observed significantly lower rates of relapse and improved OS and DFS among the treated cohort (Li et al. 2023). Similarly, a single-center prospective cohort study of azacitidine maintenance compared a cohort of 31 patients who received 32 mg/m^2 Aza on days 1–5 of a 28-day cycle, compared to 18 patients who were on a surveillance only protocol. A total of four cycles were planned, and 76% of patients were able to tolerate the full course of treatment. At a median follow up of 27 months, the relapse rate was significantly higher in the control group at 67%, median time to relapse of 4 months compared to the 26% in the treatment cohort with median time to relapse not reached. OS was 7.6 months in the control group and not reached amongst the treated participants. The authors attributed the large difference observed in this study to their inclusion of high-risk AML patients only (Keruakous et al. 2023). Similarly, a single center case-control analysis compared 93 patients who received azacitidine maintenance post-transplant for a median of 8 cycles to 357 matched controls who did not. For the entire cohort, no significant difference was observed in the 3-year cumulative incidence of disease progression but there was a small significant improvement in OS and PFS with Aza. Moreover, in subset analyses of prespecified high-risk AML and MDS patients, Aza maintenance led to improved OS and PFS, whereas no statistically significant difference was observed in the low/intermediate risk cohorts. This subgroup analysis emphasizes the importance of identifying subgroups in whom post HCT maintenance is likely to have benefit—not only for study design eligibility, but to importantly spare potential toxicity in those subgroups whom are unlikely to benefit (Pasvolsky et al. 2024). Not all analyses have suggested a benefit for maintenance HMA as a retrospective study of 185 patients undergoing HCT for high-risk AML, MDS, and CMML found no difference in 2-year CIR, OS, and PFS between the 65 patients (35%) who received single agent Aza maintenance at 32 mg/m^2 for 5 days of every 28-day cycle for 12 cycles, and those who did not (Wattebled et al. 2022).

Finally, in a prospective, phase III, open label randomized controlled trial, 185 patients were randomized 1:1 to receive IV azacitidine, at the standard dosing of 32 mg/m^2 for 5 consecutive days per 28-day cycle versus no maintenance.

The median number of treatment cycles received was 4 and only 28% of patients in the maintenance arm completed the planned 12 cycles. No statistically significant difference in median RFS, OS, CIR, or GVHD incidence was observed between the two arms, and no new safety signals were identified in the maintenance arm. Of note, there was a clinically meaningful difference in RFS of 2.07 years in the Aza arm versus 1.28 years in the control arm; however, due to fewer than expected events in the control arm, the study was underpowered for this to translate into statistical significance. This phase 3 study illustrated several challenges in conducting prospective randomized trials in the maintenance setting: (1) relatively slow accrual, (2) over performance by the control arm, and (3) treatment arm not receiving what is believed to the ideal maintenance therapy dose or duration (Oran et al. 2020a). The ongoing phase 3 randomized double-blinded AMADEUS study using the oral HMA CC486 will hopefully help answer more questions regarding the utility of maintenance HMA.

Even without a definitive trial showing benefit, maintenance HMA use after HCT for high-risk MDS or AML has become an accepted practice. The untargeted nature of HMA therapy will make it difficult to truly figure out which subsets benefit. In addition, the lack of a truly generalizable validated ultra-sensitive MRD assay for AML regardless of genetic mutations also hampers these efforts. The development of oral formulations has made therapy easier to give, however, cytopenias remain the main concern for feasibility although dose reductions or longer cycles can likely mitigate this risk. We are hopeful that future studies can be designed to incorporate modern MRD methods to better understand who truly benefits from such an approach.

11.4.2 HMA Combinations

Combinations of HMA with other effective agents are now routine in the therapy for MDS and AML. Naturally, such combinations are being investigated as maintenance therapy after HCT. A phase II trial evaluated the efficacy of Aza with donor lymphocyte infusion in 30 patients with high-risk AML (n = 20) or MDS (n = 10). At a median follow up of 49 months, OS was 65.5% with 9 patients dying of relapse. Cumulative incidence of relapse at 2 years was 27.6%. The same group, as part of a larger cohort, was analyzed in a retrospective study with 54 AML and 23 MDS patients. At 2 years, an OS of 70.8% and PFS of 68.3% were observed. The treatment appeared to be well tolerated, with half of patients completing 12 planned cycles of Aza and receiving at least one DLI (Guillaume et al. 2021). Recent abstracts have also shown promising safety and efficacy of a novel dinucleotide derivative of decitabine, termed guadecitabine, with or without DLI (Oran et al. 2020b, c).

Another phase II study evaluated the combination of Aza with granulocyte/macrophage colony stimulating factor (GM-CSF), based on the hypothesis that this combination, which in vitro has been demonstrated to promote cell differentiation, may lead to differentiation of the persistent immature leukemic stem cells (LSC) that lead to relapse. Of 24 patients who underwent HCT, 18 (75%) received at least one cycle of maintenance therapy, with a median of 7 cycles administered. The median OS and RFS for the treated patients were 29 months and 18.6 months, respectively. A higher dose of Aza, 75 mg/m^2, was well tolerated in this study compared to similar studies, presumably because the use of GM-CSF somewhat mitigated the hematologic toxicity observed in other trials. There was no clear survival benefit noted, with the authors attributing rates of RFS and OS greater than historical controls to baseline characteristics rather than treatment effect (Webster et al. 2021).

A prospective study of 20 patients examined the combination of low dose decitabine (LDEC) with venetoclax (VEN), the oral B-cell lymphoma-2 (BCL-2) inhibitor which is now routinely used in combination with HMA in the treatment of MDS/AML. All enrolled patients in this trial received LDEC with VEN starting at approximately day +100 after HCT, for a median of five 8-week cycles. The 2-year OS and EFS were both 85%, and the 2-year CIR and NRM were 15.3% and 6%, respectively. No significant impact on GVHD was noted and no new safety signals were identified (Wei et al. 2015). Another phase I trial described 22 patients who received azacitidine (36 mg/m^2/d IV on days 1–5) with venetoclax (400 mg daily on days 1–14) in either 28 or 42-day cycles after HCT. While expected cytopenias were common, significant infections were rare and investigators felt the combination to be safe and deserving of further investigation (Garcia et al. 2024). A phase II trial of Aza and VEN maintenance only enrolled 35 patients (closed early due to slow accrual) with participants receiving a median of 9 cycles of therapy. 2-year RFS was 65% and the most common adverse events were the expected cytopenias yet no conclusions regarding efficacy could be drawn due to the small sample size (Bazinet et al. 2024). A large open-label phase 3 study (VIALE-T) which randomized AML patients after HCT to best supportive care vs. maintenance Aza/VEN has finished accrual with results eagerly awaited (NCT04161885).

Gemtuzumab ozogamicin is an antibody-drug conjugate of a humanized anti-CD33 monoclonal antibody with a calicheamicin payload. A small prospective study enrolled ten patients with high-risk AML to receive a combination of azacitidine with gemtuzumab ozogamicin (AZA-GO) for a total of 4 cycles as maintenance following HCT (with AZA at 30 mg/m^2 on days 1–7 followed by GO at 3 mg/m^2 on day 8). Most patients had dose limiting myelosuppression, with median cycle length being 53 days and median number of treatment cycles received at 1.5 (Oshikawa et al. 2015). A subsequent abstract by the same group showed no statistically significant improvement in OS or DFS when compared to matched controls (Kaito et al. 2017).

TP53 mutations portend a particularly poor prognosis in MDS and AML, with estimated OS of 4–8 months after HCT. Eprenetapopt is a novel, first-in-class, p53 "refolder" theorized to restore the loss of tumor inhibition conferred by TP53 mutations. A prospective, phase II, open-label study by Mishra et al. sought to evaluate the safety and efficacy of eprenatapopt in combination with Aza as post HCT maintenance in patients with TP53 mutated AML or MDS. Thirty-three patients were ultimately enrolled and began maintenance therapy at a median of 68 days post-transplant for a median 7 cycles. The median RFS at 14.5 months was 12.5 months, with a 1-year RFS probability of 60%. At median follow up of 17 months, the estimated median OS was 20.6 months, with a 1-year OS probability of 79%. Compared to historical controls, treated patients had a compelling prognosis, yet many patients who were screened and underwent HCT ultimately did not start therapy due to competing risks of early relapse and other complications (Mishra et al. 2022). Unfortunately, development of this agent was halted due to failure to meet the primary efficacy endpoint in primary treatment trials in MDS.

11.4.3 FLT3 Inhibitors

AML with targetable mutations, such as the internal tandem duplication mutation in the FMS-like tyrosine kinase 3 gene (FLT3-ITD), provide attractive targets for maintenance due to less off-target toxicity. Initially, a phase I study of sorafenib in 22 patients with FLT3-ITD mutated AML established the dose of 400 mg twice daily as safe after HCT. Results demonstrated remarkably high OS and PFS with very few relapses observed. Toxicity in terms of GI and skin were noted, but could easily be reversed with dose reductions or interruptions (Chen et al. 2014). Subsequently, a retrospective landmark analysis from the same investigators was conducted, comparing the use of maintenance sorafenib after HCT in 26 patients with FLT3-ITD AML in CR1 with 43 control patients with FLT3-ITD AML in CR1 undergoing HCT contemporaneously who did not receive any maintenance therapy. Sorafenib was associated with significantly improved 2-year PFS (85% vs. 52%, $p = 0.0047$) and OS (83% vs. 58%, $p = 0.019$) with the difference primarily driven by a significant decrease in disease relapse in patients who received sorafenib (8.2% vs. 37.7%, $p = 0.0077$) (Brunner et al. 2016).

Two prospective randomized trials have investigated the utility of maintenance sorafenib after HCT for FLT3-ITD mutated AML. The German SORMAIN trial was a multicenter, randomized, double blind, placebo-controlled phase II study which enrolled only 83 patients in 5 years and closed early due to poor accrual, yet was able to show a significant benefit in 24-month relapse free survival (RFS) with 24 months of maintenance sorafenib (85.0% vs. 53.3%, $p = 0.002$). Sorafenib treatment was not significantly associated with more toxicity than placebo, leading the authors to conclude that it was an efficacious and well tolerated (Burchert et al. 2020). In an open-label Chinese phase III trial, 202 patients were randomized to receive sorafenib for 6 months duration vs. no therapy. Results showed 2-year leukemia-free survival to be 78.9% vs. 56.6% ($p < 0.0001$) in favor of maintenance sorafenib, very similar to what was observed in the SORMAIN study and the prior retrospective case control analysis (Xuan et al. 2020). Extended follow-up of the phase III study showed an overall survival benefit for sorafenib at 5 years (72% vs. 55.9%, HR 0.55, $p = 0.011$) driven by a lower incidence of relapse without any difference in non-relapse mortality (Xuan et al. 2023). It should be pointed out that the majority of the patients on both of these randomized trials did not receive FLT3-TKI during induction therapy as has become standard of care. In terms of real world experience, a single-center, retrospective analysis of sorafenib maintenance demonstrated similarly favorable outcomes, including 2-year PFS of 73% and OS 90%, but extremely high rates of treatment discontinuation (60% of the total cohort) due to toxicity, primarily manifesting as dermatologic, gastrointestinal, and hematologic complications (Morin et al. 2023).

As additional TKIs were developed specifically for treatment of FLT3-ITD AML, other trials were conducted for their use in maintenance therapy. A prospective, single center, phase II trial assessed 94 patients with FLT3-ITD mutated AML undergoing HCT receiving midostaurin maintenance and compared them to historical controls, concluding that midostaurin maintenance at a dose of 50 mg twice daily for a planned one year course had high rates of treatment discontinuation due to toxicity (59%) but also improved OS and PFS compared to historical controls (Schlenk et al. 2019). Subsequently, a phase 2 randomized open label trial allocated participants 1:1 to receive standard of care (SOC) versus midostaurin 50 mg twice daily. While the trial was not powered to detect significant differences in relapse or survival, there was a trend towards improved survival in the midostaurin maintenance arm, and the 2-year OS of 76% was better than historical controls. More AEs were noted in the midostaurin arm, with a 27% rate of treatment discontinuation due to AEs versus 3% in the SOC arm (Maziarz et al. 2021).

Quizartinib is a selective type 2 FLT3 inhibitor approved for use in combination with initial chemotherapy for FLT3-ITD mutated AML. This approval was based on a randomized, placebo-controlled, phase 3 trial of 559 patients adding quizartinib to chemotherapy during induction, with consolidation and afterwards as single agent quizartinib maintenance as well. An overall survival benefit was observed with quizartinib treatment in the entire study population (induction and maintenance) compared to placebo, however no benefit was observed for those patients who underwent HCT (Erba et al. 2023). Thus, the current label for quizartinib states that it is specifically not approved for single agent maintenance after HCT. Nevertheless, clinicians may choose to use quizartinib in this setting for high-risk patients especially if patients had received it during induction therapy and tolerated it well.

Gilteritinib is another potent, well-tolerated FLT3 TKI and is approved for relapsed/refractory FLT3 mutated AML. It was studied as maintenance therapy after HCT for FLT3-ITD AML patients in first CR in a randomized, placebo-controlled, international phase 3 trial which allocated patients in a 1:1 fashion to receive gilteritinib 120 mg daily versus placebo for 2 years after HCT. A significant RFS benefit was not observed in the treatment arm (HR 0.679,

$p = 0.0518$) with large differences depending on the region of the world where treated. A pre-specified analysis showed that patients with detectable FLT3-ITD MRD either prior to or after HCT before randomization derived a significant benefit in RFS from gilteritinib (HR 0.515, $p = 0.0065$). More treatment associated myelosuppression and infectious complications were observed in the treatment arm. The authors concluded that patients with detectable FLT3-ITD MRD around the time of HCT are indicated for maintenance therapy with gilteritinib for 2 years after HCT (Levis et al. 2024).

While the initial trials employing FLT3 TKIs after HCT for FLT3-ITD mutated AML were rooted in straightforward logic of targeted therapy for enhanced cytotoxicity, other mechanisms of action have been proposed given the multiple kinases these agents inhibit. Certain TKIs, specifically sorafenib, have even been shown to be of benefit when added to conventional therapy regardless of FLT3 mutational status (Röllig et al. 2015). The most interesting mechanism of action proposed for the benefit of FLT3 inhibition after HCT suggested that sorafenib induces IL-15 secretion from FLT3-ITD mutated malignant cells. IL-15 then results in metabolic reprogramming of resident cytotoxic T-cells which can more effectively eliminate the malignant cells leading to an enhanced graft vs. leukemia effect (Mathew et al. 2018). These results were recently replicated and also shown to occur with other FLT3 TKIs including quizartinib and gilteritinib (Zhang et al. 2022).

As a result of the three prospective randomized trials (one with gilteritinib, two with sorafenib) above, FLT3 TKI maintenance after HCT has become fairly routine where available. Choice of agent is mainly based on availability. In regions where a commercial FLT3-ITD MRD assay is available, presence or absence of MRD has begun to be used to not only determine candidacy for maintenance but also duration. If choosing to not use maintenance FLT3 TKI, serial measurements of FLT3-ITD MRD should be closely followed with prompt addition of preemptive TKI if MRD becomes detectable in the first 2 years after HCT.

11.4.4 IDH1/IDH2 Inhibitors

Isocitrate dehydrogenase 1 and 2 mutations (IDH1 and IDH2) are found in approximately 15% of AML and less commonly in MDS. These mutations promote oncogenesis by disrupting the epigenetic mechanisms of normal myeloid cell differentiation. The prognostic utility of an IDH1 or IDH2 mutation is uncertain and co-occurrence with more commonly prognostic mutations such as NPM1 or FLT3-ITD have made any potential maintenance studies difficult to interpret (Chen et al. 2021). A multicenter phase I trial assessed the safety of the commercially available IDH2 inhibitor enasidenib in a cohort of 19 patients where maintenance therapy was started at a median of 55 days post HCT. Enasidenib was well tolerated at the 50 mg daily and 100 mg daily doses studied, of which the latter was the recommended phase 2 dose (RP2D). No increased incidence of GVHD was noted, and only 3/19 patients (16%) discontinued treatment due to adverse events. At 2 years, the cumulative incidence of relapse was 16%, NRM was 16%, OS was 74%, and PFS was 69% (Fathi et al. 2022). The same group assessed the safety and efficacy of the commercially available IDH1 inhibitor ivosidenib in a multicenter phase I trial. Sixteen patients were initiated on treatment at a median of 66 days post-transplant, for a median of 11 out of 12 planned cycles. The RP2D was identified as 500 mg daily. Three patients (19%) experienced dose interruption due to adverse events, but only one patient had to discontinue treatment. Outcomes were similarly encouraging, as the 2-year cumulative incidence of relapse was 19%, NRM 0%, OS 88%, and PFS 81% (Fathi et al. 2023). A multicenter randomized phase II trial with ivosidenib in IDH1 mutated myeloid malignancies is planned with results eagerly awaited.

11.4.5 Other Potential Candidates

Other therapeutic targets for pharmacologic maintenance are in development. The hedgehog inhibitor glasdegib was studied in a recent phase 2 trial yet was plagued by high rates of discontinuation due to adverse effects and no compelling signal of efficacy in preventing relapse (Kent et al. 2020). Menin inhibitors are a novel class of compounds which have shown efficacy in KMT2A-mutated and NPM1-mutated AML. Trials involving both revumenib and ziftomenib in the maintenance setting are being conducted. Immunological checkpoint inhibitors were an initially attractive therapy after HCT to try and augment the graft-vs-malignancy effect, and trials used to treat relapsed disease showed some initial efficacy (Davids et al. 2016). However, subsequently, a significant risk of immunological adverse events including severe GVHD tempered such enthusiasm.

11.5 Multiple Myeloma and Lymphoma

The use of allogeneic transplant in multiple myeloma (MM) and lymphoma is decreasing due to the emergence of other therapeutic modalities including novel targeted agents, bispecific engagers, and chimeric antigen receptor T-cells (CAR-T). As such, the role of maintenance therapy after allogeneic HCT in these diseases will likely not be an area of significant investigation going forward. Nevertheless, several studies have been conducted and merit mention. A prospective, phase II study observed increased incidence of GVHD (53%) with lenalidomide maintenance for patients with MM after HCT, leading the authors to conclude that lenalidomide maintenance was not feasible in this study population (Kneppers et al. 2011). A similar study assessing 5 mg of lenalidomide daily observed a 34% incidence of GVHD and 3-year PFS and OS of 52% and 79%, respectively (Kröger et al. 2013). More recently, a multicenter, open label phase II prospective trial observed similarly high rates of GVHD (38%) and treatment discontinuation due to AE at 10 mg daily dosing in patients with MM (Alsina et al. 2014). The proteosome inhibitor bortezomib was attempted to be evaluated in maintenance after allogeneic HCT for MM. While outcomes appeared promising, cytopenias, infectious complications, and GVHD affected more than half of the study population (Kröger et al. 2006). Another proteosome inhibitor, ixazomib, was evaluated in a multicenter, randomized controlled trial with no observed difference in PFS, OS, or GVHD between the treatment and control arms and had to be closed early due to poor accrual (Bashir et al. 2023).

In patients with lymphoma, 10 patients with aggressive B-cell non-Hodgkin lymphoma were treated with maintenance rituximab after allogeneic HCT. At the time of data analysis, all ten patients remained alive, with eight continually disease free and two that were able to be rescued back into remission. Six of ten patients (60%) had severe isolated neutropenia occurring over several months after HCT which could be mitigated with IVIG administration (Shimoni et al. 2003). Despite these promising results, a subsequent randomized controlled trial of rituximab maintenance after HCT observed no difference between the treatment and placebo arms on incidence of GVHD and survival outcomes (Glass et al. 2014). Finally despite promising retrospective data (Armand et al. 2008), randomized placebo controlled studies of the mTOR inhibitor sirolimus, commonly used for GVHD prevention in the past, have also failed to show its efficacy in reducing the risk of relapse after HCT (Armand et al. 2016).

11.6 Conclusion

As increasing progress is made in the prevention of HCT-related complications, standard practices in pharmacologic maintenance treatment after allogeneic HCT remain to be determined. The advent of several targeted therapies has allowed an increasing number of prospective trials in this setting to be conducted. No agent is approved for maintenance therapy after HCT as of yet and the difficulties in carrying out these trials are well known. Nevertheless, use of maintenance therapy at many centers across the United States is routine especially BCR-ABL TKIs, hypomethylating agents, and FLT3 TKIs. Non-engrafting cellular therapy is a rapidly increasingly successful modality of therapy and will undoubtedly also be studied in the post-HCT remission setting. Some cellular therapies (i.e., CD19 CAR-T) can be used as they target unique tumor antigens or cellular subsets that can be eliminated, however, different approaches will need to be taken for the majority of myeloid diseases. Some of these are already underway including engineered T-cells for specific host antigens and allogeneic graft editing approaches to be able to deliver maintenance therapy that would then not harm the graft. Lastly, as more sensitive MRD assays are developed and validated, candidacy and duration of maintenance therapy will be further defined, and pre-emptive therapy may eventually become standard practice for certain patients.

References

Abou Dalle I, Jabbour E, Short NJ, Ravandi F (2019) Treatment of Philadelphia chromosome-positive acute lymphoblastic leukemia. Curr Treat Options Oncol [Internet] 20(1). Available from: https://doi.org/10.1007/s11864-019-0603-z

Alsina M, Becker PS, Zhong X et al (2014) Lenalidomide maintenance for high-risk multiple myeloma after allogeneic hematopoietic cell transplantation. Biol Blood Marrow Transplant 20(8):1183–1189

Armand P, Gannamaneni S, Kim HT et al (2008) Improved survival in lymphoma patients receiving sirolimus for graft-versus-host disease prophylaxis after allogeneic hematopoietic stem-cell transplantation with reduced-intensity conditioning. J Clin Oncol 26(35):5767–5774

Armand P, Kim HT, Sainvil M-M et al (2016) The addition of sirolimus to the graft-versus-host disease prophylaxis regimen in reduced intensity allogeneic stem cell transplantation for lymphoma: a multicentre randomized trial. Br J Haematol 173(1):96–104

Bachanova V, Marks DI, Zhang M-J et al (2014) Ph+ ALL patients in first complete remission have similar survival after reduced intensity and myeloablative allogeneic transplantation: impact of tyrosine kinase inhibitor and minimal residual disease. Leukemia 28(3):658–665

Bashir Q, Nishihori T, Pasquini MC et al (2023) A multicenter phase II, double-blind, placebo-controlled trial of maintenance ixazomib after allogeneic transplantation for high-risk multiple myeloma: results of the blood and marrow transplant clinical trials network 1302 trial. Transplant Cell Ther 29(6):358.e1–358.e7

Bazinet A et al (2024) Reduced dose azacitidine plus venetoclax as maintenance therapy in acute myeloid leukaemia following intensive or low-intensity induction: a single-centre, single-arm, phase 2 trial. Lancet Haematol 11(4):e287–e298

Bejanyan N, Weisdorf DJ, Logan BR et al (2015) Survival of patients with acute myeloid leukemia relapsing after allogeneic hematopoietic cell transplantation: a center for international blood and marrow transplant research study. Biol Blood Marrow Transplant 21(3):454–459

Brissot E, Labopin M, Beckers MM et al (2015) Tyrosine kinase inhibitors improve long-term outcome of allogeneic hematopoietic stem cell transplantation for adult patients with Philadelphia chromosome positive acute lymphoblastic leukemia. Haematologica 100(3):392–399

Brunner AM, Li S, Fathi AT et al (2016) Hematopoietic cell transplantation with and without sorafenib maintenance for patients with FLT3-ITD acute myeloid leukemia in CR1. Biol Blood Marrow Transplant 22(3):S198–S199

Burchert A, Bug G, Fritz LV et al (2020) Sorafenib maintenance after allogeneic hematopoietic stem cell transplantation for acute myeloid leukemia with FLT3-internal tandem duplication mutation (SORMAIN). J Clin Oncol 38(26):2993–3002

Carpenter PA, Johnston L, Fernandez HF et al (2017) Posttransplant feasibility study of nilotinib prophylaxis for high-risk Philadelphia chromosome positive leukemia. Blood 130(9):1170–1172

Chen Y-B, Li S, Lane AA et al (2014) Phase I trial of maintenance sorafenib after allogeneic hematopoietic stem cell transplantation for fms-like tyrosine kinase 3 internal tandem duplication acute myeloid leukemia. Biol Blood Marrow Transplant 20(12):2042–2048

Chen EC, Li S, Eisfeld A-K et al (2021) Outcomes for patients with IDH-mutated acute myeloid leukemia undergoing allogeneic hematopoietic cell transplantation. Transplant Cell Ther 27(6):479.e1–479.e7

Chen H, Xu L-P, Zhang X-H et al (2022) Safety and outcomes of maintenance therapy with third-generation tyrosine kinase inhibitor after allogeneic hematopoietic cell transplantation in Philadelphia chromosome positive acute lymphoblastic leukemia patients with T315I mutation. Leuk Res 121(106930):106930

Davids MS, Kim HT, Bachireddy P et al (2016) Ipilimumab for patients with relapse after allogeneic transplantation. N Engl J Med 375(2):143–153

de Lima M, Oran B, Champlin RE et al (2018) CC-486 maintenance after stem cell transplantation in patients with acute myeloid leukemia or myelodysplastic syndromes. Biol Blood Marrow Transplant 24(10):2017–2024

DeFilipp Z, Langston AA, Chen Z et al (2016) Does post-transplant maintenance therapy with tyrosine kinase inhibitors improve outcomes of patients with high-risk Philadelphia chromosome-positive leukemia? Clin Lymphoma Myeloma Leuk 16(8):466–471.e1

DeFilipp Z, Ancheta R, Liu Y et al (2020) Maintenance tyrosine kinase inhibitors following allogeneic hematopoietic stem cell transplantation for chronic myelogenous leukemia: a Center for International Blood and Marrow Transplant Research study. Biol Blood Marrow Transplant 26(3):472–479

DiNardo CD, Jonas BA, Pullarkat V et al (2020) Azacitidine and venetoclax in previously untreated acute myeloid leukemia. N Engl J Med 383(7):617–629

Erba HP, Montesinos P, Kim H-J et al (2023) Quizartinib plus chemotherapy in newly diagnosed patients with FLT3-internal-tandem-duplication-positive acute myeloid leukaemia (QuANTUM-First): a randomised, double-blind, placebo-controlled, phase 3 trial. Lancet 401(10388):1571–1583

Fathi AT, Kim HT, Soiffer RJ et al (2022) Enasidenib as maintenance following allogeneic hematopoietic cell transplantation for *IDH2*-mutated myeloid malignancies. Blood Adv 6(22):5857–5865

Fathi AT, Kim HT, Soiffer RJ et al (2023) Multicenter phase I trial of ivosidenib as maintenance treatment following allogeneic hematopoietic cell transplantation for *IDH1*-mutated acute myeloid leukemia. Clin Cancer Res 29(11):2034–2042

Gaballa MR, Banerjee P, Milton DR et al (2022) Blinatumomab maintenance after allogeneic hematopoietic cell transplantation for B-lineage acute lymphoblastic leukemia. Blood 139(12):1908–1919

Garcia J et al (2024) Prophylactic maintenance with venetoclax/azacitidine after reduced-intensity conditioning allogeneic transplant for high-risk MDS and AML. Blood Adv 8(4):978–990

Glass B, Hasenkamp J, Wulf G et al (2014) Rituximab after lymphoma-directed conditioning and allogeneic stem-cell transplantation for relapsed and refractory aggressive non-Hodgkin lymphoma (DSHNHL R3): an open-label, randomised, phase 2 trial. Lancet Oncol 15(7):757–766

Guillaume T, Thépot S, Peterlin P et al (2021) Prophylactic or preemptive low-dose azacitidine and donor lymphocyte infusion to prevent disease relapse following allogeneic transplantation in patients with high-risk acute myelogenous leukemia or myelodysplastic syndrome. Transplant Cell Ther 27(10):839.e1–839.e6

Jabbour E, Giralt S, Kantarjian H et al (2009) Low-dose azacitidine after allogeneic stem cell transplantation for acute leukemia. Cancer 115(9):1899–1905

Kaito S, Najima Y, Kishida Y et al (2017) Post-transplant maintenance therapy with azacitidine and gemtuzumab ozogamicin for high-risk hematologic malignancies. Blood 130:4517

Kebriaei P, Saliba R, Rondon G et al (2012) Long-term follow-up of allogeneic hematopoietic stem cell transplantation for patients with Philadelphia chromosome-positive acute lymphoblastic leukemia: impact of tyrosine kinase inhibitors on treatment outcomes. Biol Blood Marrow Transplant 18(4):584–592

Kent A, Vasu S, Schatz D et al (2020) Glasdegib as maintenance therapy for patients with AML and MDS patients at high risk for postallogeneic stem cell transplant relapse. Blood Adv 4(13):3102–3108

Keruakous AR, Holter-Chakrabarty J, Schmidt SA, Khawandanah MO, Selby G, Yuen C (2023) Azacitidine maintenance therapy post-allogeneic stem cell transplantation in poor-risk acute myeloid leukemia. Hematol Oncol Stem Cell Ther 16(1):52–60

Kneppers E, van der Holt B, Kersten M-J et al (2011) Lenalidomide maintenance after nonmyeloablative allogeneic stem cell transplantation in multiple myeloma is not feasible: results of the HOVON 76 Trial. Blood 118(9):2413–2419

Kröger N, Zabelina T, Ayuk F et al (2006) Bortezomib after dose-reduced allogeneic stem cell transplantation for multiple myeloma to enhance or maintain remission status. Exp Hematol 34(6):770–775

Kröger N, Zabelina T, Klyuchnikov E et al (2013) Toxicity-reduced, myeloablative allograft followed by lenalidomide maintenance as salvage therapy for refractory/relapsed myeloma patients. Bone Marrow Transplant 48(3):403–407

Levis MJ, Hamadani M, Logan B et al (2024) Gilteritinib as post-transplant maintenance for acute myeloid leukemia with internal tandem duplication mutation of FLT3. J Clin Orthod:JCO.23.02474

Li L, Zhang R, Cao W et al (2023) Comparative analysis of hypomethylating agents as maintenance therapy after allogeneic hematopoietic stem cell transplantation for high-risk acute leukemia. Leuk Lymphoma 64(13):2113–2122

Mathew NR, Baumgartner F, Braun L et al (2018) Sorafenib promotes graft-versus-leukemia activity in mice and humans through IL-15 production in FLT3-ITD-mutant leukemia cells. Nat Med 24(3):282–291

Maziarz RT, Levis M, Patnaik MM et al (2021) Midostaurin after allogeneic stem cell transplant in patients with FLT3-internal tandem duplication-positive acute myeloid leukemia. Bone Marrow Transplant 56(5):1180–1189

Metheny LL, Sobecks R, Cho C et al (2024) A multicenter study of posttransplantation low-dose inotuzumab ozogamicin to prevent relapse of acute lymphoblastic leukemia. Blood Adv 8(6):1384–1391

Mishra A, Tamari R, DeZern AE et al (2022) Eprenetapopt plus azacitidine after allogeneic hematopoietic stem-cell transplantation for TP53-mutant acute myeloid leukemia and myelodysplastic syndromes. J Clin Oncol 40(34):3985–3993

Morin S, Giannotti F, Mamez A-C et al (2023) Real-world experience of sorafenib maintenance after allogeneic hematopoietic stem cell transplantation for FLT3-ITD AML reveals high rates of toxicity-related treatment interruption. Front Oncol 13:1095870

Olavarria E, Siddique S, Griffiths MJ et al (2007) Posttransplantation imatinib as a strategy to postpone the requirement for immunotherapy in patients undergoing reduced-intensity allografts for chronic myeloid leukemia. Blood 110(13):4614–4617

Oran B, de Lima M, Garcia-Manero G et al (2020a) A phase 3 randomized study of 5-azacitidine maintenance vs observation after transplant in high-risk AML and MDS patients. Blood Adv 4(21):5580–5588

Oran B, Alousi AM, Mehta RS et al (2020b) Minimal residual disease eradication with guadecitabine (SGI-110) in the post-transplant setting. Blood 136:10–11

Oran B, Alatrash G, Alousi AM et al (2020c) Maintenance treatment with guadecitabine (SGI-110) in high risk MDS and AML patients after allogeneic stem cell transplantation. Blood 136:29–30

Oshikawa G, Kakihana K, Saito M et al (2015) Post-transplant maintenance therapy with azacitidine and gemtuzumab ozogamicin for high-risk acute myeloid leukaemia. Br J Haematol 169(5):756–759

Pasvolsky O, Saliba RM, Popat UR et al (2024) Azacitidine post-transplant maintenance improves disease progression in high-risk acute myeloid leukemia and myelodysplastic syndrome. Clin Lymphoma Myeloma Leuk [Internet]. Available from: https://doi.org/10.1016/j.clml.2024.01.012

Pfeifer H, Wassmann B, Bethge W et al (2012) Randomized comparison of prophylactic and minimal residual disease-triggered imatinib after allogeneic stem cell transplantation for BCR–ABL1-positive acute lymphoblastic leukemia. Leukemia 27(6):1254–1262

Pusic I, Choi J, Fiala MA et al (2015) Maintenance therapy with decitabine after allogeneic stem cell transplantation for acute myelogenous leukemia and myelodysplastic syndrome. Biol Blood Marrow Transplant 21(10):1761–1769

Ribera J-M, Oriol A, González M et al (2010) Concurrent intensive chemotherapy and imatinib before and after stem cell transplantation in newly diagnosed Philadelphia chromosome-positive acute lymphoblastic leukemia. Final results of the CSTIBES02 trial. Haematologica 95(1):87–95

Röllig C, Serve H, Hüttmann A et al (2015) Addition of sorafenib versus placebo to standard therapy in patients aged 60 years or younger with newly diagnosed acute myeloid leukaemia (SORAML): a multicentre, phase 2, randomised controlled trial. Lancet Oncol 16(16):1691–1699

Schlenk RF, Weber D, Fiedler W et al (2019) Midostaurin added to chemotherapy and continued single-agent maintenance therapy in acute myeloid leukemia with FLT3-ITD. Blood 133(8):840–851

Senapati J, Jabbour E, Kantarjian H, Short NJ (2023) Pathogenesis and management of accelerated and blast phases of chronic myeloid leukemia. Leukemia 37(1):5–17

Shimoni A, Hardan I, Avigdor A et al (2003) Rituximab reduces relapse risk after allogeneic and autologous stem cell transplantation in patients with high-risk aggressive non-Hodgkin's lymphoma. Br J Haematol 122(3):457–464

Shimoni A, Volchek Y, Koren-Michowitz M et al (2015) Phase 1/2 study of nilotinib prophylaxis after allogeneic stem cell transplantation in patients with advanced chronic myeloid leukemia or Philadelphia chromosome–positive acute lymphoblastic leukemia. Cancer 121(6):863–871

Wattebled K-J, Drumez E, Coiteux V et al (2022) Single-agent 5-azacytidine as post-transplant maintenance in high-risk myeloid malignancies undergoing allogeneic hematopoietic cell transplantation. Ann Hematol 101(6):1321–1331

Webster JA, Yogarajah M, Zahurak M et al (2021) A phase II study of azacitidine in combination with granulocyte-macrophage colony-stimulating factor as maintenance treatment, after allogeneic blood or marrow transplantation in patients with poor-risk acute myeloid leukemia (AML) or myelodysplastic syndrome (MDS). Leuk Lymphoma 62(13):3181–3191

Wei A, Tan P, Perruzza S, Govindaraj C (2015) Maintenance lenalidomide in combination with 5-azacitidine as post-remission therapy for acute myeloid leukaemia. Br J Haematol [Internet]. Available from: https://onlinelibrary.wiley.com/doi/abs/10.1111/bjh.13281

Xuan L, Wang Y, Huang F et al (2020) Sorafenib maintenance in patients with FLT3-ITD acute myeloid leukaemia undergoing allogeneic haematopoietic stem-cell transplantation: an open-label, multicentre, randomised phase 3 trial. Lancet Oncol 21(9):1201–1212

Xuan L, Wang Y, Yang K et al (2023) Sorafenib maintenance after allogeneic haemopoietic stem-cell transplantation in patients with FLT3-ITD acute myeloid leukaemia: long-term follow-up of an open-label, multicentre, randomised, phase 3 trial. Lancet Haematol 10(8):e600–e611

Zhang Z, Hasegawa Y, Hashimoto D et al (2022) Gilteritinib enhances graft-versus-leukemia effects against FLT3-ITD mutant leukemia after allogeneic hematopoietic stem cell transplantation. Bone Marrow Transplant 57(5):775–780

12 Challenges and Opportunities in the Care of Hematopoietic Cell Transplant Survivors in the Modern Era

Neel S. Bhatt, Alexis Richards, Kristen L. Beebe, and Nandita Khera

Abstract

With the constantly changing field of hematopoietic cell transplant (HCT) and cellular therapy and a growing number of survivors, there is an ongoing need for work to understand the changing landscape of late effects and develop interventions to optimize the long-term outcomes of survivors. In this chapter, we summarize the studies published in last 5 years describing late effects after HCT. We specifically discuss the available literature or the lack thereof for the late effects as it relates to the changes in transplant practices such as increased use of post-transplant cyclophosphamide, newer drugs to treat chronic graft vs. host disease, and chimeric antigen receptor T-cell therapy as a bridge to HCT or for a post-HCT relapse. We also summarize the recently updated recommendations for screening and management of late effects and a monitoring plan to cater to late effects specific to patients undergoing transplants for non-malignant diseases. We discuss the specific considerations such as challenges for adolescent and young adult HCT survivors, especially as they transition from pediatric to adult care. With the information discussed here, we outline challenges and opportunities for research and clinical practice highlighting the role of health care delivery models in addressing some of these issues.

Keywords

Survivorship · Adolescent and young adults · Models of care

N. S. Bhatt (✉)
Fred Hutchinson Cancer Center, Seattle, WA, USA

University of Washington, Seattle, WA, USA
e-mail: nbhatt@fredhutch.org

A. Richards · N. Khera
Mayo Clinic Arizona, Scottsdale, AZ, USA

K. L. Beebe
Mayo Clinic Arizona, Scottsdale, AZ, USA

Phoenix Children's Hospital, Phoenix, AZ, USA

12.1 Introduction

Over the last four decades, allogeneic hematopoietic cell transplantation (HCT) has been increasingly used as a curative option for patients with hematologic malignancies and non-malignant diseases. According to the Center for International Blood and Marrow Transplant Research (CIBMTR), there were approximately 8000 allogeneic HCT performed in the United States (US) in the year 2022, which is more than two times higher compared to the number of HCT performed in the year 2000 (Cusatis et al. 2023). Around 1.5 million allogeneic and autologous HCT have been performed globally between 1957 and 2019 (Niederwieser et al. 2022).

© The Author(s), under exclusive license to Springer Nature Switzerland AG 2025
K. Ballen (ed.), *Clinical Management of Challenges in Allogeneic Stem Cell Transplant*, Advances in Experimental Medicine and Biology 1475, https://doi.org/10.1007/978-3-031-84988-6_12

With changes in HCT practices such as availability of reduced intensity regimens, improvement in supportive care, and better management of post-HCT complications, overall volumes of HCT have increased as has the number of survivors. It is estimated that there will be nearly half a million HCT survivors in the United States by the year 2030 (Majhail et al. 2013). As the number of HCT survivors continue to increase, there is a burgeoning focus on assessment and management of long-term HCT survivors and development and revision of screening and preventive practice guidelines (Rizzo et al. 2006; Majhail et al. 2012; Rotz et al. 2024a, b). There has been a significant progress in the field of HCT survivorship in last few decades, which is indicated by the reduction in late mortality post-HCT (Bhatia et al. 2021). However, with the constantly changing field of HCT and cellular therapy and increasing number of survivors, there is an ongoing need for work to understand the changing landscape of late effects and develop interventions to optimize the long-term outcomes in this field.

In this chapter, we summarize the recent studies describing late effects after HCT and specifically discuss the available literature or the lack thereof for the late effects as it relates to the changes in transplant practice such as increased use of post-transplant cyclophosphamide, availability of newer drugs to treat chronic graft vs. host disease (GVHD), and increasing population of patients who may be treated with chimeric antigen receptor T-cell (CAR-T) therapy as a bridging therapy prior to HCT or for a post HCT relapse. We also discuss the summary of recently updated recommendations for screening and management of late effects as well as a monitoring plan to cater to late effects specific to patients undergoing transplants for non-malignant diseases as their numbers increase (Shenoy et al. 2018; Dietz et al. 2017a; Heimall et al. 2017a). We also discuss the specific considerations such as challenges for adolescent and young adult (AYA) HCT survivors especially as they transition from pediatric to adult care. We outline overall challenges and opportunities for research and clinical practice highlighting the role of health care delivery models in addressing some of these issues. Finally, we review the contributions of professional and patient advocacy organizations that are increasingly involved in improving the long-term outcomes of HCT.

12.2 Complications After HCT

A plethora of descriptive studies over the years have helped understand the spectrum of physical and psychological long-term complications HCT survivors that increase the mortality and morbidity of this procedure for HCT survivors. Table 12.1 summarizes the findings of studies published in the last 5 years describing late effects in the category of infections, cardiovascular complications, secondary neoplasms, and overall burden.

Recently, the international recommendations for screening and preventive practices for long-term survivors of pediatric and adult HCT and cellular therapy were updated in 2023 as a collaborative effort between 11 international societies (Rotz et al. 2024a, b). Here, we summarize key recommendations from these guidelines.

Allogeneic HCT survivors are known to be at risk for infections and should be re-vaccinated according to the most recent immunization guidelines (Carpenter and Englund 2016, 2021; Cordonnier et al. 2019). Due to the risk of early and late pulmonary dysfunction, frequent pulmonary function testing (PFT) assessments are recommended with every 3 months in the first year, every 6 months in the second year, and annually up to 5 years post-HCT in adults or until patients reach final adult height in pediatric patients. Additionally, more frequent PFT assessments are recommended for those diagnosed with chronic GVHD. Serial two-dimensional echocardiogram-based screening for cardiomyopathy is recommended at 2–5 year intervals for pediatric patients based on the doses of anthracycline-based chemotherapy and/or radiation exposing heart according to the COG long-term follow-up guidelines (www.survivorshipguidelines.org). Assessment for osteopenia or osteoporosis should be performed through dual X-ray absorptiometry (DEXA) at 1-year post-HCT timepoint with earlier assessment (at 3-month post-HCT) for those exposed to high-dose glucocorticoids. In

Table 12.1 Current evidence on physical and psychological late effects after hematopoietic cell transplant

Category	Study	Year	Main findings
Infection	Characteristics of late fatal infections after allogeneic hematopoietic cell transplantation (Norkin et al. 2019)	2019	Late fatal infections contributed to 1/3 of deaths >2 years post transplant Older age, male sex, HCT from unmatched donor, and cGVHD on immunosuppression >2 years were associated with increased risk
	Late occurring infections in a contemporary cohort of hematopoietic cell transplantation survivors (Sy et al. 2021)	2021	Nearly 1/3 of survivors developed at least one severe-fatal infection in the 5 years after 1 year from HCT date Acute leukemia and elevated HCT-CI score at HCT were independent predictors of subsequent infections 5-year cumulative incidence of late-occurring infections exceeded 45% Incremental risk associated with severity of GVHD
	Clinical significance of late CMV disease in adult patients who underwent allogeneic stem cell transplant (Shimizu et al. 2023)	2023	5-year incidence of CMV infection was 1.6% Independent risk factors were haploidentical donor, adult T-cell leukemia lymphoma, and preemptive therapy before 100 days post-transplant Late CMV was an independent risk factor for non-relapse mortality rate and overall survival rate
	Cytomegalovirus reactivations in allogeneic hematopoietic stem cell transplantation from HLA-matched and Haploidentical donors with post-transplantation cyclophosphamide (Chorao et al. 2024)	2024	46% of patients had at least 1 CMV reactive 96% of reactivations occurred before D +100 Haploidentical donor, positive recipient CMV serology, older patient age, and grade II-IV acute GVHD were risk factors Non-relapse mortality greater in patients with CMV reactivation
	Risk of cytomegalovirus infection with post-transplantation cyclophosphamide in haploidentical and HLA-matched unrelated donor transplantation (Jamy et al. 2022)	2022	Increased risk of early CMV infection in patients receiving PTCy irrespective of donor type and number of PTCy doses compared with those not receiving PTCy
	Late infectious complications in hematopoietic cell transplantation survivors: a population-based study (Foord et al. 2020)	2020	HCT survivors (≥2 years) have a significant, persistent burden of infections vs other cancer survivors and the general population Vaccine-preventable late infections were 3 times greater in HCT vs non-HCT survivors and >30 times greater than in the general population
	Incidence and impact of fungal infections in post-transplantation cyclophosphamide-based graft-versus-host disease prophylaxis and haploidentical hematopoietic cell transplantation: a Center for International Blood and Marrow Transplant Research analysis (Papanicolaou et al. 2024)	2024	PTCy was associated with higher rates of yeast FI only in haploidentical donor HCT and with mold FI in both haploidentical donor HCT and sibling donor HCT The presence of FI by day 180 was associated with increased risk for overall mortality and transplant-related mortality at 2 years regardless of donor type or PTCy use

(continued)

Table 12.1 (continued)

Category	Study	Year	Main findings
Cardiovascular	Late cardiac events after allogeneic stem cell transplant: incidence, risk factors, and impact on overall survival (Auberle et al. 2023)	2023	Cumulative incidence of a late cardiac event at 5 years after transplant was 22% Most frequent cardiac event was a decline in LVEF to < 45% Prior cardiac history was major risk factor Almost all cardiac events demonstrated a significantly increased risk of death
	Cardiovascular events after hematopoietic stem cell transplant: incidence and risk factors (Vasbinder et al. 2023)	2023	5-year cumulative incidences of CV events was 16.4% Atrial fibrillation/flutter was the most common short- and long-term CV Baseline CV comorbidities were associated with a higher risk of long-term CV events
	Atrial fibrillation in patients undergoing allogeneic hematopoietic cell transplantation (Chang et al. 2021)	2021	5-year cumulative incidence of AF was 10.6% Older (≥50 years) age, HLA-unrelated donor, dyslipidemia and pre-HCT prolonged QTc interval were independent risk factors for AF
Secondary cancers	Total body irradiation and risk of breast cancer after blood or marrow transplantation: a blood or marrow transplantation survivor study report (McDonald et al. 2020)	2020	Exposure to TBI at age <30 years was associated with a 4.4-fold higher risk of subsequent breast cancer in allogeneic BMT survivors
	Secondary neoplasms after hematopoietic cell transplant for sickle cell disease (Eapen et al. 2023)	2023	10-year incidence of leukemia/MDS was 1.7% and of any secondary neoplasm was 2.4% Low intensity regimens leading to mixed chimerism were a risk factor
	Subsequent neoplasms and late mortality in children undergoing allogeneic transplantation for nonmalignant diseases (Kahn et al. 2020)	2020	10-year survival was 93% Seventy-one patients developed SNs (1.2%); 11-fold increased risk as compared to general population Leukemia risk was highest in the first 5 years post HCT; oropharyngeal, skin, liver, and thyroid tumors primarily occurred after 5 years

Psychosocial	Psychological and physical function in allogeneic hematopoietic cell transplant survivors with chronic graft-versus-host disease (Hansen et al. 2023)	2023	Patients with more severe cGVHD and those with cGVHD manifesting in the skin, GI tract, and lungs are at risk for poorer psychological and physical outcomes
	Post-transplantation employment status of adult survivors of childhood allogeneic hematopoietic cell transplant: a report from the Center for International Blood and Marrow Transplant Research (CIBMTR) (Bhatt et al. 2019)	2019	More than 50% of the *survivors* following allo-HSCT remain unemployed 1 year after the procedure Younger age at HCT, receipt of MAC with TBI, and international survivors had a lower likelihood of being employed
	Financial hardship in childhood cancer survivors treated with hematopoietic cell transplantation: a report from the childhood cancer survivor study (Buchbinder et al. 2023)	2023	Adult survivors of childhood cancer treated with HCT do not report greater overall financial hardship compared to non-HCT survivors but do report greater overall financial hardship compared to sibling controls Household income and chronic health conditions, but not HCT, were associated with financial hardship among all survivors.
	Health-related quality of life outcomes in older hematopoietic cell transplantation survivors (Hong et al. 2023)	2023	Older survivors experienced a low level of distress and moderate-high level of confidence in survivorship information (CSI) at ≥1 year post-HCT Lower distress and higher CSI are associated with improved HRQOL
Overall	Late effects after ablative allogeneic stem cell transplantation for adolescent and young adult acute myeloid leukemia (Lee et al. 2020)	2020	Nearly 25% of AYAs with AML who were disease-free survivors at 1 year after myeloablative HCT had ≥1 late effects TBI exposure was not an independent predictor for malignant or nonmalignant late effects except for cataracts 10-year overall survival and leukemia-free survival were 73% and 70%, respectively
	Life expectancy and burden of late complications after reduced intensity conditioning allogeneic transplantation (Del Galy et al. 2022)	2022	10-year cumulative incidence of secondary neoplasms was 12.9%, 15.1% for cardiac late effects Old age was the only risk factor for cardiac late effects
	Trends in late mortality and life expectancy after allogeneic blood or marrow transplantation over 4 decades: a blood or marrow transplant survivor study report (Bhatia et al. 2021)	2021	Late mortality among recipients of allogeneic BMT has decreased over the past 40 years; however, life expectancy still lower as compared with the general US population Reduction in risk of late mortality appeared to be confined to those who underwent transplantation at a younger age or those who received bone marrow
	Late effects after allogeneic haematopoietic cell transplantation in children and adolescents with non-malignant disorders: a retrospective cohort study (Kahn et al. 2024)	2024	19% of 5790 patients had one late effect, and 7% had at least two Age 10 years or older at HCT, unrelated donor source, total-body irradiation, and GVHD were identified as risk factors for long-term effects

terms of laboratory assessments, aggressive assessment of lipid panel, hemoglobin A1c, urinalysis, blood urea nitrogen, serum creatinine, and urine protein to creatinine ratio are recommended with every 6 months frequency in the first year followed by at least annually thereafter. Similarly, liver function assessment through bilirubin, alkaline phosphatase, and alanine transferase should be performed every 1–2 months in the first year and annually thereafter. For subsequent neoplasm screening, among survivors of pediatric HCT who are <45 years of age and exposed to radiation involving abdomen are recommended to undergo colorectal cancer screening starting at age 30 or 5-year post-HCT, whichever occurs later. Survivors of pediatric HCT who are <40 years of age and exposed to radiation involving chest are recommended to undergo breast cancer screening starting at age 25 or 8-year post-HCT, whichever occurs later. Given the risk of oral GVHD and subsequent neoplasms, survivors should be seen by oral medicine specialist at least every 6 months in the first year post-HCT and annually thereafter. While these updated recommendations account for the contemporary changes in HCT practices, several questions regarding the care of HCT survivors still remain unanswered.

12.3 Challenges and Opportunities in the Care of Hematopoietic Cell Transplant Survivors

Table 12.2 highlights the key challenges and opportunities in the care of HCT survivors in the modern era.

12.3.1 Impact of Changes in Transplant Practices on Late Effects

The landscape of allogeneic HCT is constantly evolving and the last decade has seen many changes in transplant practice such as increase in the use of haploidentical and mismatched donors (Gooptu et al. 2021; Auletta et al. 2023), higher proportion of older patients coming to allogeneic HCT (Lin and Artz 2021), increasing use of maintenance therapies post-transplant (de Lima et al. 2018) and decreasing use of methotrexate and calcineurin inhibitors (CNI) for GVHD prophylaxis with availability of alternate methods of GVHD prevention such as post-transplant cyclophosphamide (PTCy) and Abatacept (Cusatis et al. 2023). As changes occur in transplant practice, late effects are expected to change as well. Of the GVHD prevention strategies, use of PTCy across different donor types has been adopted the most widely after the randomized study showing improved survival with reduced intensity regimens and PTCy based GVHD prophylaxis (Bolanos-Meade et al. 2023). The data on its impact on long-term outcomes and late effects is still maturing with two studies in the CIBMTR portfolio examining the association of PTCy with cardiomyopathy and subsequent neoplasms. While impact of PTCy on early cardiotoxicity has been reported, it is unclear how this might translate into risk for long-term adverse cardiac outcomes in these patients (Yeh et al. 2021; Dulery et al. 2021). It will also be interesting to see the impact on subsequent neoplasms for which chronic GVHD is an associated risk factor (Vajdic et al. 2016; Inamoto et al. 2018; Tichelli et al. 2019). As for impact on other late effects, studies have shown the higher risk of infections with PTCy, some of which may be delayed indicating the need for long term monitoring (Jamy et al. 2022; Chorao et al. 2024; Singh et al. 2022; Girmenia et al. 2014; Papanicolaou et al. 2024). A delay in T-cell recovery may explain the higher incidence of infections in PTCy recipients (Jamy et al. 2022; Khimani et al. 2021). As we wait for the data to mature about long term complications of PTCy, increased awareness regarding the potential occurrence of these complications is needed while following the patients who received this regimen.

As mentioned above, chronic GVHD is integrally related to a variety of other late effects (Khera et al. 2012; Lee et al. 2022, 2024). At least four agents have now been approved for treatment of steroid refractory chronic GVHD including Ibrutinib, Ruxolitnib, Belumosudil, and

Table 12.2 Challenges and potential opportunities for preventing and mitigating those challenges in care of HCT survivors

Challenges	Opportunities
Most data about late effects are derived from retrospective institutional studies or large registries with limited collection of detailed pre- and post-transplant data on health and well-being and usually lack of controls	Designing large scale longitudinal studies with a robust reference group matched on relevant variables representing the general population allowing direct comparison. Linking registries with other databases
Scarcity of information about late effects in vulnerable groups such as racial ethnic minorities (important as the numbers of these minority patients getting allogeneic HCT is increasing)	Multistakeholder initiatives to ensure equitable access to HCT with robust data collection systems to supplement clinical information with extensive information about social determinants of health
Lack of specialized personnel to provide survivorship care	System-level efforts and policy changes to enhance workforce recruitment/retention/training and infrastructure development
Lack of financial incentives to provide well rounded survivorship care	Policy changes as well as partnerships with payers to improve reimbursement and invest in novel care delivery models
Wide variation in the profiles and needs of HCT survivors across a continuum	Increased efforts to engage patients through tailored educational materials, leverage technology for care delivery and close partnerships with all stakeholders caring for the patient including community physicians
Newer and combinations of therapies, especially for GVHD, have unknown or unrealized immunologic and late complications	Future studies looking at immune reconstitution and following late effects of newer immunosuppressants are warranted
The risks associated with CART, CRS, monoclonal antibodies, or lymphodepleting chemotherapy in combination with the use of pre- or post-HCT are not well understood. Specifically the impact on renal, cardiac, immune, pulmonary, and fertility late effects	Studies need to be designed to monitor for fertility, pregnancy and/or fetal toxicity following CAR-T, and integrate the development of CRS and use of monoclonal antibodies on late effects of HCT

Axatilimab (Miklos et al. 2017; Jagasia et al. 2020; Cutler et al. 2021; Wolff et al. 2024). While most of these agents were approved based on their beneficial impact on clinical outcomes as well as patient reported outcomes, more longitudinal studies are needed to understand their impact on occurrence and course of late effects.

While survival after HCT has improved over the years, relapse still remains the predominant cause of mortality beyond 2 years in those transplanted for hematologic malignancies. With relapse as the primary cause of mortality, use of post-transplant maintenance therapy has emerged as an important area of investigation, though it adds another factor to evaluate in terms of survivorship for patients. Infection and myelosuppression are common though the risk of acute and chronic GVHD is not different or maybe sometimes lower as with use of post-transplant tyrosine kinase inhibitors (Li et al. 2023; Xuan et al. 2023; Guan et al. 2024; Levis et al. 2024). Again, the impact of these therapies on other late effects is not well known.

12.3.2 Biologic Mechanisms for Late Effects After HCT

In the past few years, there has been increased interest in mechanistic studies to elucidate the biologic mechanisms and identify biomarkers for late effects/impact on long term outcomes and non-relapse mortality. Many different pathophysiological pathways focusing on inherent susceptibility to cancer and cumulative effect of pre-HCT therapies are being explored to explain the occurrence of late effects such as infection, chronic GVHD, and other long-term complications. Delayed immune recovery after allogeneic HCT is known to be a major contributor to infection, relapse and GVHD, and other related late effects (Bejanyan et al. 2018). Our knowledge about immune reconstitution and cell functionality has increased in the recent years to underscore the need for standardized and detailed infection surveillance and prophylaxis as well as immune reconstitution–accelerating strategies, though gaps remain (Dudakov et al. 2015). Post-HCT telomere attrition has been used to explain the premature ageing seen in cancer patients and patients who undergo HCT and can be used as a biomarker to predict non relapse mortality and complications associated with HCT (Gadalla and Savage 2011; Myllymaki et al. 2020). Emerging evidence indicates the association of donor clonal hematopoiesis with clinical outcomes in transplant recipients including graft alloimmune function and potential for leukemic transformation (Gibson et al. 2022). Another area of focus has been the gut microbiome with recent studies demonstrating the higher risk of infections and GVHD with lack of diversity of gut microbiome at the time of HCT for patients undergoing HCT (Peled et al. 2020). The adverse impact of gut microbiome on psychosocial outcomes such as post HCT fatigue has also been reported (Farhadfar et al. 2021). There is need for more information on impact of early patterns of microbiota on both late microbiota and risk for other late complications. Elevated peripheral inflammatory cytokine levels along with accumulation of neurotoxic and pro-inflammatory metabolites in the central nervous system have been postulated to drive fatigue and cognitive deficits and offer attractive pharmacological targets to address these complications (Boberg et al. 2020, 2023).

12.3.3 Long-Term Follow-Up for Patients Transplanted for Non-malignant Diseases

The utilization of allogeneic HCT as a curative option for non-malignant diseases such as hemoglobinopathies, bone marrow failure syndrome, inborn error of immunity, enzymopathies, metabolic disorders, and autoimmune disorders has significantly increased over the last two decades (Cusatis et al. 2023). According to the CIBMTR, the number of allogeneic HCT performed for non-malignant diseases in the United States nearly tripled between 2000 ($N = 347$) and 2018 ($N = 992$) (Cusatis et al. 2023). Similarly, the use of gene therapy for various non-malignant diseases is also increasing (Thompson et al. 2018; Mamcarz et al. 2019; Kanter et al. 2022). Despite the increasing use of HCT as a curative therapy for this population and increasing number of survivors, the literature focusing on late complications in survivors of non-malignant diseases who underwent HCT is limited due to the assessment of selected outcomes, heterogeneity in the evaluation modalities and definitions of outcomes, short post-therapy follow-up, and reliance on self-reported or registry-based data (Allewelt et al. 2016; Stenger et al. 2019; Kahn et al. 2020, 2024; Bense et al. 2023). The limited literature focusing on late health complications in patients with non-malignant diseases after HCT has shown low incidence of malignant and non-malignant late effects and comparable or better global quality of life (QOL) compared to the general population (Kahn et al. 2020, 2024; Bense et al. 2023). However, given the limited focus on unique challenges for this patient population, it is difficult to generalize these data and make meaningful conclusions. Additionally, the published studies have not provided insights on how the variations in donor source and conditioning intensity impacts survivors' long-term correction of the underlying disease process, health complications, and QOL.

In addition to the lack of data on long-term health complications after HCT for non-malignant diseases, the data on long-term donor engraftment also remains unknown. The goal of HCT in patients with non-malignant diseases is to correct the underlying hematopoietic, enzymatic, or immunological abnormality by introducing and sustaining an allogeneic graft. Donor engraftment is measured by donor chimerism using flow cytometry sorted CD33+, CD3+, CD19+, and CD56+ subsets by polymerase chain reaction based analysis of polymorphic microsatellite regions (Boerwinkle et al. 1989; Kasai et al. 1990; Scharf et al. 1995). While post-HCT complete donor chimerism is ideal for disease-free survival, some patients display residual recipient hematopoietic cells resulting in mixed donor chimerism. Increasing recipient-origin hematopoietic cells are considered a risk factor for graft rejection and disease recurrence in patients with malignancies triggering additional therapy such as donor lymphocyte infusion (Bader et al. 2004a, b); however, the data on incidence of mixed donor chimerism and its impact on transplant outcomes among patients with non-malignant diseases is unclear (Stikvoort et al. 2013, 2016; Faraci et al. 2018; Svenberg et al. 2009). This is especially important in these patients since it could make the differentiation between post-HCT long-term health complications and recurrence of underlying disease-related manifestations unclear. Therefore, understanding the factors associated with occurrence of mixed chimerism and the impact of mixed chimerism on clinical outcomes is important in patients with non-malignant diseases, particularly as new potential novel therapies emerge.

Understanding the unmet need for better knowledge of and screening for long-term complications in survivors of HCT for non-malignant diseases, the PTCTC (formerly Pediatric Blood and Marrow Transplant Consortium) conducted an international consensus conference in 2016 entitled "Late Effects Screening and Recommendations Following Allogeneic Hematopoietic Cell Transplant for Immune Deficiency and Nonmalignant Hematologic Disease." The conference focused on survivors of hemoglobinopathies, immune deficiency, and inherited bone marrow failure syndromes and expert recommendations were provided for the late effects screening of survivors of HCT for these diseases. The findings and recommendations were presented in a series of six manuscripts (Shenoy et al. 2017, 2018; Dietz et al. 2017a, b, c; Heimall et al. 2017a, b). Similarly, the revised international screening and preventive practice guidelines for long-term survivors of HCT and cellular therapy also provided recommendations regarding the clinical follow-up of HCT survivors of hemoglobinopathies, marrow failure syndromes, inborn errors of immunity, enzymopathies, metabolic and other disorders, and autoimmune disorders (Rotz et al. 2024a).

While these manuscripts highlighted the key clinical and research gaps in this field, due to the constantly changing field of HCT and increasing use of novel therapeutic agents including gene therapies, it is important to revisit these recommendations at frequent intervals. Additionally, to bridge the current knowledge gap, it is pivotal to establish a mechanism to systematically collect donor engraftment, organ function and QOL data in patients undergoing allogeneic HCT or gene therapy for non-malignant diseases while also considering unique characteristics of each underlying disease process. Such mechanism can not only help understand the impact of current therapeutic regimens on short- and long-term health outcomes and QOL, but also subsequently inform future efforts focusing on screening, prevention, early detection, and mitigation of these complications.

12.3.4 Late Effects After CAR-T Therapy as a Bridge to HCT

Targeted CAR-T is a growing field of immunotherapy that has been shown to have excellent anti-tumor activity in patients with relapsed/refractory diseases and is often used as a bridge to HCT for patients with B-cell malignancies. It is known that all CAR-T cell products share, to varying degrees, the class-specific adverse effects of cytokine-release syndrome (CRS), immune effector cell-associated neurotoxicity syndrome, immune effector cell-associated hemophagocytic lymphohistiocytosis-like syndrome, hypogammaglobulinemia, and cytopenia (affecting all three cell lines), particularly B cell aplasia (Brudno and Kochenderfer 2019; Hines et al. 2023). These acute toxicities may also cause late organ (renal, pulmonary, cardiac, and neurologic) toxicity and might potentially impact the known risk of chemotherapy and/or radiation-related complications post-HCT. These patients can also have poor response to vaccinations and remain at a high risk for infections and subsequent adverse outcomes due to long-term B-cell aplasia, which was highlighted during the COVID-19 pandemic (Atanackovic et al. 2022; Spanjaart et al. 2021). More recently, there have been several reports of subsequent neoplasms after CAR-T cell therapy (Ghilardi et al. 2024; Hamilton et al. 2024), which highlights the need for additional work to understand the full spectrum of late complications after CAR-T therapy. Additionally, studies assessing late cognitive, memory and executive functioning, frailty and QOL are ongoing (Shalabi et al. 2021). The impact of CAR-T therapy on fertility and pregnancy currently remains unknown and requires additional work (Ligon et al. 2022). Moreover, a subset of these patients may already have received allogeneic HCT and the cumulative impact of these treatments on both malignant and non-malignant late effects will need to be studied to better characterize their latency, intensity, reversibility, and lethality especially since cGVHD may be increased in these patients (Cappell and Kochenderfer 2023). This information can eventually help develop evidence-based survivorship care guidelines for this extremely vulnerable group of patients in future.

12.3.5 Potential Challenges Associated with the Healthcare Transition for AYA HCT Survivors

Nearly 70,000 AYAs (age 15–39 years at diagnosis) are diagnosed with cancer in the United States every year (LIVESTRONG Young Adult Alliance 2006). Compared to children and older adults with cancer, their survival outcomes have not shown improvements (Bleyer et al. 2006), which has led them to be identified as a vulnerable population by the National Cancer Institute. The difference in survival is thought to be related to the difference in disease biology, and concerns related to health care access, availability of caregiver, and medication adherence (Wood and Lee 2011). AYA population has specific challenges, exposures, and needs when facing a malignant or non-malignant diagnosis and therapy, including HCT (Husson et al. 2018; Buchbinder and Khera 2021). Moreover, AYA survivors are at a higher risk of late psychosocial complications as they are going through a critical personal and professional development phase and transitioning into adult roles at the time of their illness, which could impact their educational/ vocational attainment, romantic relationships, and peer support (Cooke et al. 2011). Up to 25% AYA patients report lapse in health insurance within the first 35 months after diagnosis and they often have challenges in returning to school or work after completion of therapy (Smith et al. 2019; Bhatt et al. 2021). This underscores the need for more frequent and targeted screening for psychosocial issues impacting QOL for this population.

Moreover, due to their several expected transitions throughout their treatment course, it is also known that AYAs are at a high risk of being lost to follow-up after completion of therapy (Miyamura et al. 2019; Buchbinder et al. 2020). Transition period is an extremely vulnerable time for these patients. In the sickle cell population, a prior study has shown increased mortality rates for young adults as they transitioned from pediatric care to adult care (Quinn et al. 2010). Transition of care education and plans are necessary to prevent deterioration of health and ensure follow-up (Cupit et al. 2016). Leveraging the NMDP survivorship registry (https://bethematch.org/tcdirectory/search/advanced) for comprehensive survivorship clinics and developing plan for annual transition of care evaluation and tracking of progress, readiness, and planning can help with optimal transitions (https://gottransition.org).

12.3.6 Healthcare Delivery and Models of Care

The need for comprehensive survivorship care for HCT survivors as they face the burden of late effects has been endorsed widely. Effective models of health care delivery for HCT survivors would need to ensure multidisciplinary care for optimal post-HCT management for medical and psychosocial complications while following the recently updated guidelines (Rotz et al. 2024a). Barriers to HCT survivorship care delivery have been described (Khera 2023).

Shared care model using a collaborative approach by integrating all the relevant stakeholders such as transplant team, primary care or oncology physician, other medical specialists, social workers, and financial counselors has been described widely as an optimal approach. More recently, a risk stratified cancer survivorship care has been proposed and implemented internationally to address a major barrier of limited resources and specialized personnel who provide survivorship care (Mayer and Alfano 2019). It would be valuable to adapt this innovative care delivery model for HCT survivors since the challenges of resource constraints/lack of specialists and need for standardization have been highlighted by two recently done survey studies (Hashmi et al. 2018; Dignan et al. 2021). Such an approach would also help provide a tailored strategy based on clinical and sociodemographic factors as we know that not every survivor's needs are same (Oliver et al. 2024).

Use of technology for health care delivery has grown exponentially after the pandemic. Virtual delivery of well- rounded risk stratified survivorship care that includes psychosocial screening and interventions to HCT survivors is emerging as an important strategy to help overcome the travel needs and their financial impact for the patient to get their long term follow-up care (Lee et al. 2021). Supplementing this with survivorship care plans and online programs to provide information about healthcare guidelines, self-care tips, and national and local resources may encourage health related self-efficacy in patients, decrease distress, increase health system capacity and promote adaptive HCT survivorship (Majhail et al. 2019; Syrjala et al. 2018). Telehealth/remotely delivered interventions to encourage exercise to reduce the burden of frailty and help improve physical and mental function are being developed (Langer et al. 2024; Lee et al. 2023). Use of mobile technology applied to healthcare (mHealth) app to provide integrated, multidisciplinary care has been studied by Redondo et al. and reported to be feasible and associated with high patient satisfaction (Redondo et al. 2023). AYA population especially is more likely to seek information about specific medical conditions, health topics, and sexual health through social media. Identifying and spreading awareness about evidence-based reliable media resources is necessary. The revised international screening and preventive practices guidelines include a list of online resources which could be helpful for patients and/or caregivers (Rotz et al. 2024a).

Remote patient monitoring (RPM) is a popular technique using digital devices for collection, transmission, and analysis of individual health data outside of the traditional clinical environment, to help update new or existing care plans for more optimal disease management (Sanders et al. 2020). Early studies in HCT patients have shown feasibility of this tool in post-HCT setting both early after HCT and for monitoring for late effects (Chang et al. 2020; Mussetti et al. 2021). The role of technology with use of artificial intelligence-based devices and biosensors and telemedicine in care delivery for post-HCT patients is evolving rapidly. Larger and rigorously conducted studies are needed to understand the benefits of incorporating apps/wearables and remote patient monitoring to improve survivorship after HCT while ensuring that the disparities in care don't widen.

12.3.7 Impact of CIBMTR and Other Professional and Patient Advocacy Groups in Understanding and Improving HCT Survivorship Landscape

CIBMTR is a national resource as the registry collects patient data on most autologous and nearly all allogeneic HCT recipients in the United States. Over the years, the observational studies done using comprehensive data collected by the registry on long-term outcomes of HCT has shaped our understanding of the wide spectrum of late effects after HCT and their impact on morbidity and mortality. Recently, CIBMTR has launched an electronic patient reported outcomes (PRO) data collection using a centralized data collection system (Cusatis et al. 2022, 2023). This will bring in a wealth of PRO data that can then be put in the context of clinical complications to understand the holistic picture for HCT survivors. Including this data in future studies is critical to understanding areas of improvement such as patients' QOL and disparities in access and health outcomes based on age, race, and ethnicity. Additionally, being able to link CIBMTR data with other databases can help supplement the extensive outcomes data collected by CIBMTR.

Several other organizations have made significant contributions to further our knowledge regarding HCT survivorship. In 2016, National Cancer Institute and National Heart, Lung and Blood Institute launched an initiative to bring together the experts in the HCT survivorship area to identify barriers and knowledge gaps and to formulate research and practice recommendations (Battiwalla et al. 2017). Reports from the working groups from this initiative provide a blueprint for efforts to improve quality of medical care, patient experiences, and clinical outcomes of HCT survivors. Similarly, Blood and Marrow Transplant Clinical Trials Network is a consortium of 100+ transplant centers that conducts multi center trials in the space of HCT/cell therapy. The recent state of science symposium in 2021 included proposals from Late Effects, Quality of Life and Economics committee who had identified several priority domains in the area (Heslop et al. 2021). More recently, the PTCTC working groups on survivorship and GVHD convened a multi-disciplinary taskforce of pediatric HCT experts called *RESILIENT after Chronic GVHD* (*R*esearch and *E*ducation towards *S*olutions for *L*ate effects to *I*nnovate, *E*xcel, and *N*ur*t*ure after cGVHD). The goal of this effort was to develop recommendations regarding the long-term follow-up of patients with cGVHD, specifically on immune dysregulation, growth, metabolism, development, and quality of life. The recommendations will be published through four manuscripts in near future. Several patient-focused organizations such as BMT Infonet (www.bmtinfonet.org), National Bone Marrow Transplant Link (www.nbmtlink.org), and GVHD Alliance (www.gvhdalliance.org) have also made significant role in advancing the field of HCT survivorship through provision of information and resources, patient advocacy, and community building.

12.4 Conclusions

While the field of HCT and cellular therapy has made significant strides in last four decades by increasing the safety and efficacy of the therapy, HCT recipients continue to be at risk for developing both early and late physical and psychological health complications necessitating long-term follow-up after completion of therapy. The constantly changing field of HCT and cellular therapy poses additional challenges in the current management of HCT survivors. Efforts from clinicians, researchers, policy makers, and professional and patient organizations are needed to continue supporting these patients and address their unique needs.

References

Allewelt H, El-Khorazaty J, Mendizabal A et al (2016) Late effects after umbilical cord blood transplantation in very young children after busulfan-based, myeloablative conditioning. Biol Blood Marrow Transplant 22(9):1627–1635. https://doi.org/10.1016/j.bbmt.2016.05.024

Atanackovic D, Luetkens T, Omili D et al (2022) Vaccine-induced T-cell responses against SARS-CoV-2 and its Omicron variant in patients with B cell-depleted lymphoma after CART therapy. Blood 140(2):152–156. https://doi.org/10.1182/blood.2022016175

Auberle C, Lenihan D, Gao F, Cashen A (2023) Late cardiac events after allogeneic stem cell transplant: incidence, risk factors, and impact on overall survival. Cardiooncology 9(1):1. https://doi.org/10.1186/s40959-022-00150-1

Auletta JJ, Kou J, Chen M et al (2023) Real-world data showing trends and outcomes by race and ethnicity in allogeneic hematopoietic cell transplantation: a report from the Center for International Blood and Marrow Transplant Research. Transplant Cell Ther 29(6):346 e1–346 e10. https://doi.org/10.1016/j.jtct.2023.03.007

Bader P, Kreyenberg H, Hoelle W et al (2004a) Increasing mixed chimerism is an important prognostic factor for unfavorable outcome in children with acute lymphoblastic leukemia after allogeneic stem-cell transplantation: possible role for pre-emptive immunotherapy? J Clin Oncol 22(9):1696–1705. https://doi.org/10.1200/JCO.2004.05.198

Bader P. Kreyenberg H, Hoelle W et al (2004b) Increasing mixed chimerism defines a high-risk group of childhood acute myelogenous leukemia patients after allogeneic stem cell transplantation where pre-emptive immunotherapy may be effective. Bone Marrow Transplant 33(8):815–821. https://doi.org/10.1038/sj.bmt.1704444

Battiwalla M, Hashmi S, Majhail N, Pavletic S, Savani BN, Shelburne N (2017) National Institutes of Health hematopoietic cell transplantation late effects initiative: developing recommendations to improve survivorship and long-term outcomes. Biol Blood Marrow Transplant 23(1):6–9. https://doi.org/10.1016/j.bbmt.2016.10.020

Bejanyan N, Brunstein CG, Cao Q et al (2018) Delayed immune reconstitution after allogeneic transplantation increases the risks of mortality and chronic GVHD. Blood Adv 2(8):909–922. https://doi.org/10.1182/bloodadvances.2017014464

Bense JE, Haverman L, von Asmuth EGJ et al (2023) Late effects in pediatric allogeneic hematopoietic stem cell transplantation for nonmalignant diseases: proxy- and patient-reported outcomes. Transplant Cell Ther 29(3):186 e1–186 e10. https://doi.org/10.1016/j.jtct.2022.12.024

Bhatia S, Dai C, Landier W et al (2021) Trends in late mortality and life expectancy after allogeneic blood or marrow transplantation over 4 decades: a blood or marrow transplant survivor study report. JAMA Oncol 7(11):1626–1634. https://doi.org/10.1001/jamaoncol.2021.3676

Bhatt NS, Brazauskas R, Tecca HR et al (2019) Post-transplantation employment status of adult survivors of childhood allogeneic hematopoietic cell transplant: a report from the Center for International Blood and Marrow Transplant Research (CIBMTR). Cancer 125(1):144–152. https://doi.org/10.1002/cncr.31781

Bhatt NS, Brazauskas R, Salit RB et al (2021) Return to work among young adult survivors of allogeneic hematopoietic cell transplantation in the United States. Transplant Cell Ther 27(8):679 e1–679 e8. https://doi.org/10.1016/j.jtct.2021.04.013

Bleyer A, O'Leary M, Barr R, LAG R (eds) (2006) Cancer epidemiology in older adolescents and young adults 15 to 29 years of age, including SEER incidence and survival: 1975–2000. National Cancer Institute, NIH Pub. No. 06-5767, Bethesda

Boberg E, Kadri N, Winterling J et al (2020) Mental fatigue after allogeneic hematopoietic stem cell transplantation is associated with cognitive dysfunction, but not central nervous system inflammation. Haematologica 105(6):e310–e314. https://doi.org/10.3324/haematol.2019.225326

Boberg E, Kadri N, Hagey DW et al (2023) Cognitive impairments correlate with increased central nervous system immune activation after allogeneic haematopoietic stem cell transplantation. Leukemia 37(4):888–900. https://doi.org/10.1038/s41375-023-01840-0

Boerwinkle E, Xiong WJ, Fourest E, Chan L (1989) Rapid typing of tandemly repeated hypervariable loci by the polymerase chain reaction: application to the apolipoprotein B 3′ hypervariable region. Proc Natl Acad Sci USA 86(1):212–216. https://doi.org/10.1073/pnas.86.1.212

Bolanos-Meade J, Hamadani M, Wu J et al (2023) Post-transplantation cyclophosphamide-based graft-versus-host disease prophylaxis. N Engl J Med 388(25):2338–2348. https://doi.org/10.1056/NEJMoa2215943

Brudno JN, Kochenderfer JN (2019) Recent advances in CAR T-cell toxicity: mechanisms, manifestations and management. Blood Rev 34:45–55. https://doi.org/10.1016/j.blre.2018.11.002

Buchbinder D, Khera N (2021) Psychosocial and financial issues after hematopoietic cell transplantation. Hematology Am Soc Hematol Educ Program 2021(1):570–577. https://doi.org/10.1182/hematology.2021000292

Buchbinder D, Brazauskas R, Bo-Subait K et al (2020) Predictors of loss to follow-up among pediatric and adult hematopoietic cell transplantation survivors: a report from the Center for International Blood and Marrow Transplant Research. Biol Blood Marrow Transplant 26(3):553–561. https://doi.org/10.1016/j.bbmt.2019.11.003

Buchbinder D, Bhatt NS, Wang H et al (2023) Financial hardship in childhood cancer survivors treated with hematopoietic cell transplantation: a report from the childhood cancer survivor study. Transplant Cell Ther 29(11):705 e1–705 e9. https://doi.org/10.1016/j.jtct.2023.07.024

Cappell KM, Kochenderfer JN (2023) Long-term outcomes following CAR T cell therapy: what we know so far. Nat Rev Clin Oncol 20(6):359–371. https://doi.org/10.1038/s41571-023-00754-1

Carpenter PA, Englund JA (2016) How I vaccinate blood and marrow transplant recipients. Blood 127(23):2824–2832. https://doi.org/10.1182/blood-2015-12-550475

Carpenter PA, Englund JA (2021) Commentary: is immune recovery-based post-transplantation vaccination in children better than time-based revaccination? Transplant Cell Ther 27(4):281–283. https://doi.org/10.1016/j.jtct.2021.03.007

Chang E, Iukuridze A, Echevarria M et al (2020) Feasibility and acceptability of using a telehealth platform to monitor cardiovascular risk factors in hematopoietic cell transplantation survivors at risk for cardiovascular disease. Biol Blood Marrow Transplant 26(6):1233–1237. https://doi.org/10.1016/j.bbmt.2020.02.027

Chang EK, Chanson D, Teh JB et al (2021) Atrial fibrillation in patients undergoing allogeneic hematopoietic cell transplantation. J Clin Oncol 39(8):902–910. https://doi.org/10.1200/JCO.20.02401

Chorao P, Henriques M, Villalba M et al (2024) Cytomegalovirus reactivations in allogeneic hematopoietic stem cell transplantation from HLA-matched and haploidentical donors with post-transplantation cyclophosphamide. Transplant Cell Ther 30(5):538 e1–538 e10. https://doi.org/10.1016/j.jtct.2024.01.082

Cooke L, Chung C, Grant M (2011) Psychosocial care for adolescent and young adult hematopoietic cell transplant patients. J Psychosoc Oncol 29(4):394–414

Cordonnier C, Einarsdottir S, Cesaro S et al (2019) Vaccination of haemopoietic stem cell transplant recipients: guidelines of the 2017 European Conference on Infections in Leukaemia (ECIL 7). Lancet Infect Dis 19(6):e200–e212. https://doi.org/10.1016/S1473-3099(18)30600-5

Cupit MC, Duncan C, Savani BN, Hashmi SK (2016) Childhood to adult transition and long-term follow-up after blood and marrow transplantation. Bone Marrow Transplant 51(2):176–181. https://doi.org/10.1038/bmt.2015.228

Cusatis R, Flynn KE, Vasu S et al (2022) Adding centralized electronic patient-reported outcome data collection to an established international clinical outcomes registry. Transplant Cell Ther 28(2):112 e1–112 e9. https://doi.org/10.1016/j.jtct.2021.10.016

Cusatis R, Litovich C, Feng Z, Allbee-Johnson M, Shaw B (2023) Current uses and outcomes of cellular therapies: CIBMTR summary slides, 2023

Cutler C, Lee SJ, Arai S et al (2021) Belumosudil for chronic graft-versus-host disease after 2 or more prior lines of therapy: the ROCKstar study. Blood 138(22):2278–2289. https://doi.org/10.1182/blood.2021012021

de Lima M, Oran B, Champlin RE et al (2018) CC-486 maintenance after stem cell transplantation in patients with acute myeloid leukemia or myelodysplastic syndromes. Biol Blood Marrow Transplant 24(10):2017–2024. https://doi.org/10.1016/j.bbmt.2018.06.016

Del Galy AS, Rousseau A, Capes A et al (2022) Life expectancy and burden of late complications after reduced intensity conditioning allogeneic transplantation. Bone Marrow Transplant 57(9):1365–1372. https://doi.org/10.1038/s41409-022-01715-5

Dietz AC, Savage SA, Vlachos A et al (2017a) Late effects screening guidelines after hematopoietic cell transplantation for inherited bone marrow failure syndromes: consensus statement from the second pediatric blood and marrow transplant consortium international conference on late effects after pediatric HCT. Biol Blood Marrow Transplant 23(9):1422–1428. https://doi.org/10.1016/j.bbmt.2017.05.022

Dietz AC, Duncan CN, Alter BP et al (2017b) The second pediatric blood and marrow transplant consortium international consensus conference on late effects after pediatric hematopoietic cell transplantation: defining the unique late effects of children undergoing hematopoietic cell transplantation for immune deficiencies, inherited marrow failure disorders, and hemoglobinopathies. Biol Blood Marrow Transplant 23(1):24–29. https://doi.org/10.1016/j.bbmt.2016.10.004

Dietz AC, Mehta PA, Vlachos A et al (2017c) Current knowledge and priorities for future research in late effects after hematopoietic cell transplantation for inherited bone marrow failure syndromes: consensus statement from the second pediatric blood and marrow transplant consortium international conference on late effects after pediatric hematopoietic cell transplantation. Biol Blood Marrow Transplant 23(5):726–735. https://doi.org/10.1016/j.bbmt.2017.01.075

Dignan FL, Hamblin A, Chong A et al (2021) Survivorship care for allogeneic transplant patients in the UK NHS: changes centre practice, impact of health service policy and JACIE accreditation over 5 years. Bone Marrow Transplant 56(3):673–678. https://doi.org/10.1038/s41409-020-01067-y

Dudakov JA, Perales MA, van den Brink MRM. Immune reconstitution following hematopoietic cell transplantation. In: S.J. Forman, Negrin, R. S, J.H. Antin, F.R. Appelbaum, ed. Thomas' hematopoietic cell transplantation. 2015 Wiley-Blackwell

Dulery R, Mohty R, Labopin M et al (2021) Early cardiac toxicity associated with post-transplant cyclophosphamide in allogeneic stem cell transplantation. JACC Cardio Oncol 3(2):250–259. https://doi.org/10.1016/j.jaccao.2021.02.011

Eapen M, Brazauskas R, Williams DA et al (2023) Secondary neoplasms after hematopoietic cell transplant for sickle cell disease. J Clin Oncol 41(12):2227–2237. https://doi.org/10.1200/JCO.22.01203

Faraci M, Bagnasco F, Leoni M et al (2018) Evaluation of chimerism dynamics after allogeneic hematopoietic stem cell transplantation in children with nonmalignant diseases. Biol Blood Marrow Transplant 24(5):1088–1093. https://doi.org/10.1016/j.bbmt.2017.12.801

Farhadfar N, Gharaibeh RZ, Dahl WJ et al (2021) Gut microbiota dysbiosis associated with persistent fatigue in hematopoietic cell transplantation survivors. Transplant Cell Ther 27(6):498 e1–498 e8. https://doi.org/10.1016/j.jtct.2021.02.017

Foord AM, Cushing-Haugen KL, Boeckh MJ et al (2020) Late infectious complications in hematopoietic cell transplantation survivors: a population-based study. Blood Adv 4(7):1232–1241. https://doi.org/10.1182/bloodadvances.2020001470

Gadalla SM, Savage SA (2011) Telomere biology in hematopoiesis and stem cell transplantation. Blood Rev 25(6):261–269. https://doi.org/10.1016/j.blre.2011.06.004

Ghilardi G, Fraietta JA, Gerson JN et al (2024) T cell lymphoma and secondary primary malignancy risk after commercial CAR T cell therapy. Nat Med 30(4):984–989. https://doi.org/10.1038/s41591-024-02826-w

Gibson CJ, Kim HT, Zhao L et al (2022) Donor clonal hematopoiesis and recipient outcomes after transplantation. J Clin Oncol 40(2):189–201. https://doi.org/10.1200/JCO.21.02286

Girmenia C, Raiola AM, Piciocchi A et al (2014) Incidence and outcome of invasive fungal diseases after allogeneic stem cell transplantation: a prospective study of the

Gruppo Italiano Trapianto Midollo Osseo (GITMO). Biol Blood Marrow Transplant 20(6):872–880. https://doi.org/10.1016/j.bbmt.2014.03.004

Gooptu M, Romee R, St Martin A et al (2021) HLA-haploidentical vs matched unrelated donor transplants with posttransplant cyclophosphamide-based prophylaxis. Blood 138(3):273–282. https://doi.org/10.1182/blood.2021011281

Guan F, Yang L, Chen Y et al (2024) Comparison of long-term outcomes between imatinib and dasatinib prophylaxis after allogeneic stem cell transplantation in patients with Philadelphia-positive acute lymphoblastic leukemia: a multicenter retrospective study. Cancer 130(12):2139–2149. https://doi.org/10.1002/cncr.35232

Hamilton MP, Sugio T, Noordenbos T et al (2024) Risk of second tumors and T-cell lymphoma after CAR T-cell therapy. N Engl J Med 390(22):2047–2060. https://doi.org/10.1056/NEJMoa2401361

Hansen JL, Juckett MB, Foster MA et al (2023) Psychological and physical function in allogeneic hematopoietic cell transplant survivors with chronic graft-versus-host disease. J Cancer Surviv 17(3):646–656. https://doi.org/10.1007/s11764-023-01354-9

Hashmi SK, Lee SJ, Savani BN et al (2018) ASBMT practice guidelines committee survey on long-term follow-up clinics for hematopoietic cell transplant survivors. Biol Blood Marrow Transplant 24(6):1119–1124. https://doi.org/10.1016/j.bbmt.2018.03.023

Heimall J, Buckley RH, Puck J et al (2017a) Recommendations for screening and management of late effects in patients with severe combined immunodeficiency after allogenic hematopoietic cell transplantation: a consensus statement from the second pediatric blood and marrow transplant consortium international conference on late effects after pediatric HCT. Biol Blood Marrow Transplant 23(8):1229–1240. https://doi.org/10.1016/j.bbmt.2017.04.026

Heimall J, Puck J, Buckley R et al (2017b) Current knowledge and priorities for future research in late effects after hematopoietic stem cell transplantation (HCT) for severe combined immunodeficiency patients: a consensus statement from the second pediatric blood and marrow transplant consortium international conference on late effects after pediatric HCT. Biol Blood Marrow Transplant 23(3):379–387. https://doi.org/10.1016/j.bbmt.2016.12.619

Heslop HE, Stadtmauer EA, Levine JE et al (2021) Blood and marrow transplant clinical trials network state of the science symposium 2021: looking forward as the network celebrates its 20th year. Transplant Cell Ther 27(11):885–907. https://doi.org/10.1016/j.jtct.2021.08.016

Hines MR, Knight TE, McNerney KO et al (2023) Immune effector cell-associated hemophagocytic lymphohistiocytosis-like syndrome. Transplant Cell Ther 29(7):438 e1–438 e16. https://doi.org/10.1016/j.jtct.2023.03.006

Hong S, Zhao J, Wang S et al (2023) Health-related quality of life outcomes in older hematopoietic cell transplantation survivors. Transplant Cell Ther 29(3):202 e1–202 e8. https://doi.org/10.1016/j.jtct.2022.11.016

Husson O, Huijgens PC, van der Graaf WTA (2018) Psychosocial challenges and health-related quality of life of adolescents and young adults with hematologic malignancies. Blood 132(4):385–392. https://doi.org/10.1182/blood-2017-11-778555

Inamoto Y, Matsuda T, Tabuchi K et al (2018) Outcomes of patients who developed subsequent solid cancer after hematopoietic cell transplantation. Blood Adv 2(15):1901–1913. https://doi.org/10.1182/bloodadvances.2018020966

Jagasia M, Perales MA, Schroeder MA et al (2020) Ruxolitinib for the treatment of steroid-refractory acute GVHD (REACH1): a multicenter, open-label phase 2 trial. Blood 135(20):1739–1749. https://doi.org/10.1182/blood.2020004823

Jamy O, Hebert C, Dunn-Valadez S et al (2022) Risk of cytomegalovirus infection with post-transplantation cyclophosphamide in haploidentical and HLA-matched unrelated donor transplantation. Transplant Cell Ther 28(4):213 e1–213 e6. https://doi.org/10.1016/j.jtct.2022.01.011

Kahn JM, Brazauskas R, Tecca HR et al (2020) Subsequent neoplasms and late mortality in children undergoing allogeneic transplantation for nonmalignant diseases. Blood Adv 4(9):2084–2094. https://doi.org/10.1182/bloodadvances.2019000839

Kahn J, Brazauskas R, Bo-Subait S et al (2024) Late effects after allogeneic haematopoietic cell transplantation in children and adolescents with non-malignant disorders: a retrospective cohort study. Lancet Child Adolesc Health 8(10):740–750. https://doi.org/10.1016/S2352-4642(24)00167-6

Kanter J, Walters MC, Krishnamurti L et al (2022) Biologic and clinical efficacy of lentiglobin for sickle cell disease. N Engl J Med 386(7):617–628. https://doi.org/10.1056/NEJMoa2117175

Kasai K, Nakamura Y, White R (1990) Amplification of a variable number of tandem repeats (VNTR) locus (pMCT118) by the polymerase chain reaction (PCR) and its application to forensic science. J Forensic Sci 35(5):1196–1200

Khera N (2023) Managing survivorship after hematopoietic cell transplantation. Curr Hematol Malig Rep 18(3):75–82. https://doi.org/10.1007/s11899-023-00694-8

Khera N, Storer B, Flowers ME et al (2012) Nonmalignant late effects and compromised functional status in survivors of hematopoietic cell transplantation. J Clin Oncol 30(1):71–77. https://doi.org/10.1200/JCO.2011.38.4594

Khimani F, Ranspach P, Elmariah H et al (2021) Increased infections and delayed CD4(+) T cell but faster B cell immune reconstitution after post-transplantation cyclophosphamide compared to conventional GVHD prophylaxis in allogeneic transplantation. Transplant

Cell Ther 27(11):940–948. https://doi.org/10.1016/j.jtct.2021.07.023

Langer SL, Joseph RP, Mistretta EG et al (2024) Family-focused facilitated fitness: feasibility and acceptability of a couple-based physical activity intervention for hematopoietic cell transplant recipients and their caregiving partners. Transplant Cell Ther 30(4):450 e1–450 e17. https://doi.org/10.1016/j.jtct.2024.01.066

Lee CJ, Kim S, Tecca HR et al (2020) Late effects after ablative allogeneic stem cell transplantation for adolescent and young adult acute myeloid leukemia. Blood Adv 4(6):983–992. https://doi.org/10.1182/bloodadvances.2019001126

Lee CJ, Wickline M, Flowers MED. Telemedicine in patient care of long-term transplant survivors. In: Savani BN, Tichelli A, eds. Blood and marrow transplantation long term management: survivorship after transplant, 2nd. 2021 Wiley

Lee CJ, Wang T, Chen K et al (2022) Association of chronic graft-versus-host disease with late effects following allogeneic hematopoietic cell transplantation for children with hematologic malignancy. Transplant Cell Ther 28(10):712 e1–712 e8. https://doi.org/10.1016/j.jtct.2022.07.014

Lee K, Shamunee J, Lindenfeld L et al (2023) Feasibility of implementing a supervised telehealth exercise intervention in frail survivors of hematopoietic cell transplantation: a pilot randomized trial. BMC Cancer 23(1):390. https://doi.org/10.1186/s12885-023-10884-5

Lee CJ, Wang T, Chen K et al (2024) Severity of chronic graft-versus-host disease and late effects following allogeneic hematopoietic cell transplantation for adults with hematologic malignancy. Transplant Cell Ther 30(1):97 e1–97 e14. https://doi.org/10.1016/j.jtct.2023.10.010

Levis MJ, Hamadani M, Logan B et al (2024) Gilteritinib as post-transplant maintenance for AML with internal tandem duplication mutation of FLT3. J Clin Oncol 42(15):1766–1775. https://doi.org/10.1200/JCO.23.02474

Li L, Zhang R, Cao W et al (2023) Comparative analysis of hypomethylating agents as maintenance therapy after allogeneic hematopoietic stem cell transplantation for high-risk acute leukemia. Leuk Lymphoma 64(13):2113–2122. https://doi.org/10.1080/10428194.2023.2252948

Ligon JA, Fry A, Maher JY et al (2022) Fertility and CAR T-cells: current practice and future directions. Transplant Cell Ther 28(9):605 e1–605 e8. https://doi.org/10.1016/j.jtct.2022.06.002

Lin RJ, Artz AS (2021) Allogeneic hematopoietic cell transplantation for older patients. Hematology Am Soc Hematol Educ Program 2021(1):254–263. https://doi.org/10.1182/hematology.2021000257

LIVESTRONG Young Adult Alliance (2006) Closing the gap: research and care imperatives for adolescents and young adults with cancer. Report of the Adolescent and Young Adult Oncology Progress Review Group. No. 06-6067. Rockville, US Department of Health and Human Services, National Institutes of Health, National Cancer Institute

Majhail NS, Rizzo JD, Lee SJ et al (2012) Recommended screening and preventive practices for long-term survivors after hematopoietic cell transplantation. Biol Blood Marrow Transplant 18(3):348–371. https://doi.org/10.1016/j.bbmt.2011.12.519

Majhail NS, Tao L, Bredeson C et al (2013) Prevalence of hematopoietic cell transplant survivors in the United States. Biol Blood Marrow Transplant 19(10):1498–1501

Majhail NS, Murphy E, Laud P et al (2019) Randomized controlled trial of individualized treatment summary and survivorship care plans for hematopoietic cell transplantation survivors. Haematologica 104(5):1084–1092. https://doi.org/10.3324/haematol.2018.203919

Mamcarz E, Zhou S, Lockey T et al (2019) Lentiviral gene therapy combined with low-dose busulfan in infants with SCID-X1. N Engl J Med 380(16):1525–1534. https://doi.org/10.1056/NEJMoa1815408

Mayer DK, Alfano CM (2019) Personalized risk-stratified cancer follow-up care: its potential for healthier survivors, happier clinicians, and lower costs. J Natl Cancer Inst 111(5):442–448. https://doi.org/10.1093/jnci/djy232

McDonald AM, Chen Y, Wu J et al (2020) Total body irradiation and risk of breast cancer after blood or marrow transplantation: a blood or marrow transplantation survivor study report. J Clin Oncol 38(25):2872–2882. https://doi.org/10.1200/JCO.20.00231

Miklos D, Cutler CS, Arora M et al (2017) Ibrutinib for chronic graft-versus-host disease after failure of prior therapy. Blood 130(21):2243–2250. https://doi.org/10.1182/blood-2017-07-793786

Miyamura K, Yamashita T, Atsuta Y et al (2019) High probability of follow-up termination among AYA survivors after allogeneic hematopoietic cell transplantation. Blood Adv 3(3):397–405. https://doi.org/10.1182/bloodadvances.2018026039

Mussetti A, Salas MQ, Condom M et al (2021) Use of telehealth for domiciliary follow-up after hematopoietic cell transplantation during the COVID-19 pandemic: prospective pilot study. JMIR Form Res 5(3):e26121. https://doi.org/10.2196/26121

Myllymaki M, Redd R, Reilly CR et al (2020) Short telomere length predicts nonrelapse mortality after stem cell transplantation for myelodysplastic syndrome. Blood 136(26):3070–3081. https://doi.org/10.1182/blood.2020005397

Niederwieser D, Baldomero H, Bazuaye N et al (2022) One and a half million hematopoietic stem cell transplants: continuous and differential improvement in worldwide access with the use of non-identical family donors. Haematologica 107(5):1045–1053. https://doi.org/10.3324/haematol.2021.279189

Norkin M, Shaw BE, Brazauskas R et al (2019) Characteristics of late fatal infections after allogeneic hematopoietic cell transplantation. Biol Blood Marrow

Transplant 25(2):362–368. https://doi.org/10.1016/j.bbmt.2018.09.031

Oliver MM, Meng Q, Hageman L et al (2024) Health care utilization by long-term survivors of blood or marrow transplantation-a Bone Marrow Transplant Survivor Study report. Cancer 130(5):803–815. https://doi.org/10.1002/cncr.35076

Papanicolaou GA, Chen M, He N et al (2024) Incidence and impact of fungal infections in post-transplantation cyclophosphamide-based graft-versus-host disease prophylaxis and haploidentical hematopoietic cell transplantation: a Center for International Blood and Marrow Transplant Research analysis. Transplant Cell Ther 30(1):114 e1–114 e16. https://doi.org/10.1016/j.jtct.2023.09.017

Peled JU, Gomes ALC, Devlin SM et al (2020) Microbiota as predictor of mortality in allogeneic hematopoietic-cell transplantation. N Engl J Med 382(9):822–834. https://doi.org/10.1056/NEJMoa1900623

Quinn CT, Rogers ZR, McCavit TL, Buchanan GR (2010) Improved survival of children and adolescents with sickle cell disease. Blood 115(17):3447–3452. https://doi.org/10.1182/blood-2009-07-233700

Redondo S, De Dios A, Gomis-Pastor M et al (2023) Feasibility of a new model of care for allogeneic stem cell transplantation recipients facilitated by eHealth: the MY-medula pilot study. Transplant Cell Ther 29(6):385 e1–385 e8. https://doi.org/10.1016/j.jtct.2023.03.016

Rizzo JD, Wingard JR, Tichelli A et al (2006) Recommended screening and preventive practices for long-term survivors after hematopoietic cell transplantation: joint recommendations of the European Group for Blood and Marrow Transplantation, the Center for International Blood and Marrow Transplant Research, and the American Society of Blood and Marrow Transplantation. Biol Blood Marrow Transplant 12(2):138–151. https://doi.org/10.1016/j.bbmt.2005.09.012

Rotz SJ, Bhatt NS, Hamilton BK et al (2024a) International recommendations for screening and preventative practices for long-term survivors of transplantation and cellular therapy: a 2023 update. Transplant Cell Ther 30(4):349–385. https://doi.org/10.1016/j.jtct.2023.12.001

Rotz SJ, Bhatt NS, Hamilton BK et al (2024b) International recommendations for screening and preventative practices for long-term survivors of transplantation and cellular therapy: a 2023 update. Bone Marrow Transplant 59(6):717–741. https://doi.org/10.1038/s41409-023-02190-2

Sanders SF, Stern AD, Gordon WJ (2020) How to make remote monitoring tech part of everyday health care. Harv Bus Rev

Scharf SJ, Smith AG, Hansen JA, McFarland C, Erlich HA (1995) Quantitative determination of bone marrow transplant engraftment using fluorescent polymerase chain reaction primers for human identity markers. Blood 85(7):1954–1963

Shalabi H, Gust J, Taraseviciute A et al (2021) Beyond the storm – subacute toxicities and late effects in children receiving CAR T cells. Nat Rev Clin Oncol 18(6):363–378. https://doi.org/10.1038/s41571-020-00456-y

Shenoy S, Angelucci E, Arnold SD et al (2017) Current results and future research priorities in late effects after hematopoietic stem cell transplantation for children with sickle cell disease and thalassemia: a consensus statement from the second pediatric blood and marrow transplant consortium international conference on late effects after pediatric hematopoietic stem cell transplantation. Biol Blood Marrow Transplant 23(4):552–561. https://doi.org/10.1016/j.bbmt.2017.01.009

Shenoy S, Gaziev J, Angelucci E et al (2018) Late effects screening guidelines after hematopoietic cell transplantation (HCT) for hemoglobinopathy: consensus statement from the second pediatric blood and marrow transplant consortium international conference on late effects after pediatric HCT. Biol Blood Marrow Transplant 24(7):1313–1321. https://doi.org/10.1016/j.bbmt.2018.04.002

Shimizu H, Najima Y, Kako S et al (2023) Clinical significance of late CMV disease in adult patients who underwent allogeneic stem cell transplant. J Infect Chemother 29(12):1103–1108. https://doi.org/10.1016/j.jiac.2023.07.015

Singh A, Dandoy CE, Chen M et al (2022) Post-transplantation cyclophosphamide is associated with an increase in non-cytomegalovirus herpesvirus infections in patients with acute leukemia and myelodysplastic syndrome. Transplant Cell Ther 28(1):48 e1–48 e10. https://doi.org/10.1016/j.jtct.2021.09.015

Smith AW, Keegan T, Hamilton A et al (2019) Understanding care and outcomes in adolescents and young adult with cancer: a review of the AYA HOPE study. Pediatr Blood Cancer 66(1):e27486. https://doi.org/10.1002/pbc.27486

Spanjaart AM, Ljungman P, de La Camara R et al (2021) Poor outcome of patients with COVID-19 after CAR T-cell therapy for B-cell malignancies: results of a multicenter study on behalf of the European Society for Blood and Marrow Transplantation (EBMT) Infectious Diseases Working Party and the European Hematology Association (EHA) Lymphoma Group. Leukemia 35(12):3585–3588. https://doi.org/10.1038/s41375-021-01466-0

Stenger E, Phelan R, Shaw BE et al (2019) Excellent overall survival and low incidence of late effects in patients undergoing allogeneic hematopoietic cell transplant for sickle cell disease: a report from the Center for International Blood and Marrow Transplant Research (CIBMTR). Blood 134(Supplement_1):697

Stikvoort A, Gertow J, Sundin M, Remberger M, Mattsson J, Uhlin M (2013) Chimerism patterns of long-term stable mixed chimeras posthematopoietic stem cell transplantation in patients with nonmalignant diseases: follow-up of long-term stable mixed chimerism patients. Biol Blood Marrow

Transplant 19(5):838–844. https://doi.org/10.1016/j.bbmt.2013.02.015

Stikvoort A, Sundin M, Uzunel M et al (2016) Long-term stable mixed chimerism after hematopoietic stem cell transplantation in patients with non-malignant disease, shall we be tolerant? PLoS One 11(5):e0154737. https://doi.org/10.1371/journal.pone.0154737

Svenberg P, Mattsson J, Ringden O, Uzunel M (2009) Allogeneic hematopoietic SCT in patients with non-malignant diseases, and importance of chimerism. Bone Marrow Transplant 44(11):757–763. https://doi.org/10.1038/bmt.2009.82

Sy A, Chanson D, Berano Teh J et al (2021) Late-occurring infections in a contemporary cohort of hematopoietic cell transplantation survivors. Cancer Med 10(9):2956–2966. https://doi.org/10.1002/cam4.3896

Syrjala KL, Yi JC, Artherholt SB et al (2018) An online randomized controlled trial, with or without problem-solving treatment, for long-term cancer survivors after hematopoietic cell transplantation. J Cancer Surviv 12(4):560–570. https://doi.org/10.1007/s11764-018-0693-9

Thompson AA, Walters MC, Kwiatkowski J et al (2018) Gene therapy in patients with transfusion-dependent beta-thalassemia. N Engl J Med 378(16):1479–1493. https://doi.org/10.1056/NEJMoa1705342

Tichelli A, Beohou E, Labopin M et al (2019) Evaluation of second solid cancers after hematopoietic stem cell transplantation in European patients. JAMA Oncol 5(2):229–235. https://doi.org/10.1001/jamaoncol.2018.4934

Vajdic CM, Mayson E, Dodds AJ et al (2016) Second cancer risk and late mortality in adult Australians receiving allogeneic hematopoietic stem cell transplantation: a population-based cohort study. Biol Blood Marrow Transplant 22(5):949–956. https://doi.org/10.1016/j.bbmt.2016.01.027

Vasbinder A, Hoeger CW, Catalan T et al (2023) Cardiovascular events after hematopoietic stem cell transplant: incidence and risk factors. JACC Cardio Oncol 5(6):821–832. https://doi.org/10.1016/j.jaccao.2023.07.007

Wolff D, Cutler C, Lee SJ et al (2024) Axatilimab in recurrent or refractory chronic graft-versus-host disease. N Engl J Med 391(11):1002–1014. https://doi.org/10.1056/NEJMoa2401537

Wood WA, Lee SJ (2011) Malignant hematologic diseases in adolescents and young adults. Blood 117(22):5803–5815. https://doi.org/10.1182/blood-2010-12-283093

Xuan L, Wang Y, Yang K et al (2023) Sorafenib maintenance after allogeneic haemopoietic stem-cell transplantation in patients with FLT3-ITD acute myeloid leukaemia: long-term follow-up of an open-label, multicentre, randomised, phase 3 trial. Lancet Haematol 10(8):e600–e611. https://doi.org/10.1016/S2352-3026(23)00117-5

Yeh J, Whited L, Saliba RM et al (2021) Cardiac toxicity after matched allogeneic hematopoietic cell transplant in the posttransplant cyclophosphamide era. Blood Adv 5(24):5599–5607. https://doi.org/10.1182/bloodadvances.2021004846

Index

© The Editor(s) (if applicable) and The Author(s), under exclusive license to Springer Nature Switzerland AG 2025

K. Ballen (ed.), *Clinical Management of Challenges in Allogeneic Stem Cell Transplant*, Advances in Experimental Medicine and Biology 1475, https://doi.org/10.1007/978-3-031-84988-6

The manufacturer's authorised representative in the EU is Springer Nature Customer Service Centre GmbH, Europaplatz 3, 69115 Heidelberg, Germany. If you have any concerns regarding our products, please contact ProductSafety@springernature.com

Printed and bound by CPI Group (UK) Ltd, Croydon, CR0 4YY
01/10/2026
02225936-0019